Learning and Analytics in Intelligent Systems

Volume 49

The main aim of the series is to make available a publication of books in hard copy form and soft copy form on all aspects of learning, analytics and advanced intelligent systems and related technologies. The mentioned disciplines are strongly related and complement one another significantly. Thus, the series encourages cross-fertilization highlighting research and knowledge of common interest. The series allows a unified/integrated approach to themes and topics in these scientific disciplines which will result in significant cross-fertilization and research dissemination. To maximize dissemination of research results and knowledge in these disciplines, the series publishes edited books, monographs, handbooks, textbooks and conference proceedings.

Indexed by EI Compendex.

Dimitrios P. Panagoulias · George A. Tsihrintzis · Maria Virvou

Artificial Intelligence-Empowered Bio-medical Applications

Challenges, Solutions and Development Guidelines

Springer

Dimitrios P. Panagoulias
Piraeus, Greece

George A. Tsihrintzis
Piraeus, Greece

Maria Virvou
Piraeus, Greece

ISSN 2662-3447 ISSN 2662-3455 (electronic)
Learning and Analytics in Intelligent Systems
ISBN 978-3-031-90173-7 ISBN 978-3-031-90174-4 (eBook)
https://doi.org/10.1007/978-3-031-90174-4

This Springer imprint is published by the registered company Springer Nature Switzerland AG
The registered company address is: Gewerbestrasse 11, 6330 Cham, Switzerland

What is the essence of life? To serve others and to do good (Aristotle). Guided by this timeless principle and the spirit of Philanthropia we pledge this work, to advancing technology as a compassionate ally in the pursuit of health and equity of access.

The research for this book took place at the Department of Informatics of University of Piraeus from June 2021 to June 2024.

It was conducted under the supervision of Prof. Dr. George A. Tsihrintzis, to whom I am immensely grateful for his guidance, ethos and trust and for opening a new world of opportunities, helping me in becoming a better and hopefully more useful individual. I am also grateful to Prof. Dr. Maria Virvou, whose insights and targeted interventions enhanced my research process, adding new perspectives and viewpoints. The contributions and accurate observations of Asst. Prof. Dr. Dionysios Sotiropoulos improved my technical approach, for which I am extremely grateful. I am also extremely appreciative of all the other collaborators I have had over these years, whose contributions and expertise have greatly enhanced my research.

For everything in my life I owe much love and gratitude to my wife Nadia my children Kira and Petros, my sister Kiriaki, my parents

Petros and Sofia and my brothers Kostas and Savvas.

Dimitrios P. Panagoulias
University of Piraeus
June 2024

Preface

This book examines the advancements in personalized medicine, emphasizing the transition from a generalized "one-size-fits-all" approach to a more tailored strategy in disease diagnosis and patient management. This shift is facilitated by the use of biomarkers, which are critical in developing and training prognostic models and neural networks within the fields of machine learning and artificial intelligence. Biomarkers, defined as measurable and reproducible medical signs, serve as objective indicators of a patient's medical state.

The research highlights significant achievements in using machine learning techniques to predict various health indicator. For example, blood exams are utilized to classify body mass index with an average accuracy of 84%, while cascaded SVM-based classifiers are used to determine systolic blood pressure with a 74% accuracy. Furthermore, a new system has been proposed and implemented to predict metabolic syndrome with an 84% accuracy, relying on multiple blood parameters excluding defining factors of it, thus proposing alternative diagnostic pathways.

In addition to specific diagnostic tools, this book explores broader applications of AI-empowered systems that use machine learning models, in healthcare through microservices in distributed systems, which simplify management and scaling by segregating functions into distinct services. In addition, the regulatory and validation challenges are also considered and investigated to define optimum development paths and digital architecture solutions using the Rational Unified Process and intuitive workflows. It also addresses the necessity of explainability and interpretability in AI systems to enhance usability and trust, particularly in the medical domain. A novel framework (PINXEL) is introduced to define explainability requirements using the Technology Acceptance Model (TAM).

Following that, the capabilities and effectiveness of large language models were evaluated, particularly ChatGPT, in the medical domain. The focus was on this novel system's ability to aid in medical diagnosis through structured evaluations and domain-specific follow-up analysis. Then a methodology was outlined and implemented to further assess correctness and accuracy of GPT-4V (Generative Pre-trained Transformer 4 with vision) using multimodal multiple-choice questions in the domain of General Pathology. In addition to that using our proposed methodology, requirements can be extracted for targeted interventions and fine-tuning of state-of-the art models, for their optimisation in domain-specific tasks.

Finally, this book proposes a novel AI-powered system designed to provide diagnostic advice in primary care settings, employing all technologies explored during this research. This system enhances large language models and other machine learning tools with a rule-based approach, offering tailored medical advice through interactive AI assistant named "Med|Primary AI assistant" and "Dermacen Analytica". This systems leverage natural language processing, image analysis and segmentation, domain-specific knowledge and multi-modal LLMs to create a personalized user interaction, which is evaluated through a rigorous context and dialogue-based methodological framework. This holistic approach underscores the integration of AI into healthcare, aiming to assist medical professionals in patient diagnosis and management with high accuracy and user-specific adaptability.

Piraeus, Greece

Dimitrios P. Panagoulias
George A. Tsihrintzis
Maria Virvou

Competing Interests The authors have no competing interests to declare that are relevant to the content of this manuscript.

Book Structure

Chapter 1: "**Introduction to AI-Empowered Medical Software: Recent Advances and Challenges**"

In this chapter an introduction is provided for the various themes that are discussed in this book. A brief overview is provided about biomarkers, Neural networks, development paradigms and generative Artificial Intelligence.

Chapter 2: "**Personalized Nutrition Applications Using Nutritional Biomarkers and Machine Learning**"

In this chapter a theoretical background of basic concepts related to personalized medicine and AI are outlined. This chapter aims to formulate a general evaluation chart of nutritional biomarkers, investigate the best methods to predict body mass index and discover dietary patterns using neural networks.

Chapter 3: "**Blood Exam Classification for Predicting Defining Factors in Metabolic Syndrome Diagnosis Using Support Vector Machines**"

In this chapter, cascaded SVM-based classifiers for automated metabolic syndrome diagnosis were created. Specifically, using blood exams, an average accuracy of about 84% in correctly classifying body mass index was achieved. Similarly, cascaded SVM-based classifiers achieved a 74% accuracy in correctly classifying systolic blood pressure. Next, we propose and implement a system that achieves an 84% accuracy in metabolic syndrome prediction.

Chapter 4: "**Extreme Value Analysis Applied in Dietary Data**"

In this chapter, extreme value theory was used to examine the significance of outliers in health data, with a focus on diet and the standard biochemistry profile. Specifically, using extreme value analysis and by applying a systemisation of the analytical process, health trends can be predicted, and thus, health interventions can be (at least partially) automated.

Chapter 5: "**Iterative Microservices Approach for Explainable and Reliable AI in Medical Applications**"

The development of NuhealthSoft an AI-infused medical application, is outlined in this chapter through the RUP methodology and built using microservices.

Chapter 6: "**Challenges in Regulating and Validating AI-Driven Healthcare**"

In this chapter, the regulatory aspects of defining regulatory parameters for deploying an AI-empowered medical applications. Thus, the regulator is considered in this, as the main stakeholder, and using the RUP the deployment parameters are outlined in a comprehensive manner.

Chapter 7: "**Framework for AI Explainability Leveraging User Acceptance and Health Literacy Models**"

In this chapter, Explainability and interpretability are considered important properties of AI-empowered systems and, as such, constitute important factors for technology acceptance. The Personalized INterface for eXplainability basEd on Levels framework is outlined. PINXEL is a novel framework for explainability requirements in AI-empowered systems using the Technology Acceptance Model (TAM).

Chapter 8: "**Enhancing Domain-Specific Interactions in Large Language Models**"

In this chapter, we present a novel AI-empowered system which enhances Large Language Models and other Machine Learning tools with rules to provide Primary Care diagnostic advice to patients.

Chapter 9: "**Extracting Requirements to Fine-Tune Large Language Models in Multimodal Medical Diagnosis**"

In this chapter, an evaluation system that incorporates two independent steps of a novel methodology was engineered, for (1) evaluation via structured multimodal interactions with LLMs and (2) follow-up, domain-specific analysis based on data extracted from the previous interactions for model fine-tuning recommendations. In addition, this framework was tested with publicly available data, pertaining to General Pathology Multiple choice quizzes with associated images.

Chapter 10: "**Integrating Multi-Modal Language Models and Machine Learning in Dermatology**"

In this chapter, using LLMs, ML services and analytical tools a workflow and a system was implemented to simulate a dermatologist's workflow, as an assistive medical agent. The system is outlined and evaluated using datasets, incorporating Images depicting various skin conditions.

Chapter 11: "**Cognitive Network Evaluation Toolkit for Medical Domains**"

In this chapter, we introduce the Cognitive Network Evaluation Toolkit for Medical Domains (COGNET-MD), a pioneering benchmark designed for evaluating large language models (LLMs) within the medical domain. Our approach includes a sophisticated scoring framework with heightened difficulty levels, aimed at rigorously assessing the capability of LLMs to accurately interpret and analyze medical text.

Chapter 12: "**Advances, Challenges and Contributions to Artificial Intelligence-Empowered Bio-Medical Applications**"

In this final chapter, we summarize our various findings. We also itemize the challenges that were identified and discuss the focus of this work, in software engineering, Neural networks, personalization approaches and Large Language models.

Contents

Abbreviations

AI	Artificial Intelligence
COGNET-MD	Cognitive Network Evaluation Toolkit for Medical Domains
LLM	Large Language Model
ML	Machine Learning
NLP	Natural Language Processing
PINXEL	Personalized INterface for eXplainability basEd on Levels
RAI	Responsible Artificial Intelligence
REALM	Rapid Estimate of Adult Literacy in Medicine
TAM	Technology Acceptance Model
XAI	Explainable Artificial Intelligence

Part I
Medical Software Empowered with/by Learning Algorithms and Advanced Statistical Processing

Chapter 1
Introduction to AI-Empowered Medical Software: Recent Advances and Challenges

The efficiency of a medical system is gauged by its capability to anticipate and fulfill the healthcare needs of the population, ensuring high quality at minimal cost while optimizing all forms of capital. Both medical science and economics rely on empirical and factual data to make informed decisions and develop models that deliver socially equitable and beneficial outcomes. Technology provides optimization solutions based on analyzing different types of data to discover new pathways. These pathways need thorough evaluation and validation to ensure they align with societal requirements and needs, ultimately becoming innovations that transform the world for the better.

1.1 Problem Definition

1.1.1 Biormarkers as Tools for Artificial Intelligence

Over the past two decades, the medical industry has experienced significant transformation. Technological innovations such as machine learning, large language models, and AI-powered systems have opened new pathways in drug discovery and patient management, and have enhanced existing methodologies. A key metric that provides insights into current, future, or past health states is biomarkers.

Biomarkers are medical signs that offer "objective indications of a medical state observed from outside the patient and can be measured accurately and reproducibly." They are molecules (such as proteins or antibodies) or biochemical changes (such as gene expressions and mutations) found in patients' tissues, blood, or other bodily fluids. Biomarkers signify the presence of a specific disease, condition, or mutation and are invaluable in clinical and pharmacological applications. Biomarkers are important features in machine learning that can lead to new discoveries and empower systems with clinical usability.

D. P. Panagoulias et al., *Artificial Intelligence-Empowered Bio-medical Applications*, Learning and Analytics in Intelligent Systems 49,
https://doi.org/10.1007/978-3-031-90174-4_1

The vast knowledge space of the medical domain, combined with the low tolerance for mistakes and the need to mitigate risks, necessitates a delicate approach during the training, validation, and deployment of AI models. Regulation and expert contributions are required, and their inclusion in the development process is a complex task.

1.1.2 System Design with Explainability and Personalization Features

Patients' needs and abilities vary, and their requirements and definitions of the usability of an application are different and should be acknowledged for the application to be successful and beneficial to the patients-users [1, 2]. The significant proliferation of AI-empowered systems and machine learning (ML) across various domains underscores the vital necessity for comprehensive and customized explainability frameworks [3–5] to lead to usable and trustworthy systems. Especially in the medical domain, where validation of methodologies and outcomes, is as important as the adoption rate of such systems, the depth and level of abstraction of the explainability are particularly important and necessitate a systemic approach to ensure proper definition. Explainability and interpretability are crucial usability and trustworthiness properties [6] of AI-empowered systems and, as such, constitute important factors for technology acceptance.

1.1.3 Multi-modal Language Models, Natural Language Processing, Transformers and Evaluation Paradigms

An emergent technology showing great promise to revolutionize healthcare is the technology of large language models (LLMs) [7, 8]. Indeed, LLMs demonstrate a remarkable capability of understanding medical texts and identifying (diagnosing) a range of symptoms and health conditions.

The Transformer model [9] has been highly influential in the field of Natural Language Processing (NLP) and serves as the engine for state-of-the-art LLMs. The Transformer model is based on a mechanism, referred to as self-attention, that directly models the relationships between words in a sentence, regardless of their respective positions in the sentence.

An exemplary LLM is the Generative Pre-trained Transformer (GPT >=4) by OpenAI, which powers ChatGPT and generates accurate, human-like text responses. Other notable LLMs include Google's BERT (Bidirectional Encoder Representations from Transformers), Meta's LLaMA (Large Language Model Meta AI), Mistral and Anthropic's Claude among others. While each LLM [10] and NLP has its limitations, selectively integrating elements from various technologies can offer both efficacy and cost-effectiveness. Under different circumstances, LLMs perform better under the supervision and assistance of an expert who can evaluate the responses and provide additional instructions that can refine generated output, providing opportunities for optimisation and novel systemic approaches.

The evolution of ChatGPT and similar applications empowered by Large Language Models (LLMs) represent a landmark in Artificial Intelligence (AI), offering extensive applications and raising complex challenges. These developments span education, cultural preservation, decision-support systems, and healthcare, as explored in various recent studies. ChatGPT's ability to generate natural and coherent language has made it invaluable in educational settings including medical contexts. It facilitates innovative approaches to e-learning by creating interactive experiences, as for example, in cultural heritage education through poetry and lyrics [11]. The possible limitations can be addressed by rule augmentations. For instance, LYRICEL integrates rule-based AI and machine learning with open AI's GPT for song lyric education, bridging traditional cultural knowledge and modern AI tools [12]. This synergy underscores the potential for large-scale AI in preserving and disseminating cultural heritage. The holistic evaluation frameworks developed for ChatGPT have demonstrated its potential as an intelligent tutoring system, highlighting its role in personalized learning [13]. In healthcare, the GPT 4.0 and GPT 4.0 with vision is being tested for its ability to support diagnosis and treatment planning, including complex cases like lung cancer staging and tele-dermatology [14, 15]. Rule-augmented systems further enhance GPT's domain-specific reliability, especially in multimodal tasks combining medical images with symptom analysis [16].

Despite its promise, ChatGPT faces significant challenges in areas like user trust, ethical responsibility, and domain specificity. The VIRTSI model, for example, emphasizes the dynamics of trust in AI-human collaboration, noting the need for transparency and context-aware interactions [6]. Similarly, studies on AI responses have highlighted how varying consequences impact user trust and satisfaction [17]. Ethical concerns include biases embedded in training data and the potential for misinformation. In medical applications, frameworks like COGNET-MD have been developed to evaluate LLM performance against rigorous benchmarks, ensuring safety and accuracy [18]. Moreover, integrating expert oversight into AI systems is increasingly recognized as essential for mitigating risks [19].

To fully realize ChatGPT's potential, researchers are exploring methods to augment LLMs with rules and hybrid models. These approaches aim to enhance domain-specific accuracy while ensuring scalable, trustworthy, and ethically sound AI applications for assisting in decision making [20]. Multimodal AI integration also holds promise for advancing fields such as telemedicine and cultural preservation, offering new paradigms for AI-human interaction.

1.1.4 Software Engineering Processes and Approaches in AI—Empowered Software Engineering

The integration of Artificial Intelligence (AI) into software engineering necessitates a reevaluation of development methodologies. Rational Unified Process (RUP), Agile, Scrum, and microservices each play a critical role in addressing the unique challenges and opportunities presented by AI-empowered systems. RUP is a comprehensive, iterative framework emphasizing use-case-driven design, architecture-centric planning, and phased development cycles [21, 22]. These characteristics are essential in AI projects where complexity and interdependencies dominate. Savvopoulos et al. [23] demonstrated RUP's efficacy in e-commerce systems by using clustering techniques for user modeling, showcasing its potential for personalization. Dimitrios P Panagoulias, Virvou, et al. [24] further illustrated its value in a health data processing tool, where structured validation was vital for precision. Hirsch [25] suggested making RUP more agile to better address evolving requirements, a critical factor in AI systems where data and objectives can shift. Agile methodologies, including Scrum, bring iterative flexibility and responsiveness to development, fostering frequent feedback cycles and adaptive planning. Virvou and George A Tsihrintzis [26] highlighted Agile's capacity for seamless integration of human-AI collaboration, crucial in projects where explainability and user alignment are priorities. Similarly, Dimitrios P Panagoulias, Sarmas, et al. [27] emphasized Agile's role in developing interpretable AI decision support systems. Scrum, as a specific Agile framework, adds operational focus with its sprint-based iterations, facilitating continuous validation and improvement. Shafiee et al. [28] noted that Scrum enhances project management in dynamic environments, adding agility to projects where frequent adjustments are required.

Microservices architecture complements these methodologies and stands out as a foundational element in AI-empowered software engineering. It enables modular development, where independent components can evolve separately, making the system inherently scalable and adaptable. Schäffer et al. [29] demonstrated the effectiveness of microservices in creating scalable and integrable platforms, a necessity for AI systems handling diverse functionalities. George A Tsihrintzis, Virvou, Sakkopoulos, et al. [30] underscored the importance of modularity in intelligent systems that integrate machine learning paradigms, further highlighting microservices' pivotal role. The ability of microservices to isolate and optimize individual components ensures that AI systems can quickly adapt to technological advancements, integrate cutting-edge tools, and scale seamlessly to meet increasing computational demands.

In conclusion, while RUP provides structural depth and Agile and Scrum offer flexibility and responsiveness, microservices are the cornerstone of AI-empowered software engineering. They enable modularity, scalability, and innovation, addressing the complexity of AI systems effectively. By emphasizing microservices in conjunction with RUP and Agile methodologies, software teams can develop adaptable, scalable, and cutting-edge AI-driven solutions tailored to the dynamic needs of modern applications.

1.1.5 Support Vector Machine Classifiers and Machine Learning

Support Vector Machines (SVM) have proven to be effective in various applications, including recommendation systems, where they improve personalization through collaborative filtering and user modeling [23, 31]. SVM is also effective in addressing class imbalance challenges in anomaly detection, particularly when integrated with artificial immune system-based approaches or one-class classifiers [32, 33]. In healthcare, SVM has been applied to diagnostic imaging, such as breast cancer detection through mammogram classification, achieving significant accuracy improvements in early detection [34]. The use of interpretable SVM models is essential in healthcare applications where transparency and trust are critical [35]. Additionally, reviews of AI-based technologies have highlighted SVM's role in advancing healthcare solutions, including precision medicine [36].

1.2 Research Objectives and Goals

The objective of this book is to identify the shortcomings and opportunities of recent technologies, with a particular focus on the medical domain. It aims to propose solutions and provide a basis for cost-effective and modern improvements that will enhance the patient's journey and streamline the doctor's workflow. In the evolving space of Artificial Intelligence and software development within the medical domain, business scaling often overshadows important social parameters such as equity, safety, and validation.

This book addresses these important issues by highlighting the following:

1. The personalization of technology based on the perceived usability and usefulness as understood by its users, providing tailored Explainability functionalities
2. The identification of regulators and experts as the main stakeholders who ensure safety and equity, with design paradigms provided as part of the novel and modern software development of AI-empowered systems
3. The evaluation of medical systems, with various proposed, implemented, and open-sourced pipelines to facilitate this process.

1.3 Novelties and Contributions

As shown in Fig. 1.1 three distinct systems were implemented addressing the issues previously highlighted.

Utilizing publicly available datasets, novel machine learning models and analytical tools were developed as classification engines for conditions like obesity, metabolic syndrome, and blood pressure, investigating alternative pathways through

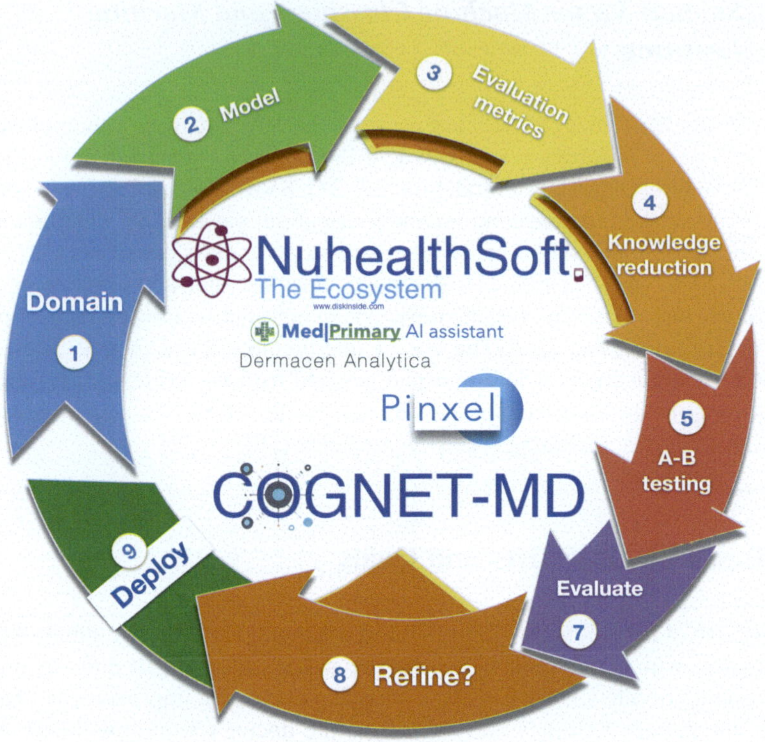

Fig. 1.1 Nuhealthosft is an ecosystem of Machine learning and analytical processes for biomedical classifications. PINXEl is a personalised explainability framework. COGNET-MD is a framework to assess LLMs in medical knowledge via a novel scoring system

a variety of features. These processes were systematically applied, leading to semantic groupings that formed the basis for explainable models of the decision-making process.

An AI-infused medical application—NuhealtSoft-, incorporating the previous, was outlined using the Rational Unified Process (RUP) and developed using microservices [37, 38], where the patients' needs and stakeholder (regulators and doctors) requirements are at the center of a continuous development process focused on improvements. Following the RUP methodology, we demonstrated how microservices are suitable as a proposed solution for a medical-health application from an architectural standpoint, emphasizing the business and economic benefits. Through the RUP methodology, we have established how key concepts of microservices evolve over time, defining an optimized system design in the elaboration phase. In addition ChatBots in the domain of pathology and dermatology were also constructed and incorporated in the NuhealtSoft ecosystem.

A framework—PINXEL- employing targeted ML (hierarchical clustering, k-means, or other methods) to acquire a user model for personalized, multi-layered explainability was also developed. It integrates a rule-based system that guides the degree of trustworthiness to be achieved based on user perception and AI literacy level.

The various limitations of multi-modal LLMs were addressed using evaluation pipelines for assessing their accuracy in assessing symptoms either unilateral or accompanied by associated images. For that purposes publicly available datasets were used and information was extracted, for optimisation and finetuning of the underlying models. Additionally, COGNET-MD a novel toolkit was created, consisting of expertly curated datasets and an intuitive framework for scoring and assessing these systems.

References

1. J. Hesketh, Personalised nutrition: how far has nutrigenomics progressed? Eur. J. Clin. Nutr. **67**(5), 430–435 (2013)
2. A. Acosta, et al., Precision medicine and obesity. Gastroenterol. Clin. **50**(1), 127–139 (2021)
3. D. Gunning, M. Stefik, J. Choi, T. Miller, S. Stumpf, G.-Z. Yang, XAI—Explainable artificial intelligence. Sci. Robot. **4**(37), eaay7120 (2019). https://doi.org/10.1126/scirobotics.aay7120. eprint: https://www.science.org/doi/pdf/10.1126/scirobotics.aay7120. https://www.science.org/doi/abs/10.1126/scirobotics.aay7120
4. K. Rasheed, A. Qayyum, M. Ghaly, A. Al-Fuqaha, A. Razi, J. Qadir, Explainable, trustworthy, and ethical machine learning for healthcare: a survey. Comput. Biol. Med. 106043 (2022)
5. D.P. Panagoulias, M. Virvou, G.A. Tsihrintzis, A novel framework for artificial intelligence explainability via the Technology Acceptance Model and Rapid Estimate of Adult Literacy in Medicine using machine learning. Expert Syst. Appl. **248**, 123375 (2024). ISSN: 0957-4174. https://doi.org/10.1016/j.eswa.2024.123375. https://www.sciencedirect.com/science/article/pii/S0957417424002409
6. M. Virvou, G.A. Tsihrintzis, E.-A. Tsichrintzi, VIRTSI: a novel trust dynamics model enhancing Artificial Intelligence collaboration with human users–Insights from a ChatGPT evaluation study. Inf. Sci. 120759 (2024)
7. Y. Labrak, A. Bazoge, E. Morin, P.-A. Gourraud, M. Rouvier, R. Dufour, BioMistral: a collection of open-source pretrained large language models for medical domains. arXiv preprint arXiv:2402.10373 (2024)
8. *GPT-4V(ision)* (2023)
9. A. Vaswani, N. Shazeer, N. Parmar, J. Uszkoreit, L. Jones, A.N. Gomez, Ł. Kaiser, I. Polosukhin, Attention is all you need. Adv. Neural Inf. Process. Syst. **30** (2017)
10. S. Locke, A. Bashall, S. Al-Adely, J. Moore, A. Wilson, G.B. Kitchen, Natural language processing in medicine: a review. Trends Anaesth. Crit. Care **38**, 4–9 (2021)
11. M. Virvou, G.A. Tsihrintzis, D.N. Sotiropouloss, K. Chrysafiadi, E. Sakkopoulos, E.-A. Tsichrintzi, ChatGPT in artificial intelligence-empowered E-learning for cultural heritage: the case of lyrics and poems, in *2023 14th International Conference on Information, Intelligence, Systems & Applications (IISA)*. IEEE (2023), pp. 1–9
12. D.P. Panagoulias, D. Sotiropoulos, K. Chrysafiadi, E.-A. Tsichrintzi, E. Sakkopoulos, G.A. Tsihrintzis, M. Virvou, LYRICEL: a rule-augmented artificial intelligence-empowered cultural E-learning with GPT and machine learning for song lyrics, in *2024 15th International Conference on Information, Intelligence, Systems & Applications (IISA)*. IEEE (2024), pp. 1–5
13. M. Virvou, G.A. Tsihrintzis, Is ChatGPT beneficial to education? A holistic evaluation framework based on intelligent tutoring systems, in *2023 14th International Conference on Information, Intelligence, Systems & Applications (IISA)*. IEEE (2023), pp. 1–8

14. D.P. Panagoulias, F.A. Palamidas, M. Virvou, G.A Tsihrintzis, Evaluation of ChatGPT-supported diagnosis, staging and treatment planning for the case of lung cancer, in *2023 20th ACS/IEEE International Conference on Computer Systems and Applications (AICCSA)*. IEEE (2023), pp. 1–8
15. D.P. Panagoulias, E. Tsoureli-Nikita, M. Virvou, G.A. Tsihrintzis, *Dermacen Analytica: A Novel Methodology Integrating Multi-Modal Large Language Models with Machine Learning in Tele-dermatology* (2024). arXiv: 2403.14243 [cs.CL]
16. D.P. Panagoulias, M. Virvou, G.A. Tsihrintzis, Augmenting large language models with rules for enhanced domain-specific interactions: the case of medical diagnosis. Electronics **13**(2), 320 (2024)
17. M. Virvou, G.A. Tsihrintzis, Impact of consequenses on human trust dynamics in artificial intelligence responses, in *2024 15th International Conference on Information, Intelligence, Systems & Applications (IISA)*. IEEE (2024), pp. 1–6
18. D.P. Panagoulias, P. Papatheodosiou, A.P. Palamidas, M. Sanoudos, E. Tsoureli-Nikita, M. Virvou, G.A. Tsihrintzis, COGNET-MD, an evaluation framework and dataset for Large Language Model benchmarks in the medical domain. arXiv: 2405.10893 [cs.CL] (2024)
19. M. Virvou, G.A. Tsihrintzis, A novel trust state-chart model for requirements engineering of trustful AI-empowered software, in *2023 14th International Conference on Information, Intelligence, Systems & Applications (IISA)*. IEEE (2023), pp. 1–6
20. G.A. Tsihrintzis, M. Virvou, H. Doukas, L.C. Jain, Introduction to advances in artificial intelligence-empowered decision support systems, in *Advances in Artificial Intelligence-Empowered Decision Support Systems: Papers in Honour of Professor John Psarras* (Springer, 2024), pp. 1–11
21. A.K. Shuja, J. Krebs, IBM Rational unified process reference and certification guide: solution designer (RUP) (Pearson Education, 2007)
22. A. Anwar, A review of rup (rational unified process). Int. J. Softw. Eng. (IJSE) **5**(2), 12–19 (2014)
23. A. Savvopoulos, M. Virvou, D.N. Sotiropoulos, G.A. Tsihrintzis, Clustering for user modeling in recommender e-commerce application: a rup-based intelligent software life-cycle, in *Knowledge-Based Software Engineering* (IOS Press, 2008), pp. 295–304
24. D.P. Panagoulias, M. Virvou, G.A. Tsihrintzis, nuhealthsoft: a nutritional and health data processing software tool from a patient's perspective, in *2022 16th International Conference on Signal-Image Technology & Internet-Based Systems (SITIS)*. (IEEE, 2022), pp. 386–393
25. M. Hirsch, Making RUP agile, in *OOPSLA 2002 Practitioners Reports*, 1–ff (2002)
26. M. Virvou, G.A. Tsihrintzis, Pre-made empowering artificial intelligence and ChatGPT: the growing importance of human AI-experts, in *2023 14th International Conference on Information, Intelligence, Systems & Applications (IISA)*. IEEE (2023), pp. 1–8
27. D.P. Panagoulias, E. Sarmas, V. Marinakis, M. Virvou, G.A. Tsihrintzis, H. Doukas, Intelligent decision support for energy management: a methodology for tailored explainability of artificial intelligence analytics. Electronics **12**(21), 4430 (2023)
28. S. Shafiee, Y. Wautelet, L. Hvam, E. Sandrin, C. Forza, Scrum versus Rational Unified Process in facing the main challenges of product configuration systems development. J. Syst. Softw. **170**, 110732 (2020)
29. E. Schäffer, L.N. Penczek, M. Bartelt, M. Brossog, B. Kuhlenkötter, J. Franke, A microservice-And AutomationML-based reference architecture for an engineering configurator web platform. Procedia CIRP **103**, 274–279 (2021)
30. G.A. Tsihrintzis, M. Virvou, E. Sakkopoulos, L.C. Jain, *Machine Learning Paradigms: Applications of Learning and Analytics in Intelligent Systems*, vol. 1 (Springer, 2019)
31. A.S. Lampropoulos, D.N. Sotiropoulos, G.A. Tsihrintzis, Cascade hybrid recommendation as a combination of one-class classification and collaborative filtering. Int. J. Artif. Intell. Tools **23**(04), 1460009 (2014)
32. D.N. Sotiropoulos, G.A. Tsihrintzis, Artificial immune system-based classification in extremely imbalanced classification problems. Int. J. Artif. Intell. Tools **26**(03), 1750009 (2017)

33. D. Sotiropoulos, C. Giannoulis, G.A. Tsihrintzis, A comparative study of one-class classifiers in machine learning problems with extreme class imbalance, in *IISA 2014, The 5th International Conference on Information, Intelligence, Systems and Applications*. IEEE (2014), pp. 362–364
34. A. Agrios, D.N. Sotiropoulos, G.A. Tsihrintzis, Computer aided breast cancer diagnosis: an SVM-based mammogram classification approach. Int. J. Comput. Intell. Stud. **7**(2), 138–163 (2018)
35. E. Alogogianni, M. Virvou, Interpretable machine learning for undeclared work prediction, in *2023 14th International Conference on Information, Intelligence, Systems & Applications (IISA)*. IEEE (2023), pp. 1–8
36. G.A. Tsihrintzis, M. Virvou, G. Phillips-Wren, Surveys in artificial intelligence-based technologies. Intell. Decis. Technol. **13**(4), 393–394 (2019)
37. *What are microservices?* (N.d.). https://microservices.io. Accessed: 2022-08-5
38. D.P. Panagoulias, M. Virvou, G.A. Tsihrintzis, A microservices-based iterative development approach for usable, reliable and explainable A.I.-infused medical applications using R.U.P., in *34th IEEE Conference on Tools with Artificial Intelligence (ICTAI 2022)*. Virtually (2022)

Chapter 2
Personalized Nutrition Applications Using Nutritional Biomarkers and Machine Learning

Abstract The traditional "one-size-fits-all" methodology in disease diagnosis and patient management has been largely replaced by a more individualized strategy known as personalized medicine. In this evolving paradigm, biomarkers play a crucial role, serving as fundamental variables (features) in the advancement of machine learning and artificial intelligence-driven prognostic models. Among the diverse categories of biomarkers, metabolites are particularly significant due to their direct association with metabolic processes. Metabolomics, a discipline dedicated to the systematic examination of chemical signatures produced by cellular activities, offers valuable insights into metabolic functions. By analyzing metabolic profiles, researchers can gain a comprehensive understanding of cellular physiology, providing a direct functional assessment of an organism's physiological state. The objective of this chapter is to develop a structured evaluation framework for nutritional biomarkers, explore predictive models for body mass index (BMI), and identify dietary patterns through the application of neural networks.

Keywords Artificial intelligence · Personalised medicine · Biomarkers · Nutrition · Obesity

2.1 Introduction

The most accurate approach to evaluating patients and tailoring personalized medical treatments, disease classification, and medication strategies is through the identification of biomarkers. These biomarkers serve as medical indicators that, in a broad sense, *"provide objective indications of the medical state observed from outside the patient and can be measured accurately and reproducibly"* [26]. Biomarkers encompass various biological components, including molecules, proteins, antibodies, and biochemical alterations detected in tissue, blood, or other bodily fluids. Their presence can signify an existing medical condition, forecast potential disease progression, or indicate genetic mutations [43, 59]. In this respect, biomarkers function either as a diagnostic tool for present conditions or as a predictive mechanism for future health outcomes. They may derive from a single measurement during an examination or

D. P. Panagoulias et al., *Artificial Intelligence-Empowered Bio-medical Applications*, Learning and Analytics in Intelligent Systems 49,
https://doi.org/10.1007/978-3-031-90174-4_2

from a combination of multiple observed metrics. For instance, a well-known example of a composite biomarker is the Body Mass Index (BMI), which is computed from an individual's mass and height.

Although the use of biomarkers is not a novel concept-they have been widely studied and applied in medicine for more than two decades-their significance continues to grow. The rationale behind their increasing importance is that the discovery of novel biomarkers is expected to refine medical diagnostics and therapeutic pathways, leading to enhanced treatment accuracy, reduced healthcare costs, and optimized patient outcomes. However, the implementation and widespread use of biomarker-based medical strategies also raise critical ethical, social, and economic concerns that warrant thorough investigation. First, the inherent complexity of human disease biology may limit the applicability of certain biomarker-driven approaches. Second, the cost of some advanced treatments, particularly in oncology, may be prohibitively high, posing challenges in resource allocation and treatment accessibility. These financial constraints necessitate objective decision-making regarding the prioritization of patient groups, selection criteria for treatment eligibility, and the anticipated impact on healthcare systems. Given the dynamic nature of both research and medical data, these concerns must be continuously reassessed.

Among the most critical biomarkers linked to metabolism are metabolites. The field of metabolomics is dedicated to the systematic analysis of chemical fingerprints (metabolites) that result from specific biochemical processes. A metabolic profile provides a real-time snapshot of cellular physiology, and consequently, metabolomics offers a direct functional interpretation of an organism's physiological state [41].

Conversely, obesity has emerged as a major global health challenge, not only in industrialized nations but also in developing regions with lower and middle-income populations. Effective weight management strategies require a deeper understanding of the molecular mechanisms driving weight fluctuations. Several biological pathways, including oxidative stress responses and insulin regulation frameworks, have been implicated in weight gain and metabolic regulation. Additionally, growing research highlights the role of gut microbiota-derived metabolites, particularly short-chain fatty acids, in metabolic balance and energy homeostasis [16, 39].

In the subsequent sections, this chapter explores fundamental domain concepts, methodological approaches for data analysis, and the construction of machine learning-driven predictive models. The primary objective of this study is to establish a comprehensive evaluation framework for nutritional biomarkers, assess their predictive capabilities for BMI estimation, and uncover dietary patterns using neural networks. The analytical approach is structured around both a theoretical exploration of the problem and a practical solution modeling process, leveraging automated machine learning algorithms and pattern recognition techniques. A thorough literature review has been conducted to build a foundational understanding of the subject. Additionally, Python and relevant scientific computing libraries have been utilized to implement and validate the proposed models.

2.2 Basic Concepts

2.2.1 AI—Driven Personalization Technologies and Personalized Medicine

Artificial intelligence (AI)-driven personalization has demonstrated its transformative potential across multiple domains, including education [12], cultural learning [46], energy management [63], emotion recognition [58, 64], assistive technologies [6], and decision-making systems [62]. By dynamically adapting to individual user needs, AI personalization proves especially valuable in the biomedical field, where personalized medicine is becoming a central focus.

In the healthcare domain, AI-powered personalization facilitates the development of customized treatment plans, patient-specific diagnostics, and adaptive health monitoring systems. AI models can process vast amounts of patient-specific data, including genetic profiles, lifestyle behaviors, and medical history, to assess disease risks and recommend individualized therapeutic strategies. Additionally, AI-enhanced assistive technologies provide cognitive and functional support to individuals with disabilities or chronic illnesses [6]. Furthermore, emotion recognition models [58, 64] can be integrated into mental health applications [45, 47], delivering real-time emotional analysis and adaptive intervention mechanisms, thereby improving patient engagement and healthcare outcomes.

The integration of AI-driven personalization in biomedical science has the potential to enhance healthcare efficiency, optimize treatment outcomes, and empower patients by delivering targeted medical solutions based on their individual profiles. These technologies contribute to greater diagnostic precision, streamlined therapeutic protocols, and ultimately, improved patient care and quality of life.

The efficiency of a healthcare system is largely determined by its ability to anticipate and address the medical needs of populations, ensuring high-quality service delivery while optimizing financial, technological, and human resources [52]. Both medical science and healthcare economics rely on empirical data and statistical models to support evidence-based decision-making, enabling the design of socially equitable and medically beneficial interventions.

Over the past two decades, the healthcare industry has undergone a substantial shift. The traditional one-size-fits-all treatment paradigm has been replaced by a more individualized approach, where treatments are tailored to subpopulations with shared genetic or biological characteristics [1]. Personalized medicine has emerged as an alternative to conventional medical practices that often led to over-treatment or under-treatment. By refining therapeutic pathways, personalized medicine enhances patient outcomes while minimizing unnecessary medical interventions.

To facilitate comparative effectiveness research, methodologies such as health outcome measurements, quality-of-life assessments, economic evaluations, decision modeling, and clinical trials are employed. These evidence-building strategies support stakeholders by providing crucial insights into treatment efficacy, healthcare guidelines, and insurance coverage decisions.

Recognizing the importance of personalized medicine, the European Commission prioritized its implementation under the Luxembourg Presidency, officially adopting relevant conclusions on December 7, 2015. According to these policy directives,

"Personalized medicine refers to a medical model that utilizes the characterization of phenotypic and genotypic profiles (e.g., molecular markers, medical imaging data, and lifestyle information) to tailor treatment strategies to the right individual at the right time. This approach also facilitates disease predisposition assessment and enables targeted preventive interventions" [50].

In alignment with this initiative, the Council of the European Union urged Member States and the European Commission to assess existing legislative frameworks through the STAMP initiative to maximize the efficiency of European Union (EU) regulatory instruments. The Council also called for ongoing evaluation of the European Medicines Agency (EMA)'s adaptive approval mechanisms to expedite access to novel therapeutics for targeted patient groups with urgent medical needs.

As a direct outcome of these efforts, the ERA PerMed organization was established, uniting 32 research and funding institutions across 23 countries with co-financing from the European Commission. The primary objective of ERA PerMed is to align research strategies, foster collaboration among European entities, and strengthen the global competitiveness of EU stakeholders in the field of personalized medicine [23].

Clinical research

The personalization of treatment necessitates the integration of diverse health data, encompassing medical records, imaging diagnostics, and lifestyle information, to enable the implementation of precise and effective interventions.

Within the domain of clinical trials, several challenges must be addressed to facilitate innovation and methodological advancements. The adoption of novel, more adaptive clinical trial frameworks, along with the development of advanced statistical methodologies, alternative data generation techniques, and enhanced analytical tools, is essential for optimizing trial efficiency and improving evidence-based medical decision-making.

Approval

The joint development of medicines and medical devices presents significant challenges, primarily due to the varied regulatory frameworks across the European Union member states. These complexities necessitate streamlined approaches to ensure effective harmonization.

Notably, several regulatory pathways and incentives exist to facilitate the approval of such medicines. These include conditional authorizations, approvals under exceptional circumstances, and accelerated evaluations for cases where conventional development methods are not feasible. Additionally, orphan drugs, developed for rare diseases, may benefit from ten-year market exclusivity. Achieving international regulatory alignment could further enhance the cross-border acceptance of medical products and should be actively promoted.

Equally crucial is the education and training of healthcare professionals in the field of personalized medicine. A precise understanding of drug characteristics, genetic testing protocols, and mutation-specific formulations is fundamental for optimizing treatment effectiveness. For instance, the molecular profiling of pathogens is essential before prescribing antibiotics, as this practice minimizes multi-drug resistance and enhances treatment precision.

A coordinated approach among pharmaceutical companies, medical institutions, and regulatory authorities is imperative. Establishing quality control standards, drug efficacy thresholds, and regulatory frameworks will facilitate the rapid implementation of novel treatments.

However, individual patient responses to treatment vary significantly. While some individuals experience positive outcomes, others may see no improvement or suffer adverse effects, as illustrated in Fig. 2.1. These differences are largely influenced by genetic composition and metabolic profiles, which play a crucial role in drug efficacy. Personalized medicine accounts for these individual variations by leveraging biological markers to categorize patients into subgroups with similar characteristics, thereby guiding the selection of optimal therapeutic interventions.

On a broader scale, personalized medicine has the potential to revolutionize healthcare by enabling more effective interventions tailored to local populations based on their socioeconomic and lifestyle characteristics. A comprehensive understanding of age, gender, ethnicity, and lifestyle factors will lead to smarter resource allocation, ensuring that limited medical resources are utilized optimally. Achieving this

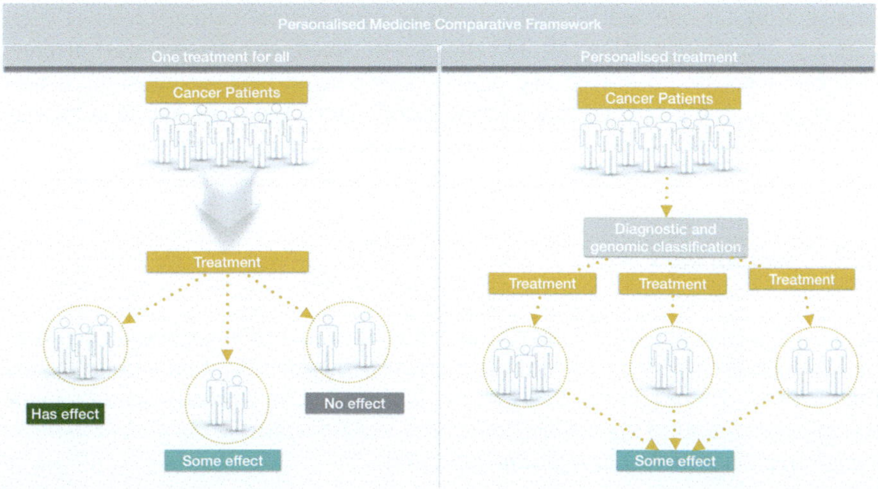

Fig. 2.1 Comparative framework

requires multidisciplinary collaborations and the active participation of patients in decision-making processes.

The rapid advancement of technology continues to provide an environment conducive to healthcare innovation. Genomic technologies, already a vital component of modern medicine, are expected to play an even more prominent role in the future. These technologies facilitate genetic profiling, allowing clinicians to identify the most effective drug regimens, minimize adverse reactions, and tailor treatments to individual patients.

The ultimate goal of personalized, predictive, and participatory medicine is to ensure that the right patient receives the right treatment at the right time. By doing so, it prevents the overprescription of expensive and ineffective medications, mitigates potentially harmful side effects, and enhances overall treatment efficacy.

At this point, it is important to establish the following assumptions:

A. Medicine is fundamentally an information-driven discipline.
B. A systematic approach is necessary to address the complex biological nature of diseases and their detailed analysis.
C. Mathematical modeling and increasing computational power enable the processing of vast datasets at speeds that facilitate real-time, actionable insights.

Despite its vast potential, the widespread implementation of personalized medical approaches is met with two primary challenges: the intricate biology of the human body and the economic complexities that govern the healthcare industry [3]. Access to basic healthcare remains a universal right, yet practical obstacles-including limited resources, high production costs of pharmaceuticals, and the extensive financial and time investment required for research-often hinder the widespread adoption of personalized medical strategies.

To successfully integrate personalized medicine into mainstream healthcare, it is essential to educate and involve patients in the decision-making process. Their awareness and understanding of the benefits associated with personalized medical treatments must be prioritized and actively encouraged [48]. Establishing a new medical paradigm, grounded in a revised social contract, will be key to achieving broader accessibility and adoption of these advanced medical solutions.

2.2.2 *Biomarkers*

The most effective approach to assessing patient health, refining disease classification, and optimizing treatment strategies involves the identification and utilization of biomarkers. These serve as biological indicators that, in a broad sense, *"provide objective indications of the medical state observed from outside the patient and can be measured accurately and reproducibly"* [59].

Biomarkers encompass a wide range of biological entities, including proteins, antibodies, gene expressions, and mutations, which can be identified in tissues, blood, or other bodily fluids. Their primary function is to indicate the presence, progression, or potential future onset of a specific medical condition.

Biomarkers are classified into two main categories: predictive and prognostic. They serve as monitoring tools, aiding in disease progression assessment, treatment evaluation, and medication response analysis [17]. The discovery of novel biomarkers holds the potential to enhance early cancer detection, refine disease management strategies, and enable more effective interventions. In recognition of their importance, the European Union and the U.S. Government have significantly increased funding for ongoing cancer biomarker research, as only a small subset of biomarkers currently have clinical applications.

In 1998, a working group at the American National Institutes of Health (NIH) formally defined biomarkers as follows:

"A characteristic that is objectively measured and evaluated as an indicator of normal biological processes, pathogenic processes, or pharmacological responses to therapeutic intervention" [40].

Cancer biomarkers play a crucial role in clinical decision-making, guiding diagnostic and therapeutic strategies based on three key principles:

1. Identifying who requires treatment (predictive biomarkers).
2. Determining how to treat (prognostic biomarkers).
3. Assessing treatment intensity (pharmacodynamic biomarkers).

In medical and paramedical fields, biomarkers answer three fundamental questions: how much, what, and for whom (Fig. 2.2). Many biomarkers are routinely measured in clinical practice, either individually or in combination, to detect early warning signs of disease in targeted populations. For instance, blood pressure, heart rate, and pulse are commonly used to assess cardiovascular health, while cholesterol and triglyceride levels provide insights into metabolic function and help evaluate the risk of coronary artery disease. Additionally, body mass index (BMI) and waist-to-hip ratio serve as indicators of obesity and metabolic disorders, whereas T-cell counts offer insights into immune system status. Similarly, cortisol levels, a steroid hormone, are frequently monitored in patients experiencing chronic stress [57].

Exceeding a specific threshold for an individual biomarker may indicate an elevated risk of developing a disease in the future [53]. Moreover, when combined with other clinical parameters from a patient's medical history, novel biomarkers can be developed to refine risk assessment models. For example, studies suggest that older men face increased health risks due to a combination of immune and neuroendocrine dysfunctions, whereas in older women, elevated systolic blood pressure, when analyzed alongside additional biomarker readings, is often linked to heightened health risks (Table 2.1).

While biomarkers play a role in nearly every medical specialty, they are particularly integral to oncology, where they have been systematically adopted in clinical applications. As illustrated in Fig. 2.3, over 50% of biomarkers in use have practical applications in cancer treatment, and a significant proportion of clinical trials incorporate biomarkers to guide therapy selection [14]. Genetic testing enables clinicians

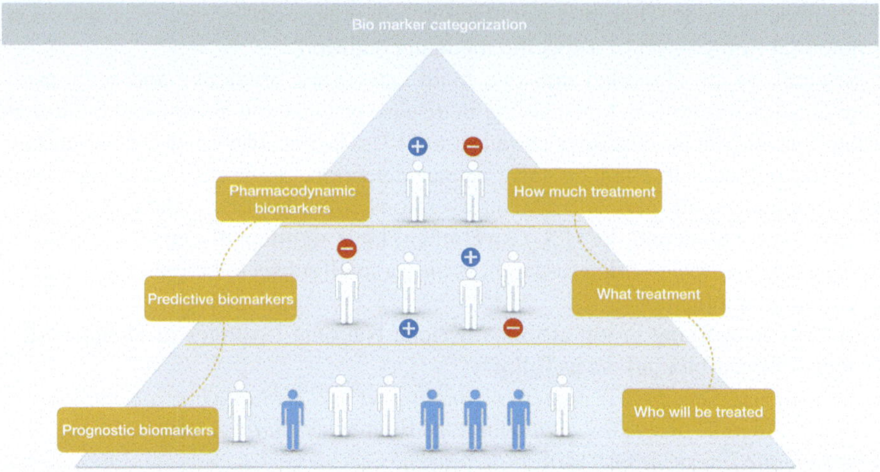

Fig. 2.2 Biomarker categories

Table 2.1 Biomarkers and health outcomes

Biomarkers	Description	Health outcome
C-reactive protein (CRP)	Marker of systemic inflammation and immune response	Increased risk of cardiovascular disease (CVD), chronic inflammatory diseases, and metabolic syndrome
Hemoglobin A1c (HbA1c)	Long-term marker of blood glucose regulation	Risk factor for diabetes, cardiovascular complications, and neuropathy
Serum creatinine	Indicator of kidney function and muscle metabolism	Associated with chronic kidney disease (CKD) and renal failure
High-density lipoprotein (HDL) cholesterol	"Good" cholesterol that helps remove excess cholesterol from the bloodstream	Lower levels associated with increased risk of cardiovascular disease (CVD) and metabolic disorders
Vitamin D (25-hydroxyvitamin D)	Essential for bone health, immune function, and metabolic balance	Deficiency linked to osteoporosis, autoimmune diseases, and increased susceptibility to infections

to predict disease susceptibility and detect specific genetic mutations. When combined with computational disease modeling, genetic analysis can enhance treatment selection and optimize therapeutic interventions [18].

The advancement of genomic medicine is expected to substantially improve cancer patient outcomes. A notable example is the Herceptin™(trastuzumab) therapy, which targets specific breast cancer biomarkers [17].

Fig. 2.3 Biomarker adoption

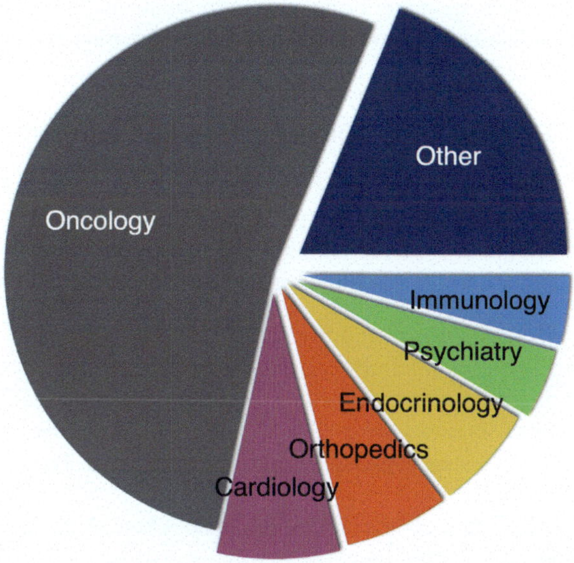

Multiple external factors, including age, ethnicity, gender, lifestyle, and socioeconomic status, influence the applicability and effectiveness of biomarkers [23].

For instance, in the United States, racial disparities contribute to increased exposure to stressful conditions such as discrimination in employment and housing. Chronic psychological stress has well-documented biological consequences, affecting immune response and overall health. Additionally, aging increases susceptibility to chronic diseases and disabilities. Long-term exposure to environmental stressors and social inequalities can negatively impact an individual's biological resilience over time.

By incorporating multi-dimensional health data, including genetic, environmental, and lifestyle factors, biomarker research can pave the way for more precise, equitable, and effective medical interventions.

Ideal Biomarker

An ideal biomarker should be defined by several key attributes, including accuracy, sensitivity, specificity, patient safety, and ease of use. A biomarker test must exhibit clinical validity, meaning it must reliably identify specific health conditions, such as determining which patients are at risk of disease recurrence, across independent patient cohorts [7, 10, 38]. The sensitivity of a biomarker reflects its ability to detect patients with a given disease [22], whereas its specificity indicates how well it can differentiate between those with and without the condition. Typically, an increase in sensitivity results in a reduction in specificity. Biomarkers should effectively classify patients based on treatment outcomes, such as distinguishing between responders and non-responders or healthy individuals and those at risk of premature mortality. Establishing the threshold values of a biomarker test is crucial, as it directly impacts treatment decisions, influencing whether a patient will (or will not) receive a particular intervention [13].

Several factors can influence biomarker sensitivity and specificity, including:

- Type of sample used (e.g., blood-based or tissue-based biomarker detection)
- Sample stability and required processing time
- Adequate implementation of negative control tests
- Patient profile (medical history and physiological characteristics)

Beyond clinical validity, biomarker tests must demonstrate clinical utility, meaning they should improve patient health outcomes compared to cases where no biomarker-guided decision-making is used. Typically, randomized clinical trials (RCTs) serve as the final phase in biomarker development, where patient assignment to treatment groups (e.g., Treatment A vs. Treatment B) is based on biomarker test results.

According to the U.S. Food and Drug Administration (FDA) [27], an ideal biomarker should meet the following criteria:

Specificity: It must correspond to a specific disease and distinguish between different physiological conditions.

Additionally, an effective biomarker should be:

- Safe and easily measurable
- Rapidly deployable, allowing for quicker diagnosis
- Cost-efficient, ensuring economic feasibility
- Highly accurate, minimizing misdiagnosis
- Consistent across diverse populations, considering ethnic and gender variations

The FDA's landmark study, titled "Innovation or Stagnation: Challenges and Opportunities for the Critical Path to New Medical Products", highlights concerns about the outdated methodologies used to assess modern biomedical advancements [41]. This issue is particularly evident in biomarker research. Janet Woodcock (FDA) remarked that while hundreds of potential biomarkers are identified each year, only a small fraction achieve clinical relevance due to the high costs and challenges in validating their utility [55]. She further emphasized that the biomarker development pipeline is fragmented, requiring a more structured bioinformatics-driven approach. By standardizing biomarker terminology and promoting computational modeling, researchers can establish quantitative disease models, facilitating simulation-based validation [55].

Evaluating biomarkers and integrating them into clinical frameworks is a complex process. Simultaneously, there is a growing need for medical innovations that provide both clinical and economic value. One of the ongoing dilemmas in clinical practice revolves around weighing costs against benefits. For instance, early detection of prostate cancer using the prostate-specific antigen (PSA) test can be life-saving for some patients (Fig. 2.4). However, the test is not universally beneficial, as most

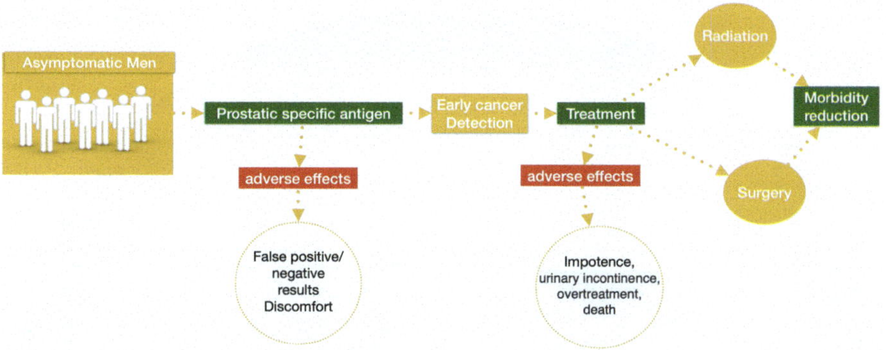

Fig. 2.4 PSA diagnosis and treatment mapping

screened individuals will not experience a substantial survival advantage. Studies indicate that to prevent a single prostate cancer-related death within an eight-year period, approximately 1,000 men must undergo PSA screening [31]. This results in numerous patients experiencing unnecessary anxiety, false positives, and potential complications such as erectile dysfunction, incontinence, and bowel dysfunction.

From a healthcare policy perspective, an ideal biomarker should fulfill the following objectives:

- Aid in clinical decision-making
- Enhance patient outcomes and quality-adjusted life years (QALY)
- Ensure sustainability within healthcare systems
- Be accessible at both national and global levels

The ACCE model (Fig. 2.5), initially developed by the Center for Disease Control and Prevention (CDC), provides a structured approach for evaluating biomarkers and genetic tests [25]. This model is widely applicable across various diagnostic tests and offers a holistic framework for assessing biomarkers prior to their integration into clinical practice. A major advantage of the ACCE framework is that it encompasses both clinical validity and utility, ensuring that biomarkers undergo rigorous evaluation before widespread adoption.

The ACCE framework consists of four fundamental elements:

- Analytical validity: Measures the accuracy and reproducibility of biomarker test results.
- Clinical validity: Determines how well the test can differentiate between individuals who have or will develop a condition and those who remain disease-free.
- Clinical utility: Assesses risks and benefits, ensuring the biomarker improves patient care and clinical outcomes.
- Ethical, legal, and social implications (ELSI): Evaluates concerns related to privacy, regulatory safeguards, and ethical considerations associated with biomarker testing.

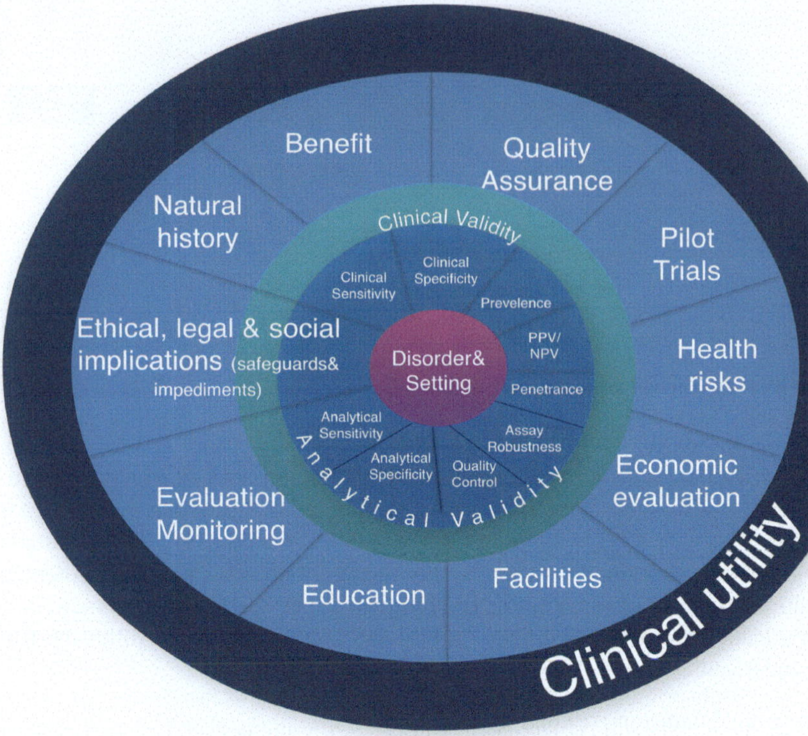

Fig. 2.5 ACCE framework

2.2.3 Validation and Evaluation of Novel Bio Markers

The discovery and validation of biomarkers is a lengthy and intricate process, which explains why only a limited number of biomarkers have successfully transitioned into clinical applications [15]. However, some biomarkers have proven highly effective, such as HER2, a genetic and protein biomarker extensively used in breast cancer diagnostics and treatment.

The development of HER2 as a biomarker dates back to 1984, when Weinberg et al. [65] identified the human epidermal growth factor receptor 2 (HER2/neu) gene. Following years of extensive research, the first HER2-based biomarker treatment was introduced in 1998, marking a significant milestone in breast cancer management.

Nevertheless, the clinical success of HER2 also highlights the complexities of tumor biology. It was observed that some patients with HER2-positive tumors exhibited primary resistance to trastuzumab, a HER2-targeted therapy. This finding indicated that simply inhibiting a tumor protein does not guarantee disease eradication, as the underlying tumor biology is more complex [50]. As a result, combination therapies incorporating trastuzumab and HER2-targeted treatments have been explored

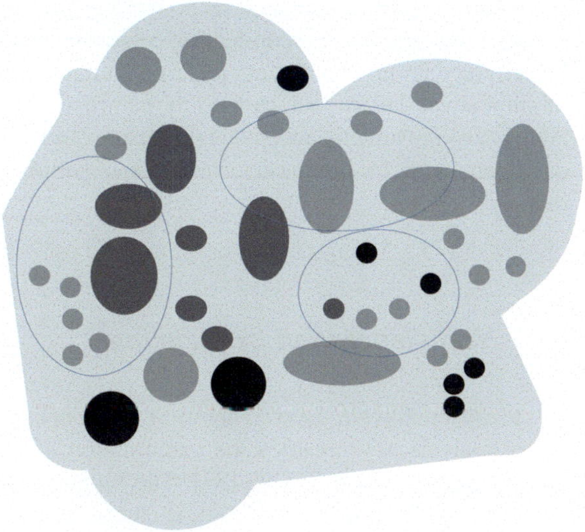

Fig. 2.6 Complexity of tumor sampling

by both the U.S. Food and Drug Administration (FDA) and the European Medicines Agency (EMA) [8]. These combination therapies are now used in both metastatic breast cancer treatment and as adjuvant therapies (administered post-surgery to prevent recurrence) in breast cancer patients.

Figure 2.6 illustrates the complexity of tumor heterogeneity, where different biomolecular elements are distributed unevenly throughout the tumor. The black elliptical circles represent various tissue sampling regions for biomarker testing. Traditional molecular biomarkers focus solely on detecting a specific biomolecular component within a designated sampling area. However, advanced composite biomarkers, which aggregate the expression patterns of multiple molecules, provide a more comprehensive analysis of tumor heterogeneity, thereby enhancing the reliability of diagnostic outcomes [21].

With the advancement of multi-omic approaches, it is now possible to transcend single-gene or single-protein analysis and instead integrate large-scale, multi-level biological datasets spanning genomics (DNA), epigenetics, and metabolomics. This shift has paved the way for collaborative research initiatives, such as The Cancer Genome Atlas Network, which aims to aggregate and analyze large-scale multi-omic data across diverse cancer types.

Despite the progress in biomarker research, several challenges remain in their evaluation and implementation [36]. Key barriers include:

- The increasing scientific complexity of biomarkers, necessitating specialized knowledge beyond standard medical training to accurately interpret results.
- Insufficient clinical performance data, as biomarker validation studies are often costly, time-intensive, and require large-scale trials to establish their efficacy and utility.

- Ambiguity regarding financial responsibility for funding biomarker studies, generating reliable data, and conducting performance evaluations.
- Uncertainty among biomarker and test manufacturers regarding the necessary level of clinical validation data, leading to inconsistencies in regulatory standards.
- Absence of universally accepted consensus standards or a structured quality assurance framework for biomarker evaluation, resulting in fragmented validation practices.

2.2.4 Next Generation Sequencing

With advancements in next-generation sequencing (NGS) technology, modern genome analysis now enables the examination of hundreds of millions of DNA fragments per sequencing cycle [33]. The sequencing of the human genome using NGS methods generates up to 100 times more data than earlier approaches, such as the Sanger method [61]. This has revolutionized genomic research, allowing scientists to analyze entire genomes and identify complex genetic mutations associated with various diseases. Given that cancer is fundamentally a genomic disease, it is not surprising that NGS technology has been widely adopted in cancer research, offering unprecedented insights into carcinogenesis and tumor evolution. Genome-based cancer diagnostics hold immense potential for improving early detection, risk assessment, and personalized treatment strategies [20].

NGS operates by fragmenting DNA into smaller sequences and performing massively parallel sequencing on these fragments. The process generates multiple overlapping sequences, known as "reads", which are assembled into a continuous genome sequence. To minimize sequencing errors, each region of the genome undergoes multiple sequencing iterations, ensuring accuracy. This sequencing approach is based on the assumption that genomic DNA fragments are randomly distributed and independent of sequence context.

NGS technology involves three primary phases:

1. Template Preparation—This step involves fragmenting genomic DNA into smaller sequences to create a representative and unbiased library. The fragments are immobilized on a solid substrate or confined within nanoparticle chambers for parallel sequencing reactions.
2. Sequence Analysis—The sequencing process employs various reaction principles, including cyclic reversible termination, single-nucleotide addition, real-time sequencing, and ligase-based sequencing [56].
3. Data Processing and Mutation Detection—Sequencing outputs undergo bioinformatics analysis to identify genetic variants, mutations, and structural alterations relevant to disease processes.

The genome of a living organism functions like a library of encoded information, where chromosomes are analogous to books, and nucleotide sequences (A, T, C, G) form the fundamental genetic instructions [11]. Early sequencing techniques required precise excision of specific DNA segments for direct sequencing analysis. In 1979, an alternative fragmentation-based sequencing method was introduced, proposing that multiple copies of an entire genome could be randomly fragmented and sequenced independently. These sequences ("reads") were then assembled based on overlapping edges.

With the rise of automated sequencing technologies and advances in computational biology, this fragmentation-based approach-now known as shotgun sequencing-became the dominant method. Modern sequencing platforms are now capable of generating hundreds of millions of sequencing reads per day, each consisting of tens to hundreds of nucleotides, making high-throughput genome analysis a reality.

There are two primary laboratory-based sequencing strategies: - Whole Genome Sequencing (WGS): This method analyzes the complete genomic structure, including coding and non-coding regions, providing a high-resolution genetic profile. - Targeted Sequencing: This technique selectively focuses on specific genomic regions of interest, offering a cost-effective alternative to WGS. Targeted sequencing requires lower DNA input, faster processing, and simpler library preparation, making it particularly useful for focused research applications [24].

The continued evolution of NGS technology has significantly influenced the development of new diagnostic and therapeutic pathways in oncology [8]. By enabling comparative genome analysis, researchers can identify key genetic variations in tumor genomes, leading to more precise cancer classifications [60].

The application of NGS in cancer genomics offers several advantages: - Rapid identification of genetic alterations in solid tumors for personalized treatment strategies [48]. - Non-invasive biomarker detection, such as tumor-specific gene rearrangements in plasma samples. - Development of digital PCR assays targeting genomic rearrangements, allowing for sensitive detection of residual disease following treatment completion [34]. - High-throughput NGS (HT-NGS) techniques facilitate the identification of mutations driving cancer metastasis, enabling more effective oncological interventions [32].

Both RNA and DNA are nucleic acids essential for genetic expression and regulation. Alongside lipids, proteins, and carbohydrates, these macromolecules are fundamental to all life forms. Ribonucleic acid (RNA) plays a pivotal role in encoding genetic instructions, regulating gene expression, and directing protein synthesis. Unlike DNA, RNA is typically single-stranded and exists in various forms, including messenger RNA (mRNA), which transmits genetic information to cellular machinery for protein production. Many viruses also utilize RNA genomes for replication.

A promising avenue in biomarker research is microRNA (miRNA), small non-coding RNA molecules that regulate protein translation and are present in blood plasma. NGS-based approaches could be leveraged to profile miRNA expression patterns in tissue and plasma samples, enabling the identification of novel biomarker signatures for disease detection.

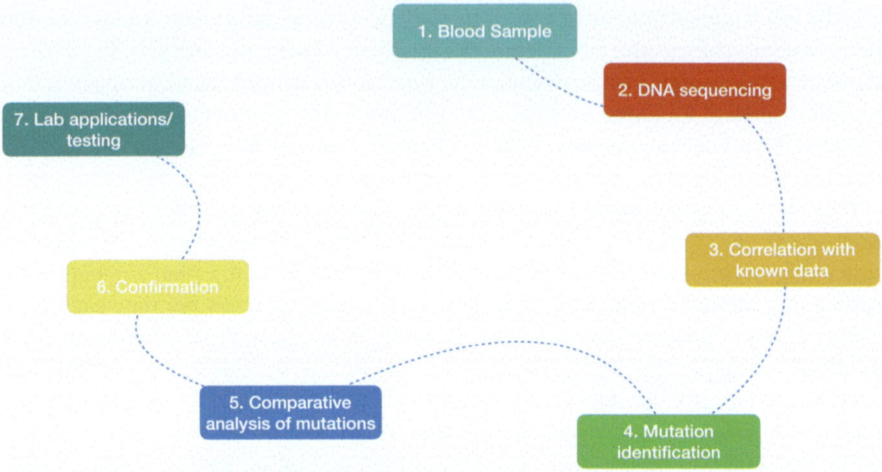

Fig. 2.7 Mapping of the DNA sequencing analysis

Figure 2.7 provides an overview of the genomic sequencing workflow, illustrating the steps involved in analyzing, validating, and integrating sequencing data with existing clinical datasets [7]. The final stage of analysis involves assessing the clinical relevance of identified mutations to determine their diagnostic and therapeutic implications [51].

2.2.5 Metabolic Syndrome and the Global Burden of Obesity

Obesity is increasingly recognized as a major global health concern, affecting not only high-income Western nations but also low- and middle-income countries. In order to formulate effective weight management strategies, it is essential to explore the molecular mechanisms underlying weight fluctuations. Several biological pathways have been implicated in weight gain and energy balance regulation, including oxidative stress responses and insulin signaling networks. Additionally, metabolites derived from gut microbiota, particularly short-chain fatty acids (SCFAs), have been identified as key contributors to metabolic homeostasis and obesity risk [19].

Obesity, or a genetic predisposition toward excessive weight gain, is widely regarded as a precursor to multiple severe health conditions. This concept is not novel; Hippocrates famously observed that "sudden death is more common in those who are naturally fat than in the lean". The strong correlation between obesity and type 2 diabetes (T2D) has been extensively documented across diverse populations, both male and female, and across ethnic groups. A pivotal discovery in diabetes research is that insulin resistance serves as a core metabolic dysfunction in individuals with T2D, a condition that is also highly prevalent among obese individuals. Evidence suggests that over 90% of diabetics are either overweight or obese [4, 37].

Further research indicates that genetic factors may play a significant role in obesity susceptibility. Genome-wide linkage analyses have identified tumor necrosis factor-alpha (TNF-a) as a key determinant of body fat percentage. Other studies have confirmed that fat cells actively produce TNF-a, linking it to obesity pathophysiology.

Obesity arises primarily from prolonged caloric excess relative to energy expenditure. While multiple biological and environmental factors contribute to body weight homeostasis, the development of obesity triggers metabolic adaptations that may persist even after weight loss, making long-term weight management challenging [13].

Metabolic syndrome, a cluster of conditions associated with obesity, insulin resistance, dyslipidemia, and hypertension, contributes to a substantial proportion of morbidity and mortality worldwide [16, 39]. While historically regarded as a disease of affluence, obesity is now an emerging epidemic in low- and middle-income countries. Nations such as India face a dual burden of malnutrition, with widespread undernutrition in rural areas and a rapidly increasing prevalence of obesity-related disorders in urban populations.

Globally, nearly 40% of adults are classified as overweight, while an additional 13% meet the criteria for obesity, highlighting the urgent need for scalable interventions to address this escalating public health challenge [29].

2.2.6 Nutritional Biomarkers

2.2.6.1 Metabolism and Metabolic Dynamics

The metabolic profile refers to the systematic investigation of metabolites, which serve as unique chemical signatures of cellular activity. By analyzing these metabolites, researchers can obtain a snapshot of cellular physiology, providing a functional assessment of an organism's physiological state.

Research in metabolomics has revealed a strong correlation between type 2 diabetes (T2D) and obesity, as evidenced by distinct metabolite patterns. Studies have identified higher levels of several metabolites in individuals with obesity and diabetes, including branched-chain amino acids (BCAAs), glutamine, proline, cysteine, tyrosine, threonine, phenylalanine, tryptophan, pantothenic acid, and choline. Conversely, asparagine, citrulline, and methionine tend to be found in lower concentrations in these individuals. A cross-sectional study involving 947 participants found that 37 metabolites were significantly associated with body mass index (BMI), including nineteen lipids, twelve amino acids, and six other metabolic compounds. Among these, eighteen metabolites had not been previously reported, including histidine and butyrylcarnitine [41].

The field of genomics focuses on DNA and genetic studies [11] (Fig. 2.9), while transcriptomics investigates mRNA expression patterns. In contrast, metabolomics examines the metabolic end products influenced by both genetic and environmental factors, providing a direct measurement of metabolite concentrations. This enables a comprehensive understanding of the molecular phenotype [54].

Metabolomics employs advanced analytical technologies [10] that facilitate:

- Metabolic fingerprinting—Diagnostic pattern recognition based on metabolite profiles.
- Absolute quantification—Precise measurement of targeted metabolites through pooled sample analysis.
- Metabolite profiling—Relative quantification of large metabolite subsets to assess metabolic shifts.
- Flux analysis—Tracking the biochemical transformations of specific metabolites.

The two most widely used techniques in metabolomics research are mass spectrometry (MS) and nuclear magnetic resonance (NMR), both of which provide high-throughput metabolite identification [28].

A metabolite is a low-molecular-weight organic compound that participates in biological processes, either as a substrate, intermediate, or final product. Metabolomic studies typically focus on small molecules within a mass range of 50–1500 daltons (Da).

Examples of common metabolite classes (Fig. 2.8) include:

- Sugars (carbohydrates)
- Lipids (fats and fatty acids)
- Amino acids (protein building blocks)

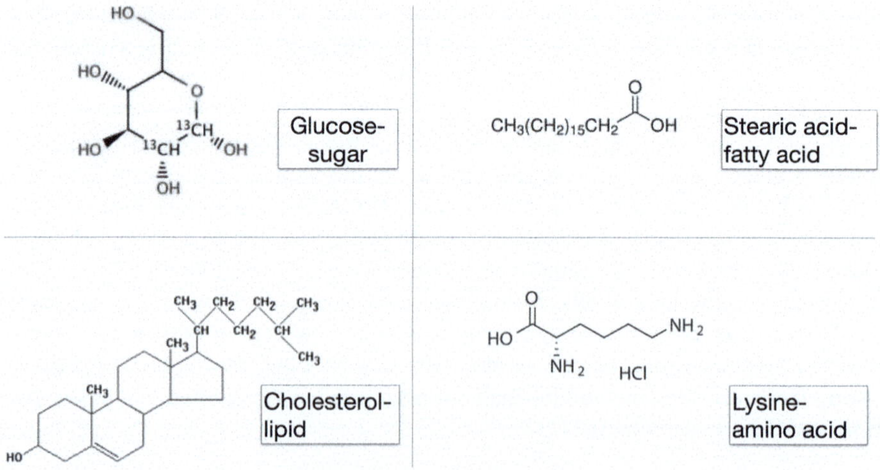

Fig. 2.8 Metabolites

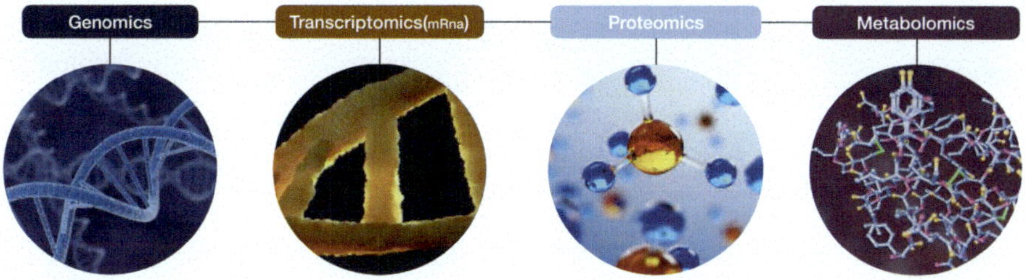

Fig. 2.9 The omics space

- Phenolic compounds (antioxidants and secondary metabolites)
- Alkaloids (bioactive nitrogen-containing compounds)

There is considerable variation in metabolite composition across different species [2]. It is estimated that the plant kingdom contains approximately 200,000 unique metabolites, with individual plant species harboring between 7,000 and 15,000 metabolites. In contrast, humans possess around 3,000 endogenous or common metabolites. However, these figures may be underestimations, as many low-abundance molecules remain undetected due to technological limitations. Notably, plants exhibit far greater biochemical diversity than most other eukaryotic organisms and often contain a larger number of genes.

Metabolism encompasses the complete set of metabolites present in a cell, tissue, or biological sample at a specific moment in time. This biochemical network is highly dynamic, as small molecules are constantly absorbed, synthesized, transformed, and degraded. These metabolites also interact with other biological systems and environmental factors, shaping overall physiological function.

Figure 2.10 presents an overview of core metabolic reactions, as adapted from the Reactome database, highlighting key pathways involved in cellular metabolism.

Nutritional Genomics

Nutritional genomics applies advanced functional genomic technologies to the field of nutrition research, allowing for a high-throughput analysis of gene-nutrient interactions. By integrating these technologies with genomic sequence databases, researchers can investigate gene expression patterns across thousands of genes simultaneously. This approach facilitates the determination of optimal dietary strategies at the population level, for specific demographic groups, or on an individual basis.

Diet plays a crucial role in chronic disease prevention and overall health, and the study of nutritional biomarkers may enable the identification of bioactive compounds present in food [38]. Understanding these bioactive interactions could lead to improvements in public health, including dietary modifications, the development of functional foods, and personalized nutrition strategies. However, significant chal-

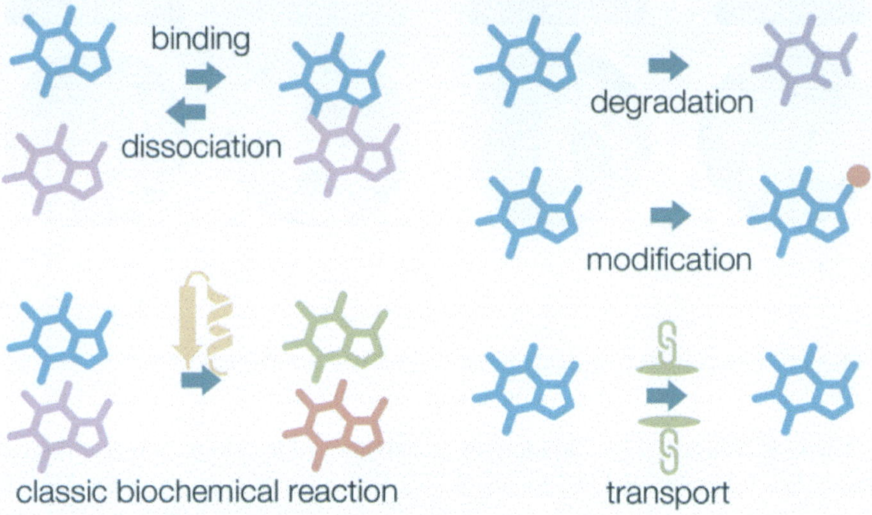

Fig. 2.10 Metabolic reactions (Reactome database)

lenges remain, particularly in the design of nutrition studies and the management of extensive genomic and metabolic datasets.

With advances in genetic analysis, it is now possible to identify polymorphisms that may predispose individuals to specific diseases, thereby allowing for dietary recommendations tailored to genetic profiles [51]. The characterization of gene polymorphisms could enable targeted nutritional interventions and personalized counseling for at-risk populations [66].

Standard Biochemistry Profile and Semantic Grouping

For this study, datasets related to biochemical profiles obtained from targeted blood tests were utilized. The Standard Biochemistry Profile (SBP) is an essential diagnostic tool that assesses the functional capacity of major organs, including the liver and kidneys. Additionally, it provides insights into protein and lipid metabolism, carbohydrate processing, enzyme production, and the body's ability to utilize key compounds (Table 2.4).

The Standard Biochemistry Profile (SBP) encompasses three primary laboratory panels, each providing targeted biochemical assessments:

- Comprehensive Metabolic Panel (CMP): Measures blood levels of albumin, blood urea nitrogen (BUN), calcium, carbon dioxide, chloride, creatinine, glucose, potassium, sodium, total bilirubin, total protein, and liver enzymes. It is frequently used to diagnose diabetes and assess metabolic and renal function.

- Liver (Hepatic) Panel: Evaluates liver functionality by measuring total protein, albumin, bilirubin, and liver enzyme levels, including aspartate aminotransferase (AST), alanine aminotransferase (ALT), and alkaline phosphatase (ALP).
- Lipid Panel: Assesses cholesterol and lipid metabolism, measuring total cholesterol, low-density lipoprotein (LDL), high-density lipoprotein (HDL), and triglycerides to evaluate cardiovascular risk factors.

By analyzing the relationships between biochemical parameters and organ functionality, a semantic grouping framework has been established, as illustrated in Fig. 2.11. Semantic similarity and relatedness are crucial in understanding how medical concepts are interconnected. These methodologies enhance biomedical data clustering, clinical variable classification, and the development of biomedical ontologies, contributing to the advancement of precision medicine and medical informatics [35].

2.3 Neural Networks, Pattern Recognition and Datasets

Neural networks address the classification problem for non-linearly separable datasets by leveraging hidden layers, where neurons are not directly connected to the output layer. These additional hidden layers can be interpreted as higher-level feature extractors, enhancing the network's capacity to separate complex patterns. By connecting artificial neurons through nonlinear activation functions, neural networks are able to generate complex decision boundaries, enabling them to effectively handle problems where categories cannot be linearly separated.

Neural network training is performed using a supervised learning approach. The fundamental concept involves presenting input vectors to the network, computing forward propagation to generate layer-wise outputs, and ultimately producing the final network output. The desired output is predefined, allowing for the adjustment of weights using optimization techniques.

In the case of the backpropagation (BP) algorithm, the gradient descent method is employed to iteratively minimize the error between predicted and actual outputs. The weight adjustments in hidden layers are determined by propagating the error backwards, ensuring that updates are applied according to the network's connection structure. This iterative process continues for each training sample, and a complete cycle through the dataset is referred to as an epoch. The total number of epochs required for convergence depends on multiple factors, particularly the magnitude of the computed error at the output level.

Figure 2.12 illustrates a schematic representation of the described neural network architecture. The configuration used in this figure reflects the settings deployed for training the dataset in this study.

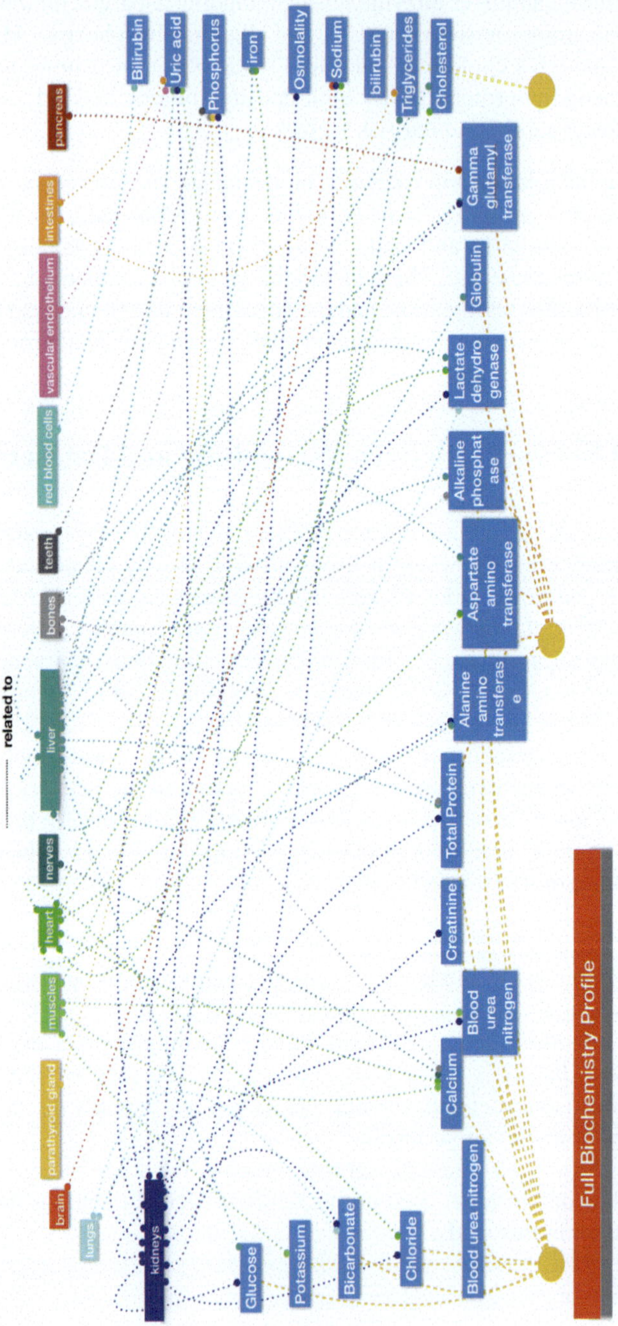

Fig. 2.11 Grouping by relation to organ

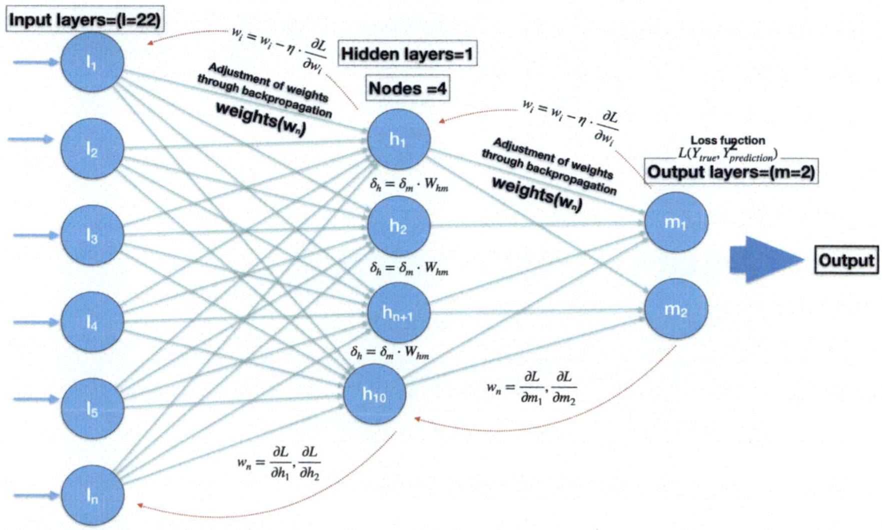

Fig. 2.12 Feedforward neural network

2.3.1 Methods

For the purposes of model training and evaluation, Python libraries were utilized, adhering to established best practices in machine learning model design. The following libraries and methodologies were employed to define and optimize the objective function of the neural network:

- Scikit-learn [49]: A widely used open-source machine learning library that provides modules for classification, regression, clustering, and dimensionality reduction. It supports various models, including support vector machines (SVMs), random forests, gradient boosting, k-means clustering, and DBSCAN, and is designed to integrate seamlessly with NumPy and SciPy.
- Cross-Validation: In classification problems, models are trained on known datasets and tested on unseen data to assess performance. The purpose of cross-validation is to evaluate the generalization capability of a model, mitigating issues such as overfitting and selection bias.
  A cross-validation cycle involves partitioning the dataset into complementary subsets, where one subset is used for training and another for validation. To improve reliability, multiple validation rounds are conducted using different partitions, and the final validation results are aggregated (averaged) to estimate the model's predictive performance (Fig. 2.13).
- StandardScaler: Normalizes features by scaling them to have a mean of 0 and a standard deviation within a range of -1 to 1, ensuring numerical stability.
- Random State: Controls the random shuffling of data before splitting into training and testing sets, ensuring reproducibility.

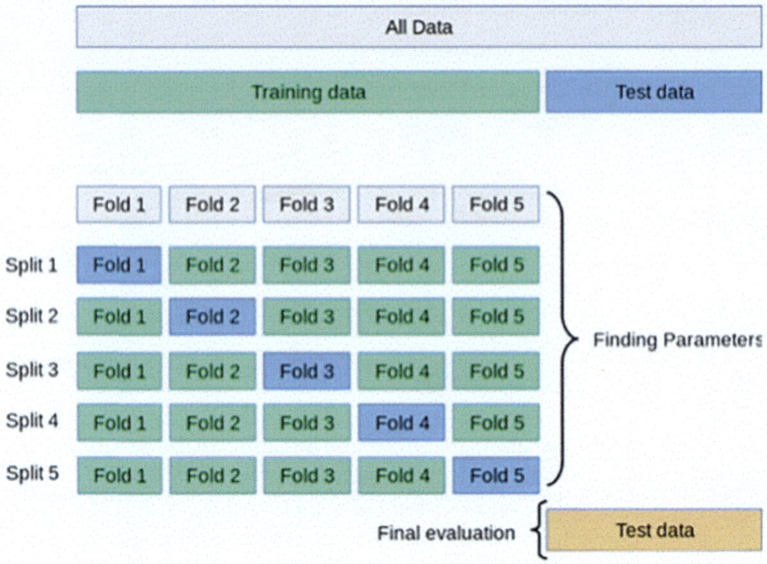

Fig. 2.13 Example of using cross validation with five-fold data separation and model evaluation

- Solver: Optimization algorithm responsible for minimizing the loss function during model training.
- Hidden Layer Size (Tuple, Length = n_layers - 2, Default = (100)): The 'n' element specifies the number of neurons in each hidden layer, allowing flexibility in network depth.
- Activation Functions: Define the non-linearity applied to neuron outputs. Common activation functions include:
  - Identity Function—A linear activation function, expressed as: $f(x) = x$ Suitable for linear regression tasks.
  - Logistic (Sigmoid) Function—Defined as: $f(x) = \frac{1}{1+e^{-x}}$ Often used for binary classification problems.
  - Tanh Function—Defined as: $f(x) = \tanh(x)$ Maps input values to a range between -1 and 1, providing advantages in gradient-based learning.
  - ReLU (Rectified Linear Unit)—Defined as: $f(x) = \max(0, x)$ A widely used activation function due to its computational efficiency and ability to mitigate the vanishing gradient problem.
- Alpha: Regularization parameter (L2 penalty) that controls overfitting by discouraging excessively large weight values.
- Max_Iter: Defines the maximum number of iterations for training. The optimization algorithm continues looping until convergence is achieved.
- RadViz Visualizer: A multi-factor data visualization algorithm that positions each feature dimension along the circumference of a circular plot and projects data

points inside the circle. The positioning of each point reflects the influence of multiple variables, making it useful for analyzing high-dimensional feature spaces.

2.3.2 *Dataset and Feature Engineering*

The datasets utilized in this study were sourced from the National Health and Nutrition Examination Survey (NHANES), a comprehensive research program designed to assess the health and nutritional status of both adults and children in the United States. NHANES is unique in that it combines both survey interviews and physical examinations, providing a holistic perspective on public health trends. This program is an essential initiative under the National Center for Health Statistics (NCHS), which operates as part of the Centers for Disease Control and Prevention (CDC), tasked with generating vital national health statistics.

The NHANES program was initially introduced in the early 1960s, structured as a series of health-focused surveys targeting specific population groups. However, in 1999, it transitioned into a continuous program, adapting to emerging health and nutrition concerns through periodic updates in data collection methodology. Each year, a nationally representative sample of approximately 5,000 individuals is selected to participate in the study [61].

For this research, all available NHANES datasets from 2001 to 2016 were retrieved. These datasets encompass a wide range of variables, including demographics, physiological tests (e.g., blood work), dietary records, laboratory test results, and medical and pharmacological histories. Further details regarding the datasets can be found in the appendix.

The proposed system integrates neural networks and machine learning techniques to establish a functional pipeline for personalized nutritional recommendations. The goal is to utilize blood biomarkers and metabolic signatures to automate the identification of nutritional associations.

The underlying assumption of our model is that observational data—such as medical history, dietary habits, and lifestyle factors—along with laboratory test results, contain valuable insights into an individual's health status. Since weight is a key health indicator, BMI (Body Mass Index) is selected as the target variable for the deep neural network. The Standard Biochemistry Profile (SBP) test results will be used as input features for model training.

The BMI dataset is formatted into an x*2 matrix, where individuals are categorized based on the widely accepted BMI classification ranges:

$if\ BMI > 30$ then Obese

$ElseIf\ BMI$ between 18.5 and 25 then Normal

$ElseIf\ BMI$ between 25 and 30 then Overweight

To systematically assess the model's performance, the study was conducted in three experimental cycles: 1. First cycle—Evaluating model performance in distinguishing normal-weight versus overweight individuals. 2. Second cycle—Evaluating model performance in distinguishing normal-weight versus obese individuals. 3. Third cycle—Evaluating model performance in distinguishing overweight versus obese individuals.

The results from each cycle were analyzed to refine model performance, leading to further adjustments in feature selection and optimization techniques. While medical literature served as a guiding reference for feature selection, the primary objective was to develop a system capable of independently identifying and evaluating patterns in health data.

Obesity is often identified as a risk factor for various chronic health conditions, while underweight status may indicate malnutrition, underlying illnesses, neurological disorders, or psychological conditions. In this study, the primary focus is on identifying the ideal weight range by classifying individuals into two categories at a time. The aim is to detect patterns in biochemical markers that differentiate these weight classes.

A correlation algorithm was applied to eliminate irrelevant features, ensuring that each input variable contributes meaningfully to classification accuracy. The objective was to streamline the model while preserving high predictive accuracy, ultimately making the system more efficient, precise, and scalable.

The proposed system employs an artificial neural network (ANN)-based model capable of classifying individuals based on biochemical markers derived from blood tests. By establishing a direct link between blood test parameters and dietary profiles, this model aims to recognize nutritional patterns across different weight categories.

The neural network design consists of two primary phases and an optimization intermediary, as illustrated in Fig. 2.14.

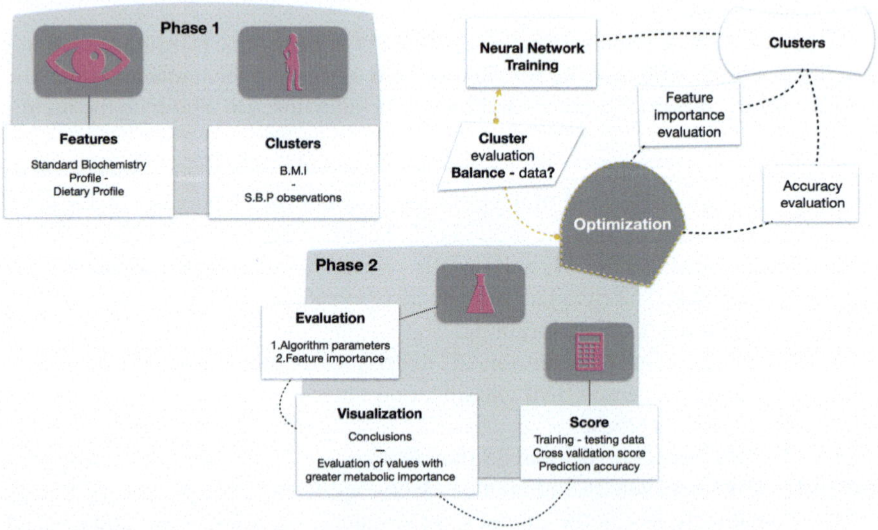

Fig. 2.14 Conceptual analysis

2.4 Application and Evaluation of Methods

2.4.1 Deep Backpropagation Neural Network (DBPNN)

A Deep Backpropagation Neural Network (DBPNN) was implemented for this study. As previously stated, Python (3.7) and the Scikit-learn (sk-learn) library were utilized to ensure scalability, reusability, flexibility, and integration with other frameworks. The architecture of the neural network is illustrated in Fig. 2.15.

For the deployment of the model and evaluation of its predictive performance, BMI data was structured into an x*2 matrix and categorized into three binary classification tasks:

- Overweight versus Normal weight (First classification cycle)
- Obese versus Normal weight (Second classification cycle)
- Obese versus Overweight (Third classification cycle)

To prevent class imbalance bias, the dataset was balanced accordingly. The NHANES dataset used in this study represents a diverse sample of the U.S. population, including men and women across different age groups and ethnic backgrounds [5].

The properties and hyperparameter settings of the most accurate trained model are listed below. During each training phase, the feature importance of input variables was evaluated, and irrelevant features were removed. This feature selection process ensured that the final model contained only the most informative features, leading to improved accuracy and predictive performance.

Several model variations were tested by incorporating different feature sets:

- Standard Biochemistry Profile (SBP)
- SBP + Smoking Status

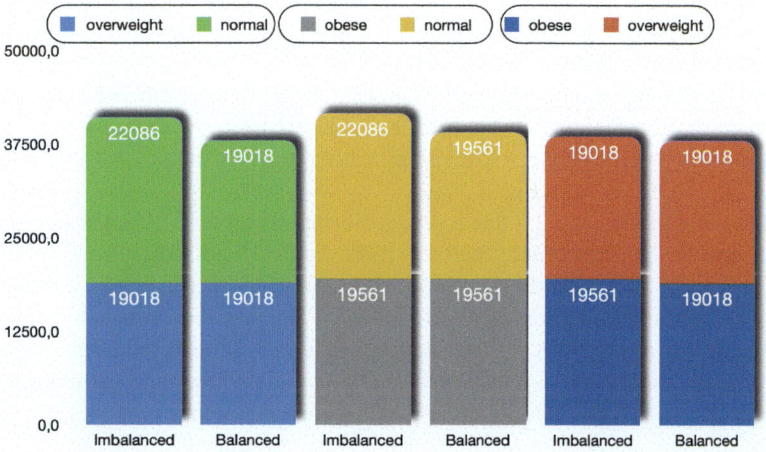

Fig. 2.15 Data analysis workflow

- SBP + Cholesterol Levels
- SBP + Cholesterol + Glycohemoglobin (HbA1c)

However, no significant improvement in predictive accuracy was observed with additional features. Therefore, the SBP-only model was chosen due to its higher accuracy and lower computational cost.

2.4.2 Standard Biochemistry Profile Neural Network (SBPNN)

Data Preprocessing and Analysis

The dataset was analyzed in binary classification groups, as depicted in Fig. 2.16, which shows the corresponding population distribution for each class. Null values were removed, and the dataset was balanced by aligning the sample size of the minority class with that of the majority.

A correlation algorithm was employed to evaluate feature relationships with BMI. This process removed duplicate features, defined as multiple measurement methods for the same biochemical parameter that exhibited identical or near-identical correlation with BMI (BMXBMI). The final set of features used for training the Dietary Profile Neural Network (DPNN) is presented in Fig. 2.16.

Model Performance and Classification Accuracy

The deep neural network was trained and evaluated across three classification cycles, with the accuracy metrics reported below:

(a) Classification Cycle 1: Overweight versus Normal weight - A 10-layer deep neural network was deployed. - The model iteratively removed less important features based on computed feature importance scores. - The final training accuracy was 0.69, and the testing accuracy was 0.68. - This accuracy was achieved on the third iteration, after eliminating four features that exhibited zero or negative contribution to classification performance. - The classification performance is illustrated in Fig. 2.17, and the ROC curve in Fig. 2.17.

(b) Classification Cycle 2: Obese versus Normal weight - A 10-layer deep neural network was used. - The model iteratively removed irrelevant features and achieved a training accuracy of 0.81 and testing accuracy of 0.80. - The true positive rate (TPR) and model evaluation are presented in Fig. 2.17. - This accuracy was achieved on the second iteration, after removing four features that had minimal predictive value.

(c) Classification Cycle 3: Obese versus Overweight - The same training methodology was applied. - The classification performance is shown in Fig. 2.17.

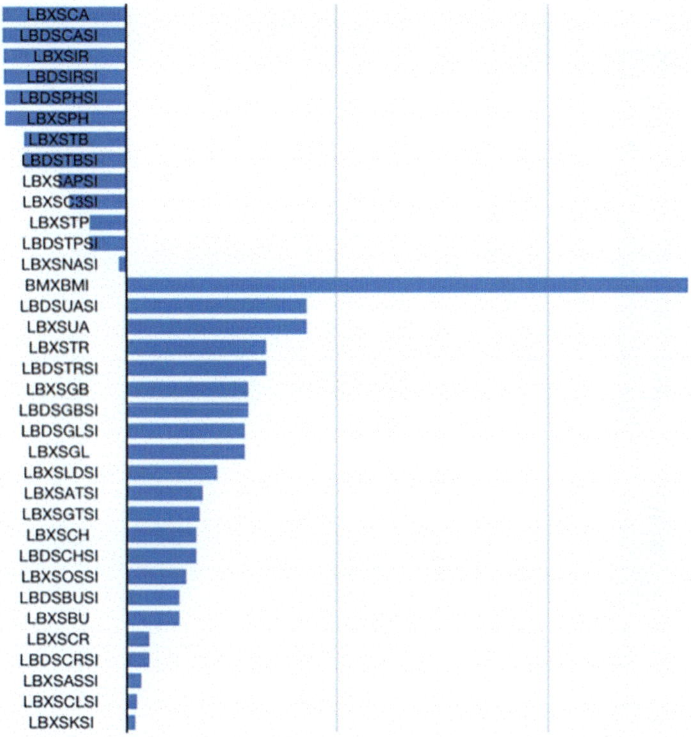

Fig. 2.16 Feature correlation with BMI

2.4.2.1 Feature Importance Analysis

Permutation feature importance is a model inspection technique useful for non-linear models, particularly opaque estimators like deep neural networks. This method assesses feature significance by randomly shuffling individual feature values and measuring the resulting drop in model accuracy. The greater the performance drop, the more critical the feature.

This technique was applied to identify the most influential biochemical features for predicting BMI-based classifications. Features were grouped into two primary categories:

- Normal Range
- Out-of-Normal Range

A comparison of feature importance scores (Fig. 2.18) revealed that three features consistently ranked among the top five predictors across all classification cycles:

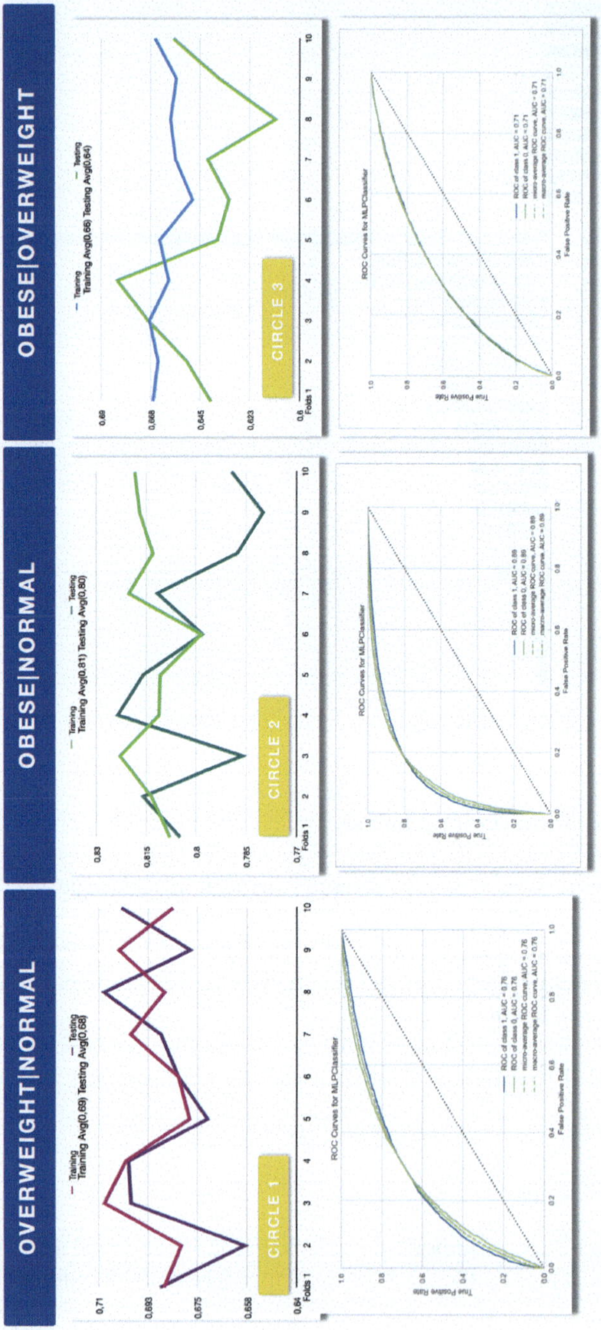

Fig. 2.17 Cross-validation and ROC curves for classification cycles 1, 2, and 3

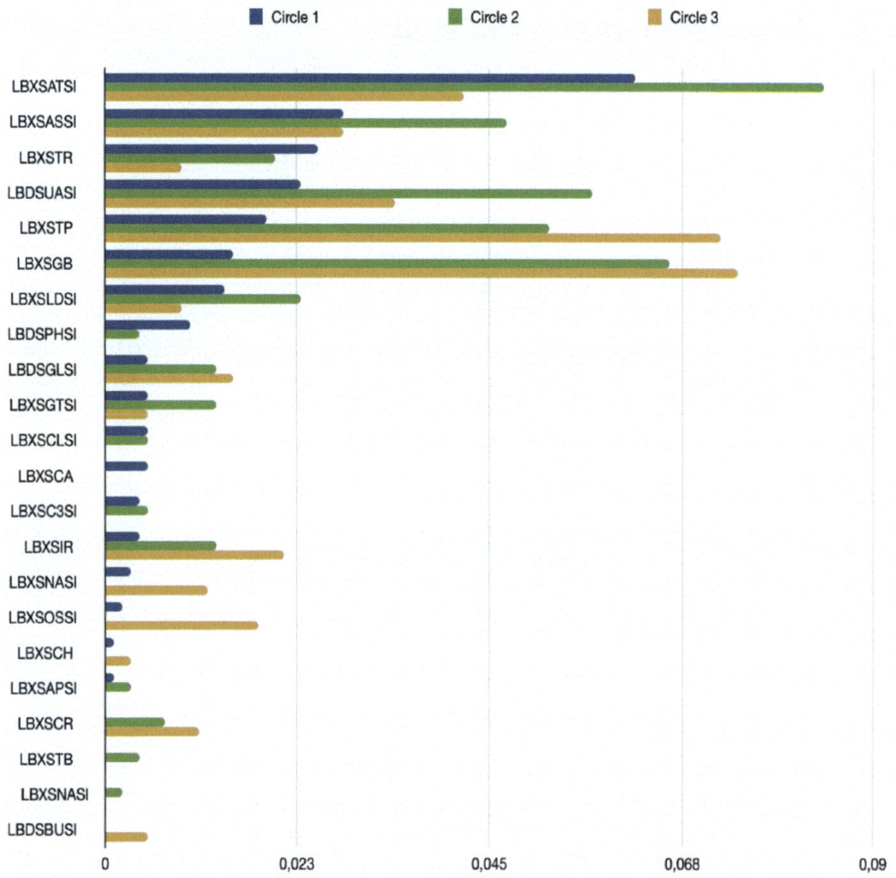

Fig. 2.18 Feature importance scores

1. Alanine Aminotransferase (ALT)—LBXSATSI
2. Aspartate Aminotransferase (AST)—LBXSASSI
3. Uric Acid—LBXDSUASI

These features exhibited a strong positive correlation with BMI.

Since uric acid levels differ by gender, ALT and AST were selected as the primary biochemical markers for training the Dietary Profile Neural Network (DPNN). Further analysis showed that:

- 95% of ALT and AST values were within normal range.
- ALT:AST ratio is a key indicator of liver health
  - A ratio below 1 is considered normal
  - A ratio above 1 is often indicative of liver disease
  - In alcoholic liver disease, 92% of patients exhibit an AST:ALT ratio greater than 1, and 70% exceed a ratio of 2

2.4.3 *Neural Network Dietary Profile*

Data Analysis

Table 2.2 presents the selected features used for training the neural network (NN) in relation to the dietary profile, with all null values removed. Each participant's daily total energy and nutrient intake from food and beverages was considered, alongside an assessment of whether their food consumption was typical, significantly above usual, or significantly below usual. These data points were obtained from the Total Nutrient Intakes files, which include daily aggregates of energy intake and 64 nutrient/food components, as computed using the USDA's Food and Nutrient Database for Dietary Studies (FNDDS) [30]. To facilitate the training process, the dataset was divided into two groups, which served as the classification targets for the neural network. This classification was based on the Alanine Aminotransferase (ALT) to Aspartate Aminotransferase (AST) ratio (ALT:AST):

- ALT:AST ratio < 1, Classified as Normal.
- ALT:AST ratio $>= 1$, Classified as Potential risk indicator.

The classification framework based on the ALT:AST ratio is illustrated in Fig. 2.19.

Classification Accuracy

A substantial improvement in accuracy was observed compared to our previous research and initial exploration of this hypothesis [42, 44]. The earlier studies were based on a smaller dataset and employed a more generalized approach in defining the neural network's target variables. In the current implementation, the model achieved a tenfold increase in accuracy, with a testing performance of approximately 0.7, representing an improvement of 16% over our previous findings, where the average

Table 2.2 Dietary profile features

Feature	Description
DRITKCAL	Energy (kcal)
DRITTFAT	Total fat (gm)
DRITSUGR	Total sugars (gm)
DRITPROT	Protein (gm)
DRITCAFF	Caffeine (mg)
DRITCARB	Carbohydrate (gm)
DRITATOC	Vitamin E as alpha-tocopherol (mg)
DRITACAR	Alpha-carotene (mcg)
DRITRET	Retinol (mcg)

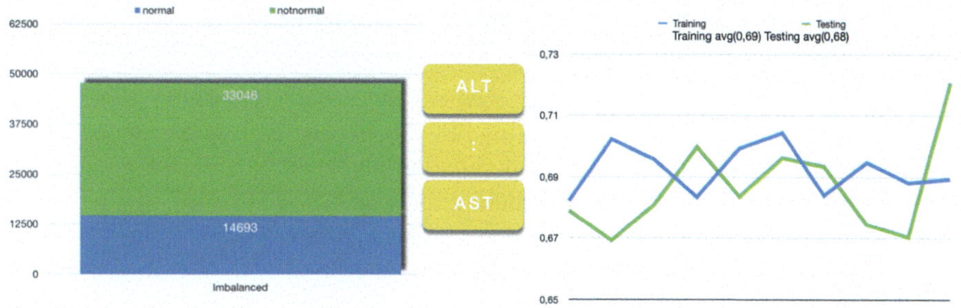

Fig. 2.19 Dataset and cross validation scores

testing accuracy was 0.6. This enhancement in predictive capability is depicted in Fig. 2.19. Initially, 27 data points were selected as input features for training. However, through an iterative optimization process, only 9 features remained active in the final model, while the rest were systematically eliminated using a feature selection methodology similar to that applied in the Standard Biochemistry Profile Neural Network (SBPNN). The final optimized feature set is summarized in Table 2.2.

2.5 Conclusions and Future Research

2.5.1 Preventive Health Strategies

Reducing the incidence of diseases is the most effective and cost-efficient approach to healthcare management. Alongside optimal treatment and patient care, preventive measures play a crucial role in maintaining public health.

Prevention can be categorized into three levels:

- Primary Prevention: Focuses on reducing exposure to risk factors through legislation, regulation, education, and behavioral interventions.
- Secondary Prevention: Involves screening, early detection, and medical interventions to identify high-risk subgroups. The use of biological markers can improve diagnostic accuracy and cost-effectiveness, preventing unnecessary treatments.
- Tertiary Prevention: Aims to prevent disease recurrence and minimize long-term health complications.

Maintaining an optimal body weight is a key indicator of overall health, and weight management is an essential component of preventive healthcare. Dietary interventions—both qualitative and quantitative—play a significant role in disease prevention and patient management.

2.5.2 AI-Based Modeling for Health Assessment

Modeling biochemical evaluation processes enhances both speed and accuracy in medical diagnostics and decision-making. Beyond individual assessments, these models can serve as powerful tools for population-level healthcare management.

Key considerations in AI-driven modeling include:

- Cost-efficiency and accessibility—Reducing healthcare costs while improving diagnostic reach.
- Data availability—Leveraging large-scale biochemical datasets to develop highly predictive AI systems.

2.5.3 Automation in Personalized Health Management

The usability of AI-driven health models can be significantly expanded through interactive web platforms and mobile applications, ensuring broader public engagement.

These systems can:

- Continuously update with new user data, refining prediction models.
- Generate personalized dietary recommendations, reinforcing nutritional awareness.
- Enhance public health education, promoting the importance of ideal body weight and balanced nutrition.

2.5.4 Future Perspectives in Metabolic Research

Interpreting complex metabolic datasets remains a challenging yet critical research area. However, significant progress has been made in integrating metabolomics with genomics and proteomics, paving the way for breakthroughs in precision medicine.

Metabolomics holds great promise for the future by enabling:

- Comprehensive biochemical analysis from genotype to phenotype.
- New insights into human physiology and disease mechanisms.
- Identification of novel biomarkers, leading to enhanced medical classification and patient management [9].

2.5.5 Challenges in Feature Selection and System Design

The intricacy of human biology, combined with the vast amount of medical and biochemical data, presents challenges in precise process mapping and AI system design.

Key research considerations include:

- Defining problem scope and aligning with medical literature.
- Developing a foundational AI framework for advanced healthcare applications.
- Refining machine learning models to improve predictive accuracy and clinical applicability.

At this stage of our research, our focus is on constructing an intelligent system that leverages available data to lay the groundwork for future advancements in AI-driven medical analysis.

Appendix

Additional Material

See Tables 2.3, 2.4, 2.5 and 2.6.

Table 2.3 Examples of FDA-approved biomarkers-markers and their conditions, limits, sensitivity, and specificity

Marker	Condition	Limit	Sensitivity (%)	Specificity (%)
CEA	Malignant pleural effusion	NA1	58	79
CEA	Peritoneal cancer	0.5 ng/ml	76	91
Her-2/neu	Stage IV breast cancer	15 ng/mL	40	98
	Bladder Tumor Antigen Urothelial cell carcinoma	NA	53	70
Thyroglobulin	Thyroid cancer metastases	2.3 ng/ml3	75	95
Alphafetoprotein	Hepatocellular carcinoma	20 ng/ml	50	70
PSA	Prostate cancer	4.0 ng/mL	46	91
CA 125	Non-small cell lung cancer	95 IU/mL	84	80
CA19.9	Pancreatic cancer	NA	75	80
Leptin, prolactin, osteopontin, and IGF-II	Ovarian cancer	NA	95	95
CD98, fascin, sPIgR4, and 14-3-3 eta	Lung cancer	NA	96	77
Troponin I	Myocardial infarction	0.1 microg/L	93	81
B-type natriuretic peptide	Congestive heart failure	8 pg/mL	98	92

Table 2.4 Elements and their descriptions

Element	Description
Alanine aminotransferase	Alanine aminotransferase measurements are used in the diagnosis and treatment of certain liver diseases (e.g., viral hepatitis and cirrhosis) and heart disease. Elevated transaminase levels may indicate myocardial infarction, liver disease, muscular dystrophy, or organ damage
Albumin	Albumin measurements are used in the diagnosis and treatment of diseases involving the liver and/or kidneys, and are often used to assess nutritional status because plasma albumin levels are dependent on protein intake
Alkaline phosphatase	Alkaline phosphatase measurements are used in the diagnosis and treatment of liver, bone, and parathyroid disease
Aspartate aminotransferase	AST measurements are used to diagnose and treat certain types of liver and heart disease. Elevated transaminase levels can signal myocardial infarction, liver disease, muscular dystrophy, or organ damage
Bicarbonate	Along with pH determination, bicarbonate measurements are used in the diagnosis and treatment of many potentially serious disorders associated with acid-base imbalance in the respiratory and metabolic systems
Blood Urea Nitrogen (BUN)	BUN measurements are used to diagnose certain kidney and metabolic diseases. Determination of serum urea nitrogen is the most widely used test to assess renal function. High BUN levels are associated with decreased renal function, increased protein catabolism, nephritis, intestinal obstruction, urinary obstruction, metal poisoning, heart failure, peritonitis, dehydration, malignancy, pneumonia, surgical shock, and Addison's disease. Low BUN levels are associated with amyloidosis, acute liver disease, pregnancy, and nephrosis
Cholesterol	Elevated cholesterol levels are associated with diabetes, nephrosis, hypothyroidism, biliary obstruction, and rare cases of idiopathic hypercholesterolemia and hyperlipidemia. Low levels are associated with hyperthyroidism, hepatitis, and sometimes severe anemia or infection
Creatinine	Creatinine measurements are useful in the diagnosis and treatment of kidney disease. Creatinine measurements are also used in the diagnosis and treatment of myocardial infarction, skeletal muscle diseases, and diseases of the central nervous system
Gammaglutamyl Transaminase (GGT)	GGT measurements are mainly used to diagnose and monitor liver disease. It is currently the most sensitive enzymatic indicator of liver disease. It is also used as a sensitive screening test for occult alcoholism. Elevated levels are found in patients taking chronic medications such as phenobarbital and phenytoin
Globulin	Globulins are a different group of proteins that carry different substances in the blood. They are also involved in various defense mechanisms within the body

(continued)

Table 2.4 (continued)

Element	Description
Glucose	Glucose measurements are used in the diagnosis and treatment of pancreatic islet carcinoma and carbohydrate metabolism disorders, including diabetes mellitus, neonatal hypoglycemia, and idiopathic hypoglycemia
Iron	Iron measurements are used in the diagnosis and treatment of diseases such as iron deficiency anemia, chronic kidney disease, and hemochromatosis
Lactate Dehydrogenase (LDH)	LDH measurements are used in the diagnosis and treatment of liver diseases such as acute viral hepatitis, cirrhosis, and metastatic liver cancer. Heart disease, such as myocardial infarction, and tumors of the lungs or kidneys
Osmolality	Serum osmolality is a measure of the number of particles dissolved in a solution. Osmolality is a calculated value in the chemistry analyzer
Phosphorus	There is a reciprocal relationship between serum calcium and inorganic phosphorus. Hyperphosphatemia is associated with vitamin D hypervitaminosis, hypoparathyroidism, and renal insufficiency. Hypophosphatemia is associated with rickets, hyperparathyroidism, and Fanconi syndrome
Potassium, Chloride, and Sodium	Low serum chloride levels are associated with salt-losing nephritis. High serum chloride levels are associated with dehydration and conditions that cause decreased renal blood flow, such as congestive heart failure. Sodium measurements are used in the diagnosis and treatment of diseases involving electrolyte imbalance
Total Bilirubin	Low bilirubin levels are associated with aplastic anemia and certain types of secondary anemia; resulting from toxic treatment for cancer and chronic nephritis
Total Calcium	Calcium measurements are used in the diagnosis and treatment of parathyroid disease, bone disease, chronic kidney disease, and tetanus. Urine calcium measurement is used in the differential diagnosis of hypercalciuria
Total Protein	Total protein measurements are used in the diagnosis and treatment of a variety of diseases including liver, kidney, or bone marrow, as well as other metabolic or eating disorders
Triglycerides	Triglyceride measurements are used in the diagnosis of diabetes, nephrosis, liver obstruction, and other diseases including lipid metabolism and various endocrine disorders
Uric Acid	Uric acid measurements are used in the diagnosis of diseases such as renal failure, gout, leukemia, psoriasis, starvation, or other wasting conditions, and in the treatment of patients receiving cytotoxic drugs

Table 2.5 Lab tests and values

Test	Normal values
Albumin, serum	3.5–5.5 g/dL
ALT	45 U/L or less
AST	40 U/L or less
Bilirubin, total	1.2–1.3 mg/dL or less
Blood group (A,B,O)	
BUN, serum	7–25 mg/dL
Calcium, serum	2.2–2.6 mmol/L
CBC with differential	Values given with report
Chloride, serum	96–109 mmol/L
CPK	Male: 17–148 U/L, female: 10–70 U/L
Creatinine, serum	0.6–6 mg/dL
CRP	0.8 mg/dL or less
Glucose, plasma	70–110 mg/dL
LDH	Less than 240 U/L
Potassium, serum	3.5–5.3 mmol/L
Prothrombin time	9–12 s
Rapid plasma reagin	Nonreactive
Sedimentation rate	male: 0–15 mm/hr, female: 0–20 mm/hr
Sodium, serum	135–147 mmol/L
Triiodothyronine (T3)	85–205 mg/mL
Thyroxine (T4)	4.5–12 mg/dL
Total protein, serum	6–8.5 g/dL
Triglycerides	Less than 150 mg/dL

Table 2.6 Standard biochemistry profile description

Test code	Description
LBDSALSI	Albumin (g/L)
LBDSBUSI	Blood urea nitrogen (mmol/L)
LBDSCASI	Total calcium (mmol/L)
LBDSCHSI	Cholesterol (mmol/L)
LBDSCRSI	Creatinine (umol/L)
LBDSGBSI	Globulin (g/L)
LBDSGLSI	Glucose, refrigerated serum (mmol/L)
LBDSIRSI	Iron, refrigerated serum (umol/L)
LBDSPHSI	Phosphorus (mmol/L)

(continued)

Table 2.6 (continued)

Test code	Description
LBDSTBSI	Total bilirubin (umol/L)
LBDSTPSI	Total protein (g/L)
LBDSTRSI	Triglycerides, refrigerated (mmol/L)
LBDSUASI	Uric acid (umol/L)
LBXSAL	Albumin (g/dL)
LBXSAPSI	Alkaline phosphatase (10/L)
LBXSASSI	Aspartate aminotransferase AST (IU/L)
LBXSATSI	Alanine aminotransferase ALT (IU/L)
LBXSBU	Blood urea nitrogen (mg/dL)
LBXSC3SI	Bicarbonate (mmol/L)
LBXSCA	Total calcium (mg/dL)
LBXSCH	Cholesterol (mg/dL)
LBXSCK	Creatine Phosphokinase(CPK) (IU/L)
LBXSCLSI	Chloride (mmol/L)
LBXSCR	Creatinine (mg/dL)
LBXSGB	Globulin (g/dL)
LBXSGL	Glucose, refrigerated serum (mg/dL)
LBXSGTSI	Gamma glutamyl transferase (U/L)
LBXSIR	Iron, refrigerated serum (ug/dL)
LBXSKSI	Potassium (mmol/L)
LBXSLDSI	Lactate dehydrogenase (U/L)
LBXSNASI	Sodium (mmol/L)
LBXSOSSI	Osmolality (mmol/Kg)
LBXSPH	Phosphorus (mg/dL)
LBXSTB	Total bilirubin (mg/dL)
LBXSTP	Total protein (g/dL)
LBXSTR	Triglycerides, refrigerated (mg/dL)
LBXSUA	Uric acid (mg/dL)
SEQN	Respondent sequence number

References

1. A. Acosta, et al., Precision medicine and obesity. Gastroenterol. Clin. **50**(1), 127–139 (2021)
2. W.E. Bechtold, R.S. Michael , I.-Y. Chang, J.B. Ward, R.F. Henderson, Species differences in urinary butadiene metabolites: comparisons of metabolite ratios between mice, rats, and humans. Toxicol. Appl. Pharmacol. **127**(1), 44–49 (1994)
3. A. Blanchard, Mapping ethical and social aspects of cancer biomarkers. New Biotechnol. **33**(6), 763–772 (2016)

4. A. Boles, R. Kandimalla, P.H. Reddy, Dynamics of diabetes and obesity: epidemiological perspective. Biochimica et Biophysica Acta (BBA)-Molecular Basis of Disease **1863**(5), 1026–1036 (2017)
5. N. Bosanquet, K. Sikora, The economics of cancer care in the UK. Lancet Oncol. **5**(9), 568–574 (2004)
6. N.G. Bourbakis, A. Esposito, G.A. Tsihrintzis, M. Virvou, L.C. Jain, Introduction to advances in assistive technologies, in *Advances in Assistive Technologies: Selected Papers in Honour of Professor Nikolaos G. Bourbakis–Vol. 3*, pp. 1–7 (2022)
7. O.A. Brekke, T. Sirnes, Biosociality, biocitizenship and the new regime of hope and despair: interpreting "Portraits of Hope" and the "Mehmet Case" ". New Genet. Soc. **30**(4), 347–374 (2011)
8. A.J. Brown, C.L. Trimble, New technologies for cervical cancer screening. Best Pract. Res. Clin. Obstet. Gynaecol. **26**(2), 233–242 (2012)
9. R.M. Califf, Biomarker definitions and their applications. Exp. Biol. Med. **243**(3), 213–221 (2018)
10. D. Callahan, *What Price Better Health?: Hazards of the Research Imperative*, vol. 9. University of California Press (2003)
11. J.D. Carver, W.A. Walker, The role of nucleotides in human nutrition. J. Nutr. Biochem. **6**(2), 58–72 (1995)
12. K. Chrysafiadi, M. Virvou, G.A. Tsihrintzis, I. Hatzilygeroudis, Evaluating the user's experience, adaptivity and learning outcomes of a fuzzy-based intelligent tutoring system for computer programming for academic students in Greece. Educ. Inf. Technol. **28**(6), 6453–6483 (2023)
13. E.T. Cirulli, L. Guo, C.L. Swisher, N. Shah, L. Huang, L.A. Napier, E.F. Kirkness, T.D. Spector, C.T. Caskey, B. Thorens, et al., Profound perturbation of the metabolome in obesity is associated with health risk. Cell Metab. **29**(2), 488–500 (2019)
14. Europen Council, Council conclusions on personalised medicine for patients. Off. J. Eur. Union [Internet] **431**, 1–4 (2015)
15. M.G.M. De Lecea, M. Rossbach, Translational genomics in personalized medicine–scientific challenges en route to clinical practice. HUGO J. **6**, 1–9 (2012)
16. J.L. Denson, A.S. Gillet, Y. Zu, M. Brown, T. Pham, Y. Yoshida, F. Mauvais-Jarvis, I.S. Douglas, M. Moore, K. Tea et al., Metabolic syndrome and acute respiratory distress syndrome in hospitalized patients with COVID-19. JAMA Netw. Open **4**, e2140568 (2021)
17. E.P. Diamandis, The failure of protein cancer biomarkers to reach the clinic: why, and what can be done to address the problem?. BMC Med. **10**, 1–5 (2012)
18. A. Eichholz, S. Merchant, A.M. Gaya, Anti-angiogenesis therapies: their potential in cancer management. OncoTargets Ther. 69–82 (2010)
19. P. Fan, L. Li, A. Rezaei, S. Eslamfam, D. Che, X. Ma, Metabolites of dietary protein and peptides by intestinal microbes and their impacts on gut. Curr. Protein Peptide Sci. **16**(7), 646–654 (2015)
20. J. Gagan, E.M. Van Allen, Next-generation sequencing to guide cancer therapy. Genome Med. **7**, 1–10 (2015)
21. E. Giannini, F. Botta, A. Fasoli, P. Ceppa, D. Risso, P.B. Lantieri, G. Celle, R. Testa, Progressive liver functional impairment is associated with an increase in AST/ALT ratio. Dig. Dis. Sci. **44**, 1249–1253 (1999)
22. D.A. Goldstein, J. Clark, Y. Tu, J. Zhang, F. Fang, R. Goldstein, S.M. Stemmer, E. Rosenbaum, A global comparison of the cost of patented cancer drugs in relation to global differences in wealth. Oncotarget **8**(42), 71548 (2017)
23. D. Hammerl, M. Smid, A.M. Timmermans, S. Sleijfer, J.W.M. Martens, R. Debets, Breast cancer genomics and immuno-oncological markers to guide immune therapies, in *Seminars in Cancer Biology*, vol. 52. Elsevier, pp. 178–188 (2018)
24. O. Harismendy, P.C. Ng, R.L. Strausberg, X. Wang, T.B. Stockwell, K.Y. Beeson, N.J. Schork, S.S. Murray, E.J. Topol, S. Levy et al., Evaluation of next generation sequencing platforms for population targeted sequencing studies. Genome Biol. **10**, 1–13 (2009)

25. R.Z. Hayeems, D. Dimmock, D. Bick, J.W. Belmont, R.C. Green, B. Lanpher, V. Jobanputra, R. Mendoza, S. Kulkarni, M.E. Grove, et al., Clinical utility of genomic sequencing: a measurement toolkit. NPJ Genomic Med. **5**(1), 56 (2020)
26. J. Hesketh, Personalised nutrition: how far has nutrigenomics progressed?. Eur. J. Clin. Nutr. **67**(5), 430–435 (2013)
27. F.R. Hirsch, K. Suda, J. Wiens, P.A. Bunn, New and emerging targeted treatments in advanced non-small-cell lung cancer. The Lancet **388**(10048), 1012–1024 (2016)
28. L. Hood, D. Galas, The digital code of DNA. Nature **421**(6921), 444–448 (2003)
29. P.T. James, R. Leach, E. Kalamara, M. Shayeghi, The worldwide obesity epidemic. Obes. Res. **9**(S11), 228S–233S (2001)
30. C.L. Johnson, S.M. Dohrmann, V.L. Burt, L.K. Mohadjer, *National Health and Nutrition Examination Survey: Sample Design, 2011–2014* (US Department of Health and Human Services, Centers for Disease Control and Prevention, 2014)
31. D.A. Levine, Cancer Genome Atlas Research Network Genome sequencing centres: Broad Institute Getz Gad 1 Gabriel Stacey B. 1 Cibulskis Kristian 1 Lander Eric 1 Sivachenko Andrey 1 Sougnez Carrie 1 Lawrence Mike 1, Washington University in St Louis Kandoth Cyriac 2 Dooling David 2 Fulton Robert 2 Fulton Lucinda 2 Kalicki-Veizer Joelle 2 McLellan Michael D. 2 O'Laughlin Michelle 2 Schmidt Heather 2 Wilson Richard K. 2 Ye Kai 2 Ding Li 2 Mardis Elaine R. 2, University of Southern California & Johns Hopkins Baylin Stephen B. 21 Bootwalla Moiz S. 22 Lai Phillip H. 22 Triche Jr Timothy J. 22 Van Den Berg David J. 22 Weisenberger Daniel J. 22 Laird Peter W. 22 Shen Hui 22, Institute for Systems Biology Reynolds Sheila M. 23 Shmulevich Ilya 23, et al., Integrated genomic characterization of endometrial carcinoma. Nature **497**(7447), 67–73 (2013)
32. Q.Y. Lian, A. Chen, J.H. Zhang, W.J. Guan, X. Xu, B. Wei, D.X. Huang, J.X. He, C.R. Ju, High-throughput next-generation sequencing for identifying pathogens during early-stage post-lung transplantation. BMC Pulm. Med. **21**, 1–9 (2021)
33. L. Liu, Y. Li, S. Li, N. Hu, Y. He, R. Pong, D. Lin, L. Lu, M. Law, Comparison of next-generation sequencing systems. J. Biomed. Biotechnol. **2012**(251364), 251364 (2012)
34. M.R. Luskin, M.A. Murakami, S.R. Manalis, D.M. Weinstock, Targeting minimal residual disease: a path to cure?. Nat. Rev. Cancer **18**(4), 255–263 (2018)
35. B.T. McInnes, T. Pedersen, Evaluating semantic similarity and relatedness over the semantic grouping of clinical term pairs. J. Biomed. Inform. **54**, 329–336 (2015)
36. C.M. Micheel, C.M. Lovly, M.A. Levy, My cancer genome. Cancer Genet. **207**(6), 289 (2014)
37. A.H. Mokdad, E.S. Ford, B.A. Bowman, W.H. Dietz, F. Vinicor, V.S. Bales, J.S. Marks, Prevalence of obesity, diabetes, and obesity-related health risk factors, 2001. Jama **289**(1), 76–79 (2003)
38. P.J. Monaghan, S. Robinson, D. Rajdl, P.M.M. Bossuyt, S. Sandberg, A. St John, M. O'Kane, L. Lennartz, R. Röddiger, S.J. Lord, et al., Practical guide for identifying unmet clinical needs for biomarkers. Ejifcc **29**(2), 129 (2018)
39. P.M. Nilsson, J. Tuomilehto, L. Ryden, The metabolic syndrome–What is it and how should it be managed? Eur. J. Prev. Cardiol. **26**, 33–46 (2019)
40. E. Nimmesgern, I. Norstedt, R. Draghia-Akli, Enabling personalized medicine in Europe by the European Commission's funding activities. Personalized Med. **14**(4), 355–365 (2017)
41. R. Pallares-Méndez, C.A. Aguilar-Salinas, I. Cruz-Bautista, L. del Bosque-Plata, Metabolomics in diabetes, a review. Ann. Med. **48**(1-2), 89–102 (2016)
42. D.P. Panagoulias, D.N. Sotiropoulos, G.A. Tsihrintzis, Biomarker-based deep learning for personalized nutrition, in *2021 IEEE 33rd International Conference on Tools with Artificial Intelligence (ICTAI)*, Washington, DC, USA, Nov. 2021, pp. 306–313
43. D.P. Panagoulias, D.N. Sotiropoulos, G.A. Tsihrintzis, Nutritional biomarkers and machine learning for personalized nutrition applications and health optimization, in *Proceedings of the Twelfth IEEE International Conference on Information, Intelligence, Systems and Applications*. Chania, Greece, pp. 731–733, July 2021

44. D.P. Panagoulias, D.N. Sotiropoulos, G.A. Tsihrintzis, Nutritional biomarkers and machine learning for personalized nutrition applications and health optimization, in *2021 12th International Conference on Information, Intelligence, Systems & Applications (IISA)*, Chania Crete, Greece, July 2021, pp. 1–6
45. D.P. Panagoulias, P. Papatheodosiou, A. Bonakis, D. Dikeos, M. Virvou, G.A. Tsihrintzis, Mathematical representation of memory and schema for improving human-generative AI interactions, in *2024 15th International Conference on Information, Intelligence, Systems & Applications (IISA)*. IEEE, pp. 1–5 (2024)
46. D.P. Panagoulias, D. Sotiropoulos, K. Chrysafiadi, E.-A. Tsichrintzi, E. Sakkopoulos, G.A. Tsihrintzis, M. Virvou, LYRICEL: a rule-augmented artificial intelligence-empowered cultural E-learning with GPT and machine learning for song lyrics, in *2024 15th International Conference on Information, Intelligence, Systems & Applications (IISA)*. IEEE, pp. 1–5 (2024)
47. P. Papatheodosiou, D.P. Panagoulias, M. Virvou, G.A. Tsihrintzis, A. Bonakis, D. Dikeos, Leveraging artificial intelligence for personalised insomnia-sleep calibration via the big five personality traits. Procedia Comput. Sci. **246**, 2539–2548 (2024)
48. C.S. Pareek, R. Smoczynski, A. Tretyn, Sequencing technologies and genome sequencing. J. Appl. Genet. **52**(4), 413–435 (2011)
49. F. Pedregosa, G. Varoquaux, A. Gramfort, V. Michel, B. Thirion, O. Grisel, M. Blondel, P. Prettenhofer, R. Weiss, V. Dubourg, J. Vanderplas, A. Passos, D. Cournapeau, M. Brucher, M. Perrot, E. Duchesnay, Scikit-learn: machine learning in Python. J. Mach. Learn. Res. **12**, 2825–2830 (2011)
50. B. Peng, H. Li, X.-X. Peng, Functional metabolomics: from biomarker discovery to metabolome reprogramming. Protein Cell **6**(9), 628–637 (2015)
51. B.A. Perkins, C.T. Caskey, P. Brar, E. Dec, D.S. Karow, A.M. Kahn, Y.-C.C. Hou, N. Shah, D. Boeldt, E. Coughlin, et al., Precision medicine screening using whole-genome sequencing and advanced imaging to identify disease risk in adults, in *Proceedings of the National Academy of Sciences*, vol. 115(14), pp. 3686–3691 (2018)
52. T. Pham, T. Tran, D. Phung, S. Venkatesh, Predicting healthcare trajectories from medical records: a deep learning approach. J. Biomed. Inform. **69**, 218–229 (2017)
53. N. Potischman, J.L. Freudenheim, Biomarkers of nutritional exposure and nutritional status: an overview. J. Nutr. **133**(3), 873S–874S (2003)
54. R. Ramautar, R. Berger, J. van der Greef, T. Hankemeier, Human metabolomics: strategies to understand biology. Current Opinion Chem. Biol. **17**(5), 841–846 (2013)
55. W.J. Scott, J. Howington, S. Feigenberg, B. Movsas, K. Pisters, Treatment of non-small cell lung cancer stage I and stage II: ACCP evidence-based clinical practice guidelines. Chest **132**(3), 234S–242S (2007)
56. S. Serrati, S. De Summa, B. Pilato, D. Petriella, R. Lacalamita, S. Tommasi, R. Pinto, Next-generation sequencing: advances and applications in cancer diagnosis. OncoTargets Ther. 7355–7365 (2016)
57. H. Starkhammar, Aspects on priority settings in cancer treatment and care. Acta Oncologica **44**(7), 667–672 (2005)
58. I.-O. Stathopoulou, E. Alepis, G.A. Tsihrintzis, M. Virvou, On assisting a visual-facial affect recognition system with keyboard-stroke pattern information. Knowl.-Based Syst. **23**(4), 350–356 (2010)
59. K. Strimbu, J.A. Tavel, What are biomarkers? Curr. Opin. HIV AIDS **5**, 463 (2010)
60. Y. Sun, S. Zhu, K. Ma, W. Liu, Y. Yue, G. Hu, H. Lu, W. Chen, Identification of 12 cancer types through genome deep learning. Sci. Rep. **9**(1), 17256 (2019)
61. T. Davenport, R. Kalakota, National health and nutrition examination survey: sample design. Future Healthc. J. **6**, 94 (2019)
62. G.A Tsihrintzis, G. Phillips-Wren, L.C. Jain, *Intelligent Decision Technologies: An International Journal (IDT) Heading into its 18th Year: A Short Note from the Editors-in-Chief* (2024)
63. G.A. Tsihrintzis, E. Sarmas, V. Marinakis, D. Panagoulias, E.-A. Tsihrintzi, M. Virvou, Evaluating non-expert stakeholder interaction with artificial intelligence on energy urban domain using VIRTSI: the case of ChatGPT, in *2024 15th International Conference on Information, Intelligence, Systems & Applications (IISA)*. IEEE, pp. 1–10 (2024)

64. M. Virvou, G.A. Tsihrintzis, E. Alepis, I.-O. Stathopoulou, K. Kabassi, Emotion recognition: empirical studies towards the combination of audio-lingual and visual-facial modalities through multi-attribute decision making. Int. J. Artif. Intell. Tools **21**(02), 1240001 (2012)
65. F. Weinberg, D.B. Peckys, N. de Jonge, EGFR expression in HER2-driven breast cancer cells. Int. J. Mol. Sci. **21**(23), 9008 (2020)
66. A.N. Winn, D.U. Ekwueme, G.P. Guy, P.J. Neumann, Cost-utility analysis of cancer prevention, treatment, and control: a systematic review. Am. J. Prev. Med. **50**(2), 241–248 (2016). ISSN: 0749-3797. https://doi.org/10.1016/j.amepre.2015.08.009. https://www.sciencedirect.com/science/article/pii/S0749379715004833

Chapter 3
Blood Exam Classification for Predicting Defining Factors in Metabolic Syndrome Diagnosis Using Support Vector Machines

Abstract Biomarkers have been extensively explored as robust classification features in training neural network-based and other machine learning and artificial intelligence-driven prognostic models, particularly within the domain of personalized nutrition. In this chapter, we develop and analyze cascaded support vector machine (SVM)-based classifiers for the automated diagnosis of metabolic syndrome. Specifically, utilizing blood test data, we achieve an average classification accuracy of approximately 84% for body mass index (BMI). Likewise, cascaded SVM-based classifiers demonstrate a 74% accuracy in classifying systolic blood pressure. Furthermore, we design and implement a predictive system that attains an accuracy of 84% in forecasting metabolic syndrome. This system leverages not only BMI classification but also predictive insights from blood test parameters, including total cholesterol, triglycerides, and glucose levels. To ensure the self-containment of this chapter, fundamental concepts pertaining to metabolic syndrome are summarized, and a review of prior relevant studies is provided. Lastly, conclusions are drawn, and potential directions for future research in this area are discussed.

Keywords Biomarkers · Metabolic syndrome (MetS) · Body mass index (BMI) · Systolic blood pressure (SBP) · Personalized medicine · Support vector machine (SVM) · Support vector classifier (SVC) · Neural network · Data analytics

3.1 Introduction

In recent years, the integration of artificial intelligence (AI) technologies-including machine learning and deep learning-into healthcare and medicine has gained substantial attention. These technologies are transforming patient treatment and disease management by automating routine tasks, streamlining medical workflows, reducing manual interventions, and enhancing efficiency across healthcare systems [35].

At the core of many AI-driven healthcare advancements are biomarkers, which serve as objective indicators derived from blood, other bodily fluids, or tissues, reflecting a person's health status or the presence of disease. Biomarkers can represent a single measurement or a combination of multiple indicators, providing critical insights

D. P. Panagoulias et al., *Artificial Intelligence-Empowered Bio-medical Applications*, Learning and Analytics in Intelligent Systems 49,
https://doi.org/10.1007/978-3-031-90174-4_3

Table 3.1 Examples of biomarkers

Name	Description	Health results
Waist to hip ratio	Abdominal obesity index	Hypertension, CHD, insulin-dependent diabetes, and stroke
Total cholesterol	Essential for bile acid and steroid hormone synthesis	At middle age: Coronary heart disease (CHD) and overall mortality; At older age: U-shaped relation to mortality
Blood pressure (SBP)	Cardiovascular activity index: maximum arterial pressure when the heart pumps blood	Cardiovascular death (CVD), stroke, coronary heart disease (CHD)
Fasting glucose	Measures blood sugar levels as an indicator of diabetes risk	Diabetes, CHD, mortality, impaired cognitive function

into disease progression, diagnosis, and treatment responses [34]. Table 3.1 presents commonly used biomarkers, including waist-to-hip ratio, total cholesterol, systolic blood pressure (SBP), and fasting glucose. These biomarkers play a fundamental role in personalized medicine, helping clinicians tailor treatment strategies, disease prevention plans, and drug development processes.

This chapter presents recent findings from our research on the relationship between an individual's biochemical profile (derived from blood tests), body mass index (BMI), metabolic health status, and systolic blood pressure (SBP). Our work builds upon previous studies that employed deep neural networks and other machine learning techniques to explore the connections between BMI, metabolic function, and nutrition [24–27]. These advancements highlight the potential of AI-driven approaches in identifying metabolic patterns and enhancing predictive modeling for early disease detection and risk assessment.

The motivation for this study stems from the need to streamline health assessments by leveraging widely available biomarkers to predict health states, thus optimizing the patient journey by reducing response times between tests, results, and medical decisions. Another key motivation is to improve healthcare operational efficiency through machine learning methodologies that can be seamlessly integrated into telemedicine applications. AI and Big Data analytics can enhance decision-making, support medical interventions [20], and expand the scope of healthcare analytics [39]. As AI and machine learning become integral to modern medical pipelines, they are expected to serve as foundational tools for diagnostics and pathology [28].

Childhood obesity remains a pressing concern, influenced by both genetic and behavioral factors. According to the CDC, recent studies indicate that obesity prevalence reaches approximately 13% among children aged 2–5 years, 20% in the 6–11-year-old group, and 21.2% among adolescents aged 12–19 years [40]. Moreover, childhood obesity disproportionately affects specific communities and populations [40]. The distribution of age groups in our dataset is depicted in Fig. 3.1, illustrating that most subjects are young adults aged 18 and above, with only a small subset

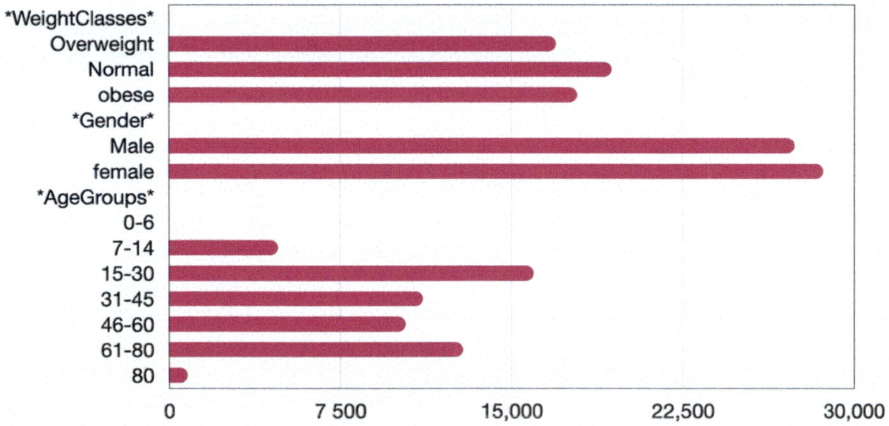

Fig. 3.1 Characteristics of the sample population

belonging to the 7–14 age range. Consequently, additional datasets are required to derive reliable conclusions regarding childhood obesity. Incorporating demographic and lifestyle determinants alongside laboratory data could enhance machine learning classifiers for childhood obesity prediction and enable automated targeted interventions. This research direction is actively being pursued, with findings to be presented in future studies.

This study focuses on the metabolic syndrome (MetS) spectrum, as illustrated in Fig. 3.2. Our approach is holistic, emphasizing automation-driven engineering shortcuts. By building upon existing literature [14] and expanding on prior findings [5], we propose a more comprehensive framework that links health factors to specific

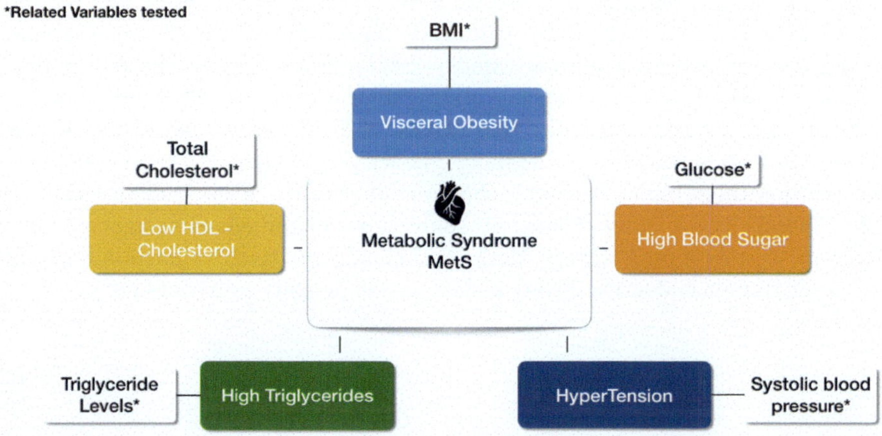

Fig. 3.2 MetS predictors

states, using these states to derive actionable recommendations. Given that MetS represents a distinct health state and weight is closely associated with MetS, we integrate triglyceride, glucose, and cholesterol levels to provide a conclusive assessment, calculating relevant risk factors and treatment recommendations.

In this chapter, we implement and evaluate various machine learning classifiers that link an individual's biochemical profile (derived from blood tests) with BMI, metabolism as a health state, and SBP. Our findings indicate that support vector machine-based (SVM) classifiers demonstrate strong predictive performance. Additionally, we compare our results with previous studies based on deep neural networks [24–27]. A detailed review of related literature is provided in Sect. 3.3.

The structure of this chapter is as follows: Sect. 3.2 presents key concepts related to MetS, establishing the theoretical foundation of our study. Section 3.3 highlights prior research in this field. Section 3.4 discusses the datasets used in our study, along with key statistical measures characterizing them. Section 3.5 explores the development and evaluation of different machine learning classifiers for BMI prediction based on blood test data, including neural network-based and SVM-based models. Our findings indicate that cascaded SVM-based classifiers exhibit the most promising performance, achieving an average classification accuracy of approximately 84%. In Sect. 3.6, we propose and implement a system for MetS prediction, which incorporates BMI predictions alongside biomarker-based predictions of total cholesterol, triglycerides, and blood pressure (Fig. 3.2). Section 3.7 summarizes our key findings and their implications. Finally, Sect. 3.8 presents our conclusions and outlines future research directions in this domain.

3.2 Metabolic Syndrome

Noncommunicable diseases (NCDs) are a major cause of morbidity and mortality worldwide, with a particularly significant impact in underdeveloped countries. Among these, metabolic syndrome (MetS)—also referred to as syndrome X or dysmetabolic syndrome—has emerged as a critical health concern. MetS encompasses a cluster of metabolic conditions that increase the risk of cardiovascular disease, among other complications. Its defining characteristics include insulin resistance, hypertension (high blood pressure), abnormal cholesterol levels, and an increased risk of blood clot formation. Numerous studies have identified overweight and obesity as the strongest predictors of MetS. Additionally, thyroid dysfunction is one of the most common endocrine disorders observed in patients with MetS [15].

3.2.1 Metabolic Syndrome Markers

MetS is recognized as a significant risk factor for atherosclerotic cardiovascular disease (ASCVD), ranking alongside cigarette smoking, hypertension, hypercholesterolemia, and diabetes [29]. Key markers of MetS include atherogenic dyslipidemia, high blood pressure, elevated glucose levels, a prothrombotic state, and chronic inflammation. Clinical hyperglycemia, particularly Type 2 diabetes, is frequently observed in advanced stages of MetS. Even in the absence of diabetes, MetS is associated with a twofold increase in relative cardiovascular disease risk.

MetS has also been recognized as a growing global health issue, particularly as urbanization and lifestyle changes contribute to increased obesity rates. Recent studies have further linked MetS to a higher risk of acute respiratory distress syndrome (ARDS) and increased mortality in COVID-19 patients [9]. Current medical guidelines define MetS based on the presence of at least three of the biomarkers illustrated in Fig. 3.2, which we discuss in further detail below.

A Large Waist

Although body mass index (BMI) is commonly used in clinical and research settings to measure adiposity, waist circumference is an important alternative metric for assessing metabolic risk [8]. While multiple factors contribute to MetS, it is rare to diagnose the syndrome in individuals who do not exhibit excess body fat. The prevalence of MetS increases proportionally with obesity. In obese individuals, excessive adipose tissue releases metabolic factors that contribute to an increased risk of metabolic disorders. In particular, the excessive release of non-esterified fatty acids promotes the accumulation of ectopic fat deposits in the liver, muscles, and visceral adipose tissue, further exacerbating metabolic risk factors [11].

A High Triglyceride Level

Triglyceride levels are determined through blood tests that measure the amount of this specific type of fat circulating in the bloodstream. Elevated triglyceride levels occur when fat intake exceeds the body's metabolic requirements, contributing to an increased risk of cardiovascular disease and metabolic dysfunction.

Reduced HDL ("Good" Cholesterol)

The body contains both high-density lipoprotein (HDL) cholesterol, known as "good" cholesterol, and low-density lipoprotein (LDL) cholesterol, or "bad" cholesterol. LDL cholesterol tends to accumulate in arteries, leading to plaque formation and an increased risk of cardiovascular disease. In contrast, HDL cholesterol helps remove LDL cholesterol from the bloodstream, preventing its accumulation in arteries.

Total cholesterol levels are calculated using the following equation: Total Cholesterol = HDL + LDL + 0.2 × Triglyceride Level.

Higher total cholesterol levels are associated with an increased risk of ischemic stroke [7]. In contrast, elevated HDL cholesterol levels are linked to a reduced risk of cardiovascular disease, whereas high LDL cholesterol levels contribute to an elevated stroke risk.

Elevated Fasting Blood Sugar (Fasting Glucose)

Glucose, commonly referred to as blood sugar, is derived from dietary carbohydrates and serves as the body's primary energy source. Insulin is essential for glucose transport into cells, where it is utilized for energy production. When insulin production is insufficient or its function is impaired, glucose levels in the bloodstream rise, leading to the development of diabetes [22].

Increased Blood Pressure

Hypertension, or high blood pressure, occurs when the force of blood against the arterial walls remains consistently elevated. Over time, this condition significantly increases the risk of major health complications, including heart disease. Blood pressure is measured in two values: • Systolic pressure: The peak pressure in arteries when the heart beats. • Diastolic pressure: The minimum pressure when the heart rests between beats.

Normal blood pressure is defined as a systolic pressure below 120 mm Hg and a diastolic pressure below 80 mm Hg. Hypertension is diagnosed when systolic pressure reaches 140 mm Hg or higher, and/or diastolic pressure remains at or above 90 mm Hg over an extended period [3]. Blood pressure classifications are further detailed in Fig. 3.3.

3.2.2 Genetics—Cardiac Effects of Obesity and MetS

The genetic basis of MetS is complex, involving numerous markers associated with various biological pathways. Hundreds of genetic variants have been linked to MetS-related traits, including factors affecting adipose tissue metabolism, insulin signaling, and regulatory mechanisms within the endocrine system. However, no single genetic marker has been identified as a universal determinant of MetS [23].

The well-documented association between obesity and hypertension has prompted extensive research into the effects of obesity on cardiovascular health [19]. Chronic

Blood Pressure Estimator

Category	Systolic(mm HG-upper#)		Diastolic(mm HG-lower#)
Normal	Less than 120	and	Less than 80
Elevated	120-129	and	Less than 80
High Blood Pressure(Hypertension) Stage 1	130-139	or	80-89
High Blood Pressure(Hypertension) Stage 2	140 or higher	or	90 or higher
Hypertensive Crisis	Higher than 180	and /or	Higher than 120

Fig. 3.3 Blood pressure classification

weight gain and increased adiposity can induce significant neurohormonal adaptations within the cardiovascular system, leading to structural and functional changes that elevate disease risk [4].

The ratio of total cholesterol to HDL cholesterol, as well as the ratio of LDL cholesterol to HDL cholesterol, are commonly used predictors of ischemic heart disease risk [18]. These ratios provide a more comprehensive assessment of cardiovascular health than total cholesterol levels alone, further emphasizing the importance of lipid profile management in reducing MetS-related complications.

3.3 Related Work and SVM Classification

In our previous study [27], we evaluated the effectiveness of neural networks in classifying individuals based on their BMI using a basic biochemical profile derived from routine blood exams (Fig. 3.4).

Support Vector Machines (SVMs) have emerged as a powerful and widely adopted machine learning technique, with applications spanning various domains, including healthcare, recommendation systems, and decision support. Recognized for their capability to handle both classification and regression tasks, SVMs are particularly effective in managing high-dimensional and nonlinear datasets by identifying optimal hyperplanes for data separation [36]. These strengths have established SVM as a leading methodology for tackling complex machine learning challenges. One notable application of SVM is in personalized recommendation systems. A. S. Lampropoulos et al. [17] introduced a hybrid model that integrates SVM with collaborative filtering, significantly improving the ability of e-commerce platforms to

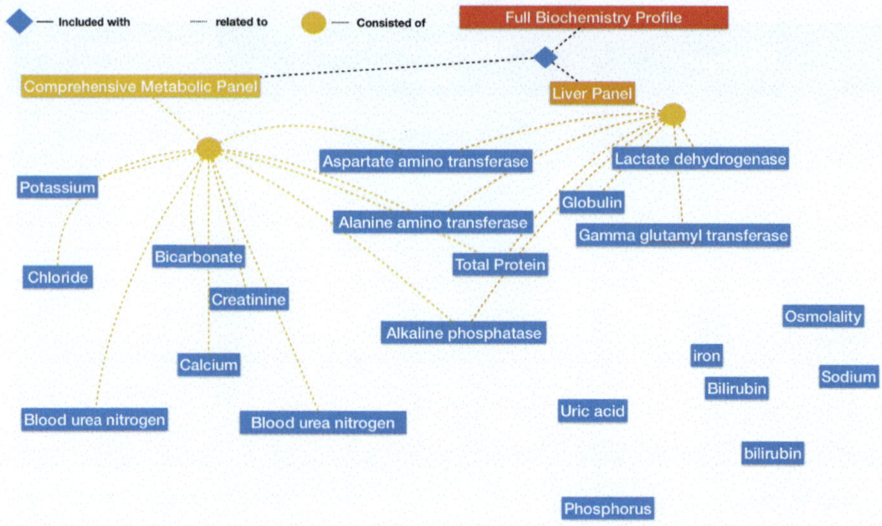

Fig. 3.4 Comprehensive biochemistry profile derived from blood exams

provide tailored recommendations. Additionally, clustering techniques combined with SVMs have demonstrated effectiveness in user modeling, enabling more intelligent personalization in recommender systems [30].

SVMs have also proven useful in handling extreme class imbalance issues, a common challenge in many machine learning datasets. For instance, D. N. Sotiropoulos et al. [32] proposed an artificial immune system-inspired approach that enhances classification accuracy when working with imbalanced datasets. Similarly, Dionysios Sotiropoulos et al. [33] conducted a comparative study of one-class classifiers, including SVM, to address such challenges effectively. These studies highlight the adaptability of SVM in applications such as anomaly detection and fraud prevention.

In healthcare, SVM has demonstrated remarkable contributions, particularly in diagnostic imaging. Agrios et al. [1] showcased SVM's utility in mammogram classification for breast cancer detection, emphasizing its potential to improve diagnostic precision. Given the increasing demand for interpretable machine learning models in healthcare, Alogogianni et al. [2] examined how interpretable SVM models enhance trust and transparency, which are critical for real-world adoption.

Advancements in machine learning paradigms have further expanded the potential of SVM. Tsihrintzis, Virvou, Doukas, et al. [37] discussed the integration of SVM in AI-powered decision support systems, illustrating its role in building modular and scalable solutions for complex applications. Additionally, ongoing research continues to explore interpretable SVM models for decision-making systems, improving the reliability of predictions in critical sectors [37]. Surveys of AI-based technologies have also emphasized the pivotal role of SVM in advancing intelligent systems [38]. In summary, SVM's adaptability and precision make it an indispensable tool in applications ranging from recommendation systems to healthcare and decision support.

As machine learning research evolves, SVM continues to play a central role in the development of innovative, transparent, and scalable AI-driven solutions.

The classification process was initially conducted using four classes, followed by three, and finally reduced to two, allowing for a deeper understanding of the relationships between variables. Specifically, the four widely recognized BMI classes—obese, overweight, normal, and underweight—were used. At each stage, accuracy improved as the classification was refined into smaller groups, as illustrated in Fig. 3.5. This led to the conception of a cascaded classification approach, inspired by similar methodologies applied in recommendation systems [16, 31]. Additionally, the findings underscored the potential of blood exam data in facilitating the classification of a broader range of health states. To ensure the integrity of our study, data analysis was conducted meticulously to prevent bias in the classification system. Each classifier's performance was cross-validated using comparison matrices. In this chapter, we report on our study of binary and one-class SVM-based classifiers and compare their performance against other machine learning classifiers.

Other related studies have suggested a correlation between electrocardiogram (ECG) patterns and MetS through the application of neural network-based classifiers [19]. Furthermore, previous research has linked MetS with demographic factors and specific blood test parameters, employing neural networks as well as traditional statistical methods for comparison [21]. The prediction of MetS using artificial neural networks and clinical data has also been explored, with BMI, age, HDL, and LDL identified as key predictive factors [12]. Additionally, an extreme learning machine approach was implemented to identify the overweight category based on blood exam data [5], using a sample of 500 male and female participants. Another study sought to diagnose MetS by integrating clinical symptoms with physio-chemical indices, analyzing 586 cases—450 of which had MetS, while 136 did not [42].

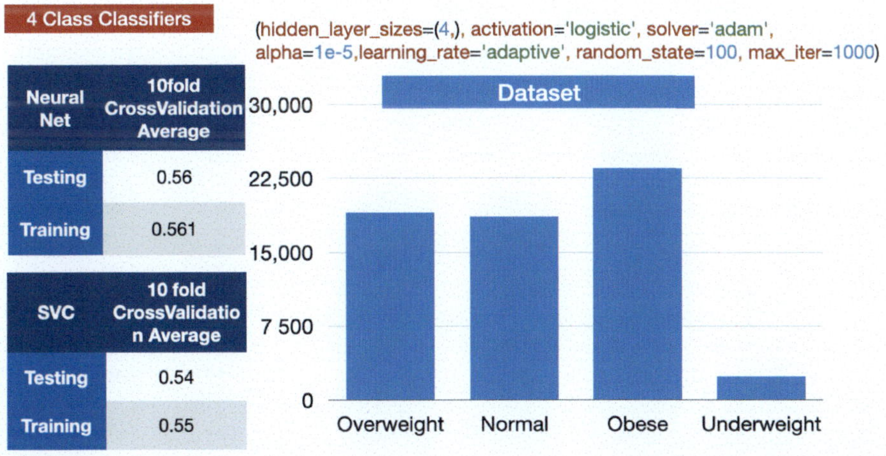

Fig. 3.5 Four-class classifier

A noteworthy contribution to the field is a recent literature review that extensively examines statistical approaches and machine learning methodologies applied to MetS classification [14]. A more detailed comparison of machine learning methodologies for predicting MetS and BMI can be found in Tables 3.2 and 3.3, respectively. A key distinction between our proposed methodology and previous approaches is that many existing studies, while methodologically sound, lack strong medical validation. Our comparison focuses on studies that follow a similar feature selection strategy to ours.

Crucial differences between our work and prior studies include sample sizes and class distribution imbalances, as outlined in Sect. 3.4. Such imbalances in prior research could introduce bias into results. Another critical factor contributing to the usability of a classification system is the feature selection process. In this study, we have ensured that our classification approach provides novel insights for medical

Table 3.2 Comparison of methodologies—MetS

Metrics	Cascaded SVC	QPHR1	ANN
Accuracy (%)	84	>95	>95
Size of sample	5000	15000	410
Balanced sample	Yes	No	No
Features	Standard Biochemistry profile. Mets defining factors not included	Uses defining factors to predict MetS (triglycerides,Bmi)	Uses defining factors to predict MetS (BMI, diastolic blood pressure, HDL-cholesterol, LDL-cholesterol)
		[41]	[12]

Table 3.3 Comparison of methodologies—BMI

Metrics	4 Class Neural Network	4 Class SVC	3 Class Neural Network	3 Class SVC	Cascaded SVM	Extreme LM
Accuracy (%)	56	55	58	62	85	90.54
Size of sample	75,000	15,000	33,000	15,000	10,000	500
Balanced sample	No	No	Yes	Yes	Yes	No
All Classes?	Yes	Yes	No	No	Yes	No (Overweight class)
Features	Full Biochemistry Profile	Full Biochemistry Profile	Full Biochemistry Profile	Full Biochemistry Profile	Without defining factors of MetS (15 features)	Blood Indexes (39 features) + age
	[25]		[24]			[5]

prognosis by excluding major defining factors that have been extensively studied-such as triglycerides, cholesterol, BMI, and glucose-as seen in previous works detailed in Tables 3.2 and 3.3.

Our objective is to develop a streamlined machine learning pipeline that encompasses a significant portion of the patient's journey, minimizing manual interventions while enhancing predictive accuracy. By optimizing the required input data, we aim to improve efficiency and outcomes without compromising classification performance. To that end, we leveraged a comprehensive dataset consisting of approximately 70,000 data points, analyzed in both balanced and imbalanced configurations. This represents a substantial increase in sample size compared to previous related studies, which typically relied on smaller datasets averaging around 500 imbalanced data points. The use of such an extensive dataset not only enhances the robustness of our findings but also allows for a more generalizable and reliable classification framework.

3.4 Exploratory Data Analysis

Initially, the data were analyzed as per the BMI via deploying frequency histograms and probability density distributions to define how the sample used is represented. Data were also analyzed via deploying the same techniques as per each variable (blood exams—standard biochemistry profile) to determine average values, correlation between each variable and other statistical metrics, as in Fig. 3.6. Variables

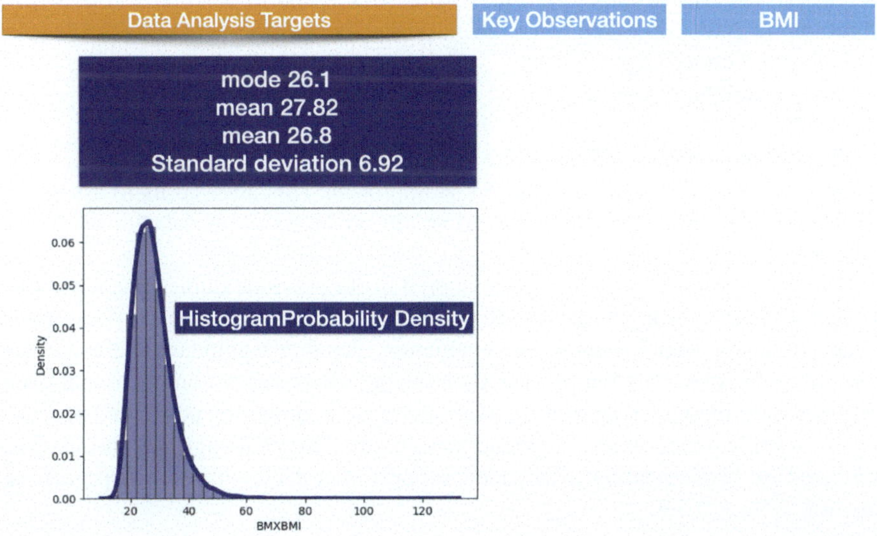

Fig. 3.6 Key observations BMI

were tested in both a generic way (complete data-set) and a more precise way as per weight category.

Key observations were drawn via examining important statistical metrics (e.g., mean, median, mode or standard deviation) to better understand the sample set under examination, to create the basis of available tools for future research endeavors and to develop a tool for missing value prediction through entity alignment. The results of this analysis can be seen in Fig. 3.1. Clearly, the sample is well-balanced among the two genders and the age classes appear normally distributed. Thus, we can safely conclude that the network fed with this particular data is less likely to be biased, since gender is equally represented and a representative group is used for all age classes.

When examining the biochemistry profile, we observe mostly normally distributed values among all variables, as can be seen in Fig. 3.18. The long tails observed in some variables suggest that more investigation could lead to the retrieval of valuable information.

3.5 Predicting BMI Based on Blood Exams

3.5.1 Neural Networks Versus SVCs, Challenges and Objectives

We begin by comparing the classification accuracy of Support Vector Classification (SVC) and neural network-based models in predicting BMI using blood exam data. The experiments were conducted by splitting the dataset into four and three weight classes. The accuracy of both models was found to be comparable, with a slight improvement observed when the classification was limited to three classes by excluding the underweight category, which had fewer available data points. When using three weight classes, the dataset was also balanced to ensure a fair comparison (Fig. 3.7).

The limited representation of the underweight category due to data scarcity prompted us to explore a one-class SVM approach. This allowed us to assess the system's ability to define classes based solely on blood exam data for each case under study.

One of the primary challenges encountered in our previous studies was the imbalance of data across weight classes. To address this, we ensured that the dataset was balanced during model training and evaluation. Another significant challenge was optimizing the classification process by reducing unnecessary computational overhead and minimizing resource costs (both in terms of computing power and processing time) while maintaining classification accuracy. This challenge was successfully mitigated by implementing a cascaded support vector classification approach, as detailed in Sect. 3.5.4.

3 Class Classifiers

Neural Net	10 fold CrossValidation Average
Testing	0.57
Training	0.58

SVC	10 fold CrossValidation Average
Testing	0.58
Training	0.62

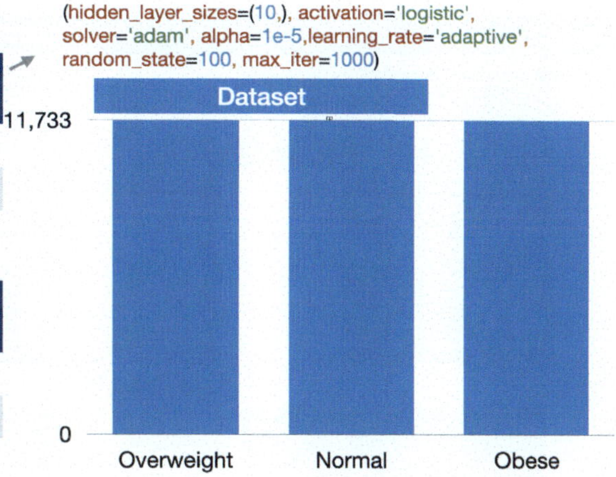

Fig. 3.7 Three-class classifier

3.5.2 *One-Class SVM Identifiers*

The dataset, consistent with earlier analyses, was further evaluated based on weight categories. The results are summarized in terms of the number of correct and incorrect predictions for both known (training) and unknown (testing) data. As illustrated in Fig. 3.8, the classifier effectively identifies the examined class even when presented with unseen data (testing sample). To assess classification accuracy, the test sample included randomly selected data points from all weight categories. This approach

OneClass SVM Confusion matrix ClassOnly Data

Training	FN	TP	Precision	MAE
Normal	222	4287	0.951	0.098
Obese	234	4266	0.948	0.104
Underweight	102	1866	0.948	0.104
Overweight	229	4271	0.949	0.102

Testing	FN	TP	Precision	MAE
Normal	23	477	0.954	0.092
Obese	38	462	0.924	0.152
Underweight	32	468	0.936	0.082
Overweight	31	469	0.938	0.124

OneClassSVM(gamma='scale', nu=0.05)

Fig. 3.8 One-class SVM

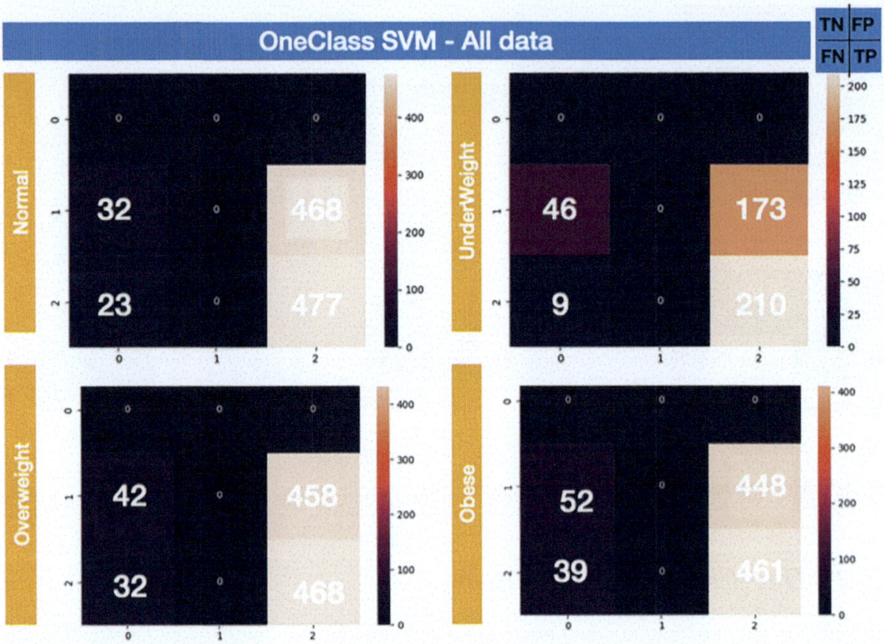

Fig. 3.9 Confusion matrices

ensured high-quality results and closely simulated real-world conditions. The test and evaluation outcomes, presented in Fig. 3.9, confirm that the blood exam dataset is well-defined, allowing the classifier to efficiently detect patterns and accurately assign classifications according to the predefined task. In all cases, the precision of the classifier was estimated to be approximately 92–95%, with a nu parameter set to 0.05, further demonstrating the robustness and reliability of the model.

3.5.3 One-vs-One SVC

An ensemble machine learning model aggregates predictions from multiple individual models to enhance overall performance. This approach is designed to outperform any single model within the ensemble by leveraging diverse perspectives on the data. A voting ensemble operates by combining predictions from multiple models and can be applied to both classification and regression tasks. In regression, the final prediction is obtained by averaging the outputs of all models. In classification, each model's predicted labels are aggregated, and the label with the majority vote is returned as the final prediction [10]. As illustrated in Fig. 3.10, the implementation of an ensemble model led to a significant improvement in classification accuracy, achieving a total average of 82% across all tested cases. The one-vs-one SVC classification scheme,

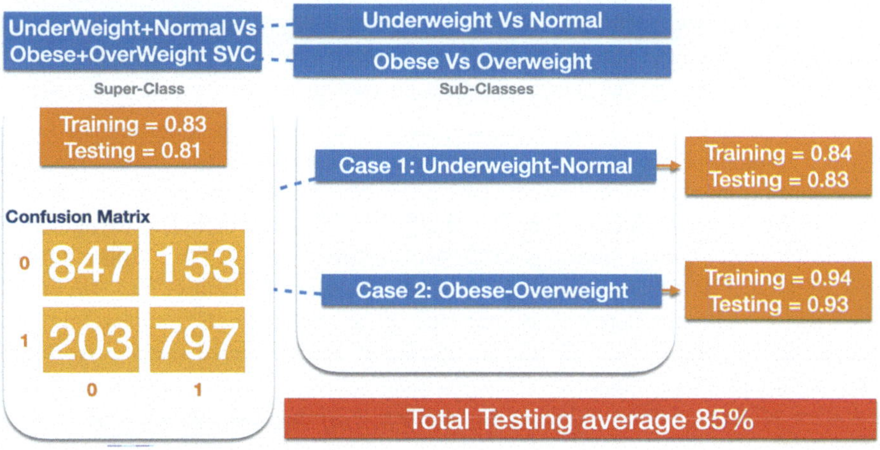

Fig. 3.10 Cascaded SVC system

OneVsOne NeuralNet classifier Accuracy 10 fold Cross Validation

Testing	OverWeight	Normal	UnderWeight	SubTotal Average
Obese	0.93	0.80	0.90	0.88
OverWeight		0.60	0.90	0.75
Normal			0.85	0.85
			TotalAverage	0.82

Training	OverWeight	Normal	UnderWeight	SubTotal Average
OBESE	0.95	0.79	0.95	0.89
OverWeight		0.64	0.90	0.77
Normal			0.86	0.86
			TotalAverage	0.84

Fig. 3.11 One-vs-one classifier

shown in Fig. 3.11, provides a foundation for a hard voting ensemble in categorizing inputs based on weight classification. However, as demonstrated in Sect. 3.5.4, a cascaded classifier was ultimately selected, as it delivered superior performance compared to alternative ensemble methods.

3.5.4 Cascaded SVC

Given the promising results of the one-vs-one SVC approach, we further developed and tested cascaded variations of it. Notably, a one-vs-one classifier cannot function as a standalone classification system; rather, it serves as a foundational testing framework upon which a more structured system can be built. Since a voting-based classification system could be computationally expensive and complex, we opted for a cascaded methodology utilizing SVCs, as illustrated in Fig. 3.10. To implement this cascaded classification scheme, we initially grouped the dataset into two broad super-classes. The first super-class consisted of samples from the underweight and normal weight categories, while the second super-class included samples from the overweight and obese categories. This grouping strategy combined extreme classes with their adjacent classes, reducing the number of primary classifications from four to two.

The classification process proceeded in two stages. The first binary classifier categorized a sample into one of the two major super-classes. Subsequently, a second classifier further refined the classification by distinguishing between the two constituent categories within the assigned super-class. Specifically:

- If the first classifier assigned a sample to the underweight-normal super-class, the second classifier further classified it as either underweight or normal.
- If the first classifier assigned a sample to the overweight-obese super-class, the second classifier further classified it as either overweight or obese.

This two-level cascaded classification approach ultimately assigned one of the four original BMI classes, similar to the initial one-vs-one classifier depicted in Fig. 3.11. However, empirical results demonstrated that the cascaded classifier significantly outperformed the direct one-vs-one approach. Specifically, the overall testing accuracy achieved with the cascaded system reached 85%, as shown in Fig. 3.10. For all experiments, approximately 4,000 data points were used, evenly distributed between weight categories (i.e., 2,000 samples from the overweight category and 2,000 from the normal weight category). A 10% subset of the data was reserved for validation (testing). The experiments were implemented in Python. All input data were normalized to a scale between 0 and 1, and a radial basis function (RBF) kernel was utilized.

To optimize classification performance, a grid search was conducted to determine the best values for the regularization parameter C and the kernel parameter γ. Initial experiments revealed that classification accuracy improved when C approached 1 and γ was closer to 0. Consequently, the search was refined in a second stage, where C was tested within the range of 0.9 to 1, and γ was varied between 0.025 and 0.225. These results are visualized in Fig. 3.12. The optimal accuracy was achieved when $C = 1$ and $\gamma = 0.05$ or $1/n_{\text{features}}$, where the number of features was 17.

Additionally, Table 3.4 presents a comparison of different kernel functions tested in the classification process along with their respective accuracy rates.

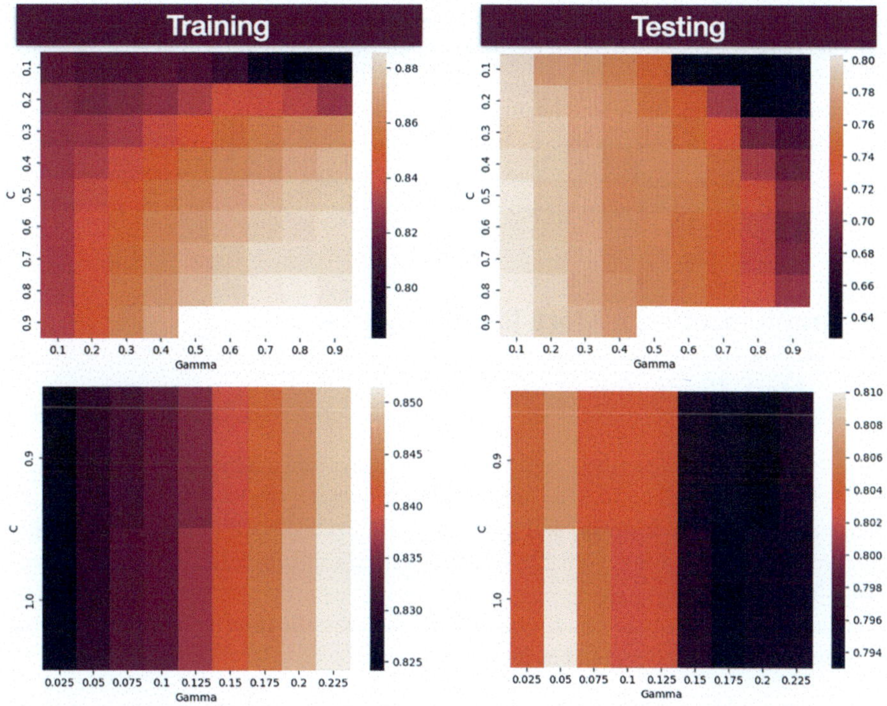

Fig. 3.12 Grid search heatmap

Table 3.4 Kernels in cascaded SVC and BMI super groups

Set/Kernels	rbf	Linear	Poly	Sigmoid
Training	0.83	0.804	0.8	0.7
Testing	0.81	0.803	0.785	0.7

While Support Vector Classifiers (SVCs) served as the primary classification approach in this research and were extensively analyzed, other machine learning classifiers were also examined and evaluated. Specifically, Gaussian Naïve Bayes, the Random Forest Classifier, and AdaBoost were tested, with their results summarized in Table 3.5. It is worth noting that the Random Forest Classifier demonstrated

Table 3.5 Kernels in cascaded SVC and BMI super groups

Set/Classifiers	Gaussian Naive Bayes	Random forest classifier	AdaBooster
Training	0.7511	0.83	0.805
Testing	0.732	0.72	0.781

promising results, with accuracy improving as the test sample size increased. However, due to its tendency to overfit the data, we ultimately opted for SVCs, which provided better class separability and generalization. Cascaded SVCs have been shown to exhibit high adaptability and accuracy in similar classification problems [31] and have also proven to be effective as a foundational approach in recommender systems [16]. Given their robustness and superior performance in our experiments, they were selected as the preferred classification method.

3.6 Implementing a More Precise System

3.6.1 MetS Prediction Model

Using the previously described methodology, we developed a model to classify individuals for MetS based on blood exam data. As shown in Fig. 3.13, MetS is identified when triglyceride, cholesterol, and glucose levels exceed normal ranges, following the "rule of three." Various MetS classification criteria and guidelines are summarized in Fig. 3.14 [13].

To enhance the precision and reliability of the classification system, we adopted the following four fundamental assumptions:

- While classifying BMI, blood exams did not include triglyceride, glucose, or cholesterol levels.

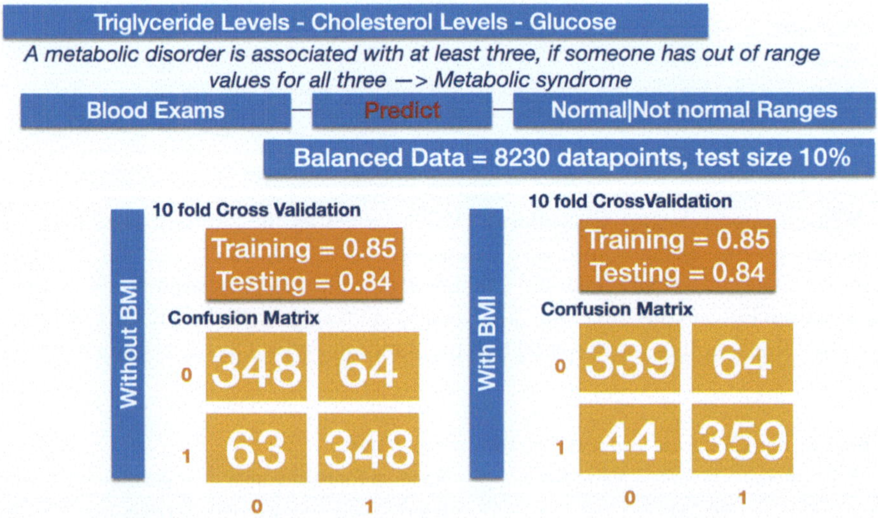

Fig. 3.13 MetS classifier-total cholesterol

	NCEP ATP III	WHO	EGIR	IDF
Required	None	Insulin Resistnace	Hyperinsulinemia	Central obesity
Criteria	**Any three of the five criteria**	**Insulin resistance or diabetes plus two of the five criteria below**	**Hyperinsulinemia plus two of the four criteria below**	**Obesity plus two of the four criteria below**
Obecity	Waist circumference	Waist to hip ratio	Waist circumference	Central obesity
Hyperglycemia	Fasting glucoce >=100 mg/dl or Rx	Insulin resistance	Insulin resistance	Fasting glucoce>=100mg/dl
Dyslipidemia	Triglycerides>=150mg/dl	Triglycerides>=150mg/dl	Triglycerides>177mg/dl	Triglycerides>150mg/dl
Dyslipidemia As a separate critereia	HDL cholesterol <40mgl/dl(M), <50mg/dl(F)	HDL cholesterol <35mgl/dl(M), <39 mg/dl(F)	HDL <39 mg/dl	HDL cholesterol <40mgl/dl(M), <50mg/dl(F)
Hypertension	>130 mmHg systolic|| 85>mmHg diastolic	>140/90 mmHg	>140/90 mmHg	>130 mmHg systolic|| 85>mmHg diastolic
Other		Microalbuminuria		

Fig. 3.14 MetS guidelines and classification criteria

- Total cholesterol is computed as a combination of LDL (bad cholesterol), HDL (good cholesterol), and 20% of triglycerides. When both total cholesterol and triglycerides are outside normal ranges, it suggests that LDL is more dominant than HDL, and HDL is more likely to be low. To validate this hypothesis, we compared classification results using total cholesterol versus only HDL as a feature.
- Since BMI was shown to be accurately predicted using blood exams—and given that BMI is a key factor in MetS—there is reason to believe that the classifier can identify metabolic patterns, making it applicable to more refined calculations.
- The tests were not gender-specific, assuming that any gender-related differences would be captured by the model. While the dataset was balanced, future experiments could investigate gender-specific variations further.

The SVC-based system achieved an average 10-fold classification accuracy of 84% in the tested sample. As shown in Fig. 3.15, our findings confirmed that when total cholesterol and triglycerides exceed normal ranges, LDL tends to dominate, while HDL is more likely to be low. Notably, BMI predictions remained stable whether total cholesterol or only HDL was used as a feature, as illustrated in Fig. 3.13. As shown in Fig. 3.15, using only HDL instead of total cholesterol as a classification feature yielded nearly identical results. This suggests that either variable could be utilized for classification. Given that HDL levels vary based on gender [6], we tested the hypothesis separately for male and female participants. The results were confirmed for both genders. A 10-fold cross-validation was performed on a 10% testing sample, drawn from the approximately 3,500 total samples. The classification accuracy was found to be 86% for males and 82% for females, with an overall accuracy of 84%, as shown in Fig. 3.15.

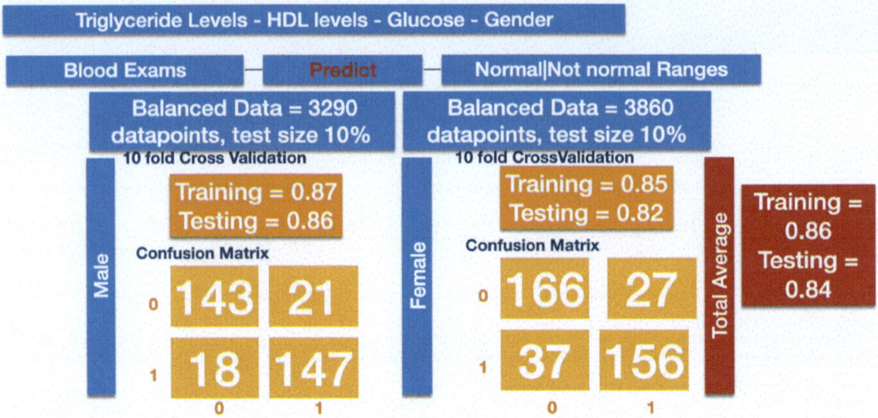

Fig. 3.15 MetS classifier-HDL

To further enhance our system, we incorporated a preliminary blood pressure prediction model, with plans to refine and expand it in future iterations. The corresponding results are displayed in Fig. 3.16. Specifically, the system was used to predict systolic blood pressure (SBP), distinguishing between values within and outside the normal range. Blood exam data, triglycerides, total cholesterol, and BMI were used as classification features. The 10-fold cross-validation accuracy on the testing sample was 74%, a promising result that can be further improved.

To evaluate the predictive capability of the system, three different scenarios were tested:

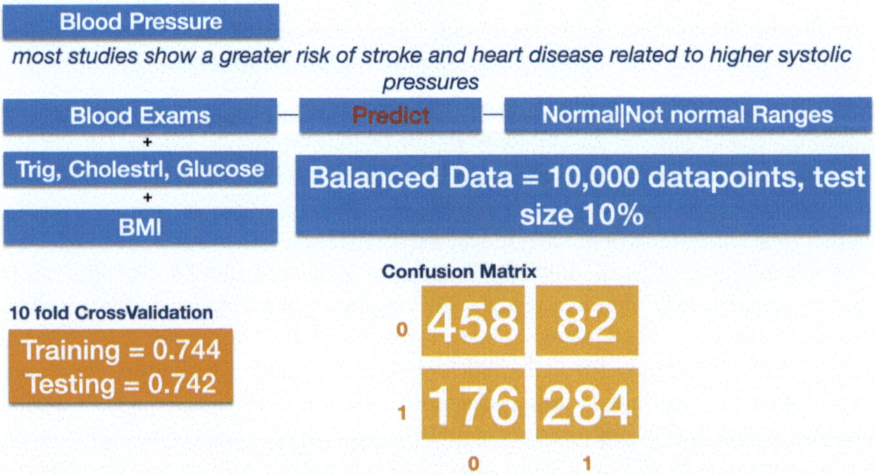

Fig. 3.16 Blood pressure classifier

1. Blood exams only—excluding key MetS indicators (triglyceride, glucose, HDL, or total cholesterol).
2. Full biochemistry profile—including all blood exam features.
3. Full biochemistry profile + BMI—incorporating BMI as an additional feature.

The third scenario consistently outperformed the other two, improving classification accuracy by approximately 7–8% in the 10-fold cross-validation process under otherwise identical conditions. Since the third scenario proved to be the most effective for predicting SBP, it was selected for the final system implementation. By first predicting MetS and BMI per class, as described in previous sections, the proposed methodology provides a valuable tool for estimating missing values and refining health assessments based on available patient data.

3.7 Discussion of Results

In this chapter, we introduced a high-performance system for predicting metabolic syndrome (MetS) based on blood exam data, as illustrated in Fig. 3.17. Our approach is designed to minimize the number of input parameters required while maintaining high classification accuracy. The system classifies individuals into BMI categories ("underweight," "normal," "overweight," and "obese") and determines MetS status (MetS "present" or MetS not "present") without directly relying on traditionally associated factors.

Beyond BMI and MetS classification, the system extends its functionality to predict additional health states, such as blood pressure status, using blood exam parameters. It also provides an assessment of related risk factors and generates recommendations, including dietary modifications, lifestyle changes, or medical interventions commonly adopted in clinical practice.

To evaluate the system's ability to infer body weight categories from blood exam data, we implemented one-class support vector machines (SVMs). By employing cascaded classifiers, we achieved an average classification accuracy of 85%, demonstrating the model's capability to optimize BMI categorization across all weight

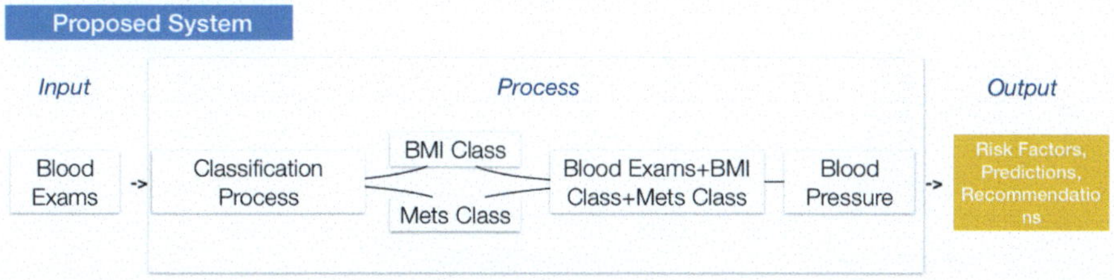

Fig. 3.17 Integration of classification methodologies

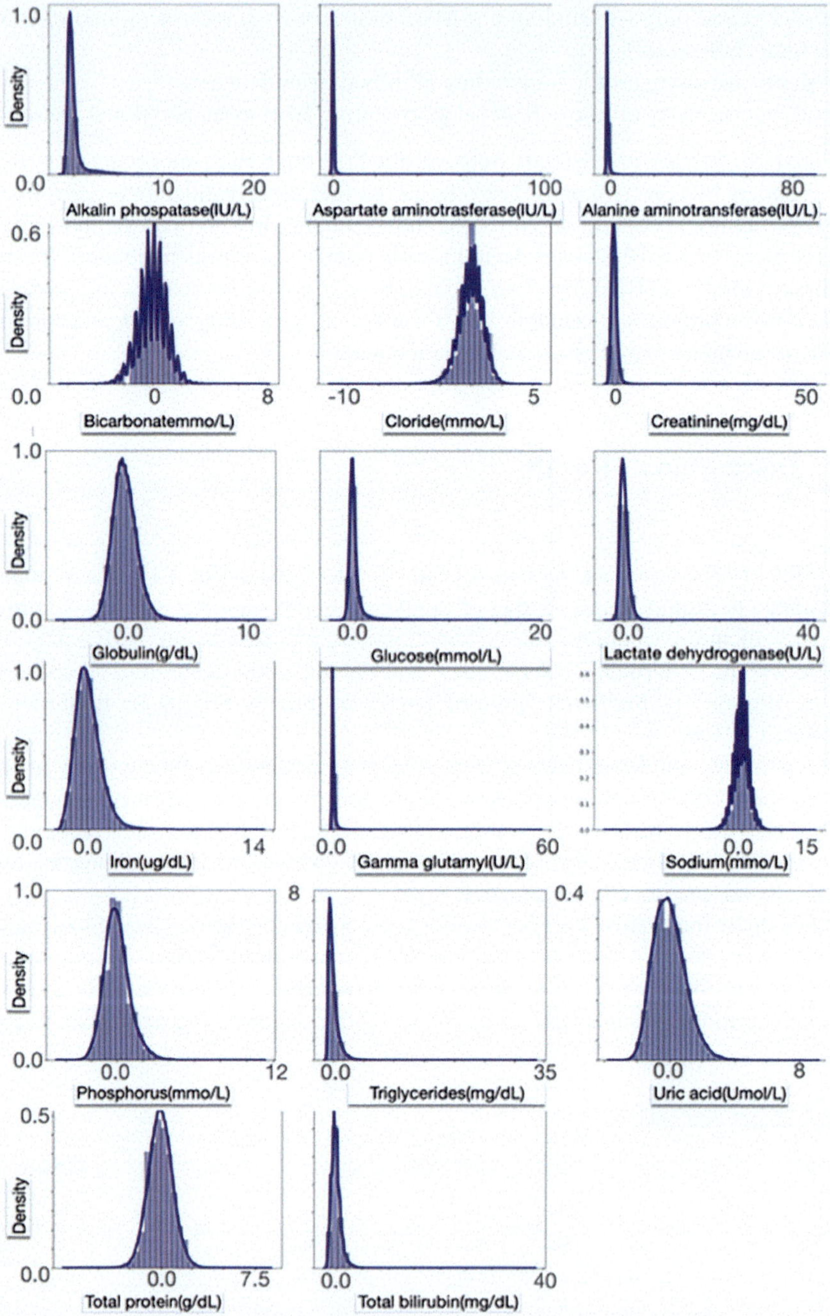

Fig. 3.18 Histograms and densities of blood features

classes using a standard biochemistry profile. This validation allowed us to extend our methodology to additional classification tasks.

Given the well-established relationship between body weight and metabolism, we applied the same approach to classify MetS status. The resulting system successfully identified MetS with an accuracy of 84%. Notably, total cholesterol and HDL levels produced similar classification results when used interchangeably as features, despite medical literature emphasizing HDL as a critical biomarker for MetS when combined with triglycerides and glucose.

Additionally, since high blood pressure serves as both a contributing factor and an outcome of MetS, we employed the same classification methodology to predict systolic blood pressure (SBP). Using a full biochemistry profile and BMI, the system achieved an accuracy of 74%, highlighting its potential as an effective predictive tool.

Through systematic experimentation with different feature selection scenarios, as discussed in previous sections, we determined that incorporating all parameters from Fig. 3.17 improved classification accuracy by approximately 8% compared to using only a partial biochemistry profile. Furthermore, since the system effectively predicts BMI and MetS status based on blood exam data, it can also function as a missing value imputation tool, ensuring robust classification even when some input values are unavailable.

Future work will continue to refine and expand this approach. Additionally, as shown in Fig. 3.17, a conceptualized system interface has been developed, and an optimized classifier will be deployed as part of its implementation.

3.8 Conclusions and Future Research Avenues

This study presents a significant advancement in the application of artificial intelligence in healthcare, introducing a streamlined, data-efficient system for health state classification. By leveraging AI-driven methodologies to reduce reliance on extensive diagnostic testing, this approach has the potential to reduce time and costs associated with medical assessments while maintaining high predictive accuracy. The integration of risk stratification and predictive modeling enables clinicians to use basic diagnostic exams to infer more complex health conditions. Furthermore, advanced pattern recognition and machine learning techniques applied to population health analysis could lead to the discovery of novel biomarkers. From a broader perspective, implementing policies based on AI-driven system recommendations could facilitate more effective and personalized healthcare interventions, guided by data-driven insights and predictive analytics.

Appendix

Abbreviations

The following abbreviations are used in this manuscript:

SBP	Systolic blood pressure
METS	Metabolic Syndrome
BMI	Body mass index
ASCVD	Atherosclerotic cardiovascular disease
HDL	High-density lipoprotein
LDL	Low-density lipoprotein
NCD	Noncommunable disease
SVM	Support vector machine
SVC	Support vector classifier
Trig	Triglycerides
Cholestrl	Cholesterol

References

1. A. Agrios, D.N. Sotiropoulos, G.A. Tsihrintzis, Computer aided breast cancer diagnosis: an SVM-based mammogram classification approach. Int. J. Comput. Intell. Stud. **7**(2), 138–163 (2018)
2. E. Alogogianni, M. Virvou, Interpretable machine learning for undeclared work prediction, in *2023 14th International Conference on Information, Intelligence, Systems & Applications (IISA)*. IEEE, pp. 1–8 (2023)
3. American Heart Association, *What Is High Blood Pressure?* (South Carolina State Library, Columbia, SC, USA, 2017), pp. 565–5698
4. S.M. Artham, C.J. Lavie, R.V. Milani, H.O. Ventura, Obesity and hypertension, heart failure, and coronary heart disease–Risk factor, paradox, and recommendations for weight loss. Ochsner J. **9**, 124–132 (2009)
5. H. Chen, B. Yang, D. Liu, W. Liu, Y. Liu, X. Zhang, L. Hu, Using blood indexes to predict overweight statuses: an extreme learning machine-based approach. PloS One **10**(11), e0143003 (2015)
6. M.T. Cooney, A. Dudina, D. De Bacquer, L. Wilhelmsen, S. Sans, A. Menotti, G. De Backer, P. Jousilahti, U. Keil, T. Thomsen et al., HDL cholesterol protects against cardiovascular disease in both genders, at all ages and at all levels of risk. Atherosclerosis **206**, 611–616 (2009)
7. R. Cui, H. Iso, K. Yamagishi, I. Saito, Y. Kokubo, M. Inoue, S. Tsugane, JPHC Study Group, High serum total cholesterol levels is a risk factor of ischemic stroke for general Japanese population: the JPHC study. Atherosclerosis **221**, 565–569 (2012)
8. S.S. Dagan, S. Segev, I. Novikov, R. Dankner, Waist circumference vs. body mass index in association with cardiorespiratory fitness in healthy men and women: a cross sectional analysis of 403 subjects. Nutr. J. **12**, 1–8 (2013)
9. J.L. Denson, A.S. Gillet, Y. Zu, M. Brown, T. Pham, Y. Yoshida, F. Mauvais-Jarvis, I.S. Douglas, M. Moore, and K. Tea et al., Metabolic syndrome and acute respiratory distress syndrome in hospitalized patients with COVID-19. JAMA Netw. Open **4**, e2140568 (2021)
10. T.G. Dietterich, Ensemble methods in machine learning, in *Archives of Internal Medicine* (Springer, Berlin/Heidelberg, Germany, 2000)

11. S.M. Grundy, Metabolic syndrome: a multiplex cardiovascular risk factor. J. Clin. Endocrinol. Metab. **92**, 399–404 (2007)
12. H. Hiroshi, T. Tetsuro, H. Shigenari, H. Toshifumi, S. Ikuo, Prediction of metabolic syndrome using artificial neural network system based on clinical data including insulin resistance index and serum adiponectin. Comput. Biol. Med. **41**, 1051–1056 (2011)
13. P.L. Huang, A comprehensive definition for metabolic syndrome. Dis. Model. Mech. **2**, 231–237 (2009)
14. H.A. Kakudi, L.C. Kiong, F.M. Moy, L.C. Kau, K. Pasupa, Diagnosis of metabolic syndrome using machine learning, statistical and risk quantification techniques: a systematic literature review. Malays. J. Comput. Sci. **34**, 221–241 (2021)
15. S. Khatiwada, S.K. Sah, R. Kc, N. Baral, M. Lamsal, Thyroid dysfunction in metabolic syndrome patients and its relationship with components of metabolic syndrome. Clin. Diab. Endocrinol. **2**, 1–5 (2016)
16. A.S. Lampropoulos, D.N. Sotiropoulos, G.A. Tsihrintzis, Cascade hybrid recommendation as a combination of one-class classification and collaborative filtering. Int. J. Artif. Intell. Tools **23**, 2685–2692 (2014)
17. A.S. Lampropoulos, D.N. Sotiropoulos, G.A. Tsihrintzis, Cascade hybrid recommendation as a combination of one-class classification and collaborative filtering. Int. J. Artif. Intell. Tools **23**(04), 1460009 (2014)
18. I. Lemieux, B. Lamarche, C. Couillard, A. Pascot, B. Cantin, J. Bergeron, G.R. Dagenais, J.P. Després, Total cholesterol/HDL cholesterol ratio vs LDL cholesterol/HDL cholesterol ratio as indices of ischemic heart disease risk in men: the Quebec Cardiovascular Study. Arch. Intern. Med. **161**, 2685–2692 (2001)
19. C. Lim, J.Y. Kim, Y. Nam, ECG signal analysis for patient with metabolic syndrome based on 1D-convolution neural network, in *2020 International Conference on Computational Science and Computational Intelligence (CSCI)*, Las Vegas, NV, USA, Dec 2020, pp. 731–733
20. A. Massaro, G. Ricci, S. Selicato, S. Raminelli, A. Galiano, Decisional support system with artificial intelligence oriented on health prediction using a wearable device and big data, in *2020 IEEE International Workshop on Metrology for Industry 4.0 IoT*. Roma, Italy, June 2020, pp. 718–723
21. S. Morteza, M. Yadollah, K. Anoshirvan, H. Farzad, Comparison of artificial neural network, logistic regression and discriminant analysis methods in prediction of metabolic syndrome. Iran. J. Endocrinol. Metab. **11**, 645–653 (2010)
22. D.M. Nathan, M.B. Davidson, R.A. DeFronzo, R.J. Heine, R.R. Henry, R. Pratley, B. Zinman, Impaired fasting glucose and impaired glucose tolerance: implications for care. Diab. Care **30**, 753–759 (2007)
23. P.M. Nilsson, J. Tuomilehto, L. Ryden, The metabolic syndrome–what is it and how should it be managed?. Eur. J. Prev. Cardiol. **26**, 33–46 (2019)
24. D.P. Panagoulias, D.N. Sotiropoulos, G.A. Tsihrintzis, Biomarker-based deep learning for personalized nutrition, in *Proceedings of the 33rd IEEE International Conference on Tools with Artificial Intelligence (ICTAI)*. Virtually, Nov 2021, pp. 73–122
25. D.P. Panagoulias, D.N. Sotiropoulos, G.A. Tsihrintzis, Nutritional biomarkers and machine learning for personalized nutrition applications and health optimization, in *Proceedings of the Twelfth IEEE International Conference on Information, Intelligence, Systems and Applications*. Chania, Greece, July 2021, pp. 731–733
26. D.P. Panagoulias, D.N. Sotiropoulos, G.A. Tsihrintzis, Nutritional biomarkers and machine learning for personalized nutrition applications and health optimization. Intell. Decis. Technol. **15**, 645–653 (2021)
27. D.P. Panagoulias, D.N. Sotiropoulos, G.A. Tsihrintzis, Towards personalized nutrition applications with nutritional biomarkers and machine learning, in *Advances in Assistive Technologies: Selected Papers in Honour of Professor Nikolaos G. Bourbakis - Vol. 3*, ed. by G.A. Tsihrintzis, M. Virvou, A. Esposito, L.C. Jain, vol. 28 (Springer, Cham, Switzerland, 2022), pp. 73–122
28. H. Ralf, G. Michael, *Artificial Intelligence Applications in Human Pathology* (World Scientific Pub Co Inc., Singapore, 2022), pp.1–216

29. S.Y. Sanisoglu, C. Oktenli, A. Hasimi, M. Yokusoglu, M. Ugurlu, Prevalence of metabolic syndrome-related disorders in a large adult population in Turkey. BMC Public Health **6**, 1–6 (2006)
30. A. Savvopoulos, M. Virvou, D.N. Sotiropoulos, G.A. Tsihrintzis, Clustering for user modeling in recommender e-commerce application: a rup-based intelligent software life-cycle, in *Knowledge-Based Software Engineering* (IOS Press, 2008), pp. 295–304
31. D.N. Sotiropoulos, G.A. Tsihrintzis, *Machine Learning Paradigms—Artificial Immune Systems and their Applications in Software Personalization*. Intelligent Systems Reference Library, vol. 118. (Springer, Berlin/Heidelberg, Germany, 2017), p. 330
32. D.N. Sotiropoulos, G.A. Tsihrintzis, Artificial immune system-based classification in extremely imbalanced classification problems. Int. J. Artif. Intell. Tools **26**(03), 1750009 (2017)
33. D. Sotiropoulos, C. Giannoulis, G.A. Tsihrintzis, A comparative study of one-class classifiers in machine learning problems with extreme class imbalance, in *IISA 2014, The 5th International Conference on Information, Intelligence, Systems and Applications*. IEEE, 2014, pp. 362–364
34. K. Strimbu, J.A. Tavel, What are biomarkers?. Curr. Opin. HIV AIDS **5**, 463 (2010)
35. D. Thomas, R. Kalakota, National health and nutrition examination survey: sample design. Future Healthc. J. **6**, 94 (2019)
36. G.A. Tsihrintzis, D.N. Sotiropoulos, L.C. Jain, *Machine Learning Paradigms: Advances in Data Analytics* (Springer, 2019)
37. G.A. Tsihrintzis, M. Virvou, H. Doukas, L.C. Jain, Introduction to advances in artificial intelligence-empowered decision support systems, in *Advances in Artificial Intelligence-Empowered Decision Support Systems: Papers in Honour of Professor John Psarras* (Springer, 2024), pp. 1–11
38. G.A. Tsihrintzis, M. Virvou, G. Phillips-Wren, Surveys in artificial intelligence-based technologies. Intell. Decis. Technol. **13**(4), 393–394 (2019)
39. A. Usman, L.J. Won, M.B.H. Syed, A. Taqdir, K.W. Ali, L. Sungyoung, The impact of big data in healthcare analytics, in *2020 International Conference on Information Networking (ICOIN)*. Barcelona, Spain, Jan. 2020, pp. 61–63
40. WHO (2021). *Obesity and Overweight*. https://www.who.int/news-room/fact-sheets/detail/obesity-and-overweight. Accessed: 2021-06-09
41. A. Worachartcheewan, C. Nantasenamat, C. Isarankura-Na-Ayudhya, V. Prachayasittikul, Quantitative population-health relationship (QPHR) for assessing metabolic syndrome. EXCLI J. **12**, 569 (2013)
42. S.-J. Xia, B.-Z. Gao, S.-H. Wang, D.S. Guttery, C.-D. Li, Y.-D. Zhang, Metabolic syndrome, machine learning, diagnosis model, symptoms, traditional Chinese medicine, physicochemical indexes. Biomed. Pharmacother. **137**, 111–367 (2021)

Chapter 4
Extreme Value Analysis Applied in Dietary Data

Abstract Biomarkers are quantifiable biological indicators that reflect an individual's health status. These measurements may involve a single variable or a combination of variables that collectively provide insights into physiological conditions. Biomarkers serve as early warning signals for potential health issues, enabling timely interventions at both individual and population levels. Nutritional biomarkers, in particular, assess the biological impact of dietary intake. In our previous work, we employed machine learning techniques to predict weight, metabolic syndrome, and blood pressure using blood-exam-based biomarkers. In this study, we leverage extreme value theory to assess the significance of outliers in health data, with a particular emphasis on dietary factors and standard biochemical profiles. Specifically, we demonstrate that extreme value analysis, combined with a systematic approach, can facilitate the prediction of health trends and partially automate health interventions. To validate our approach, we utilized publicly available datasets from the National Health and Nutrition Examination Survey (NHANES), a comprehensive program that evaluates the health and nutritional status of the U.S. population. Our analysis encompassed approximately 70,000 data points spanning a decade of observations, providing a robust basis for our findings.

Keywords Biomarkers · Bioinformatics · Extreme value theory · Extreme value analysis · Health prediction · Machine learning · Artificial intelligence · Information systems

4.1 Introduction

Predictive medicine relies on forecasting diseases through biomarker analysis, enabling the implementation of preventive measures to mitigate or altogether prevent disease progression. This approach is widely employed in cancer diagnosis and treatment. Biomarkers provide precise evaluation pathways and facilitate more effective treatment strategies by tailoring interventions to individuals or groups with similar biological characteristics. A well-known example of a biometric and biomarker is the resting heart rate, an indicator of cardiovascular function and overall physical

D. P. Panagoulias et al., *Artificial Intelligence-Empowered Bio-medical Applications*, Learning and Analytics in Intelligent Systems 49,
https://doi.org/10.1007/978-3-031-90174-4_4

fitness. An elevated resting heart rate is a recognized risk factor for coronary heart disease. Another critical biomarker associated with cardiovascular conditions-including heart attacks, congenital heart defects, coronary artery disease, and pancreatitis-is the triglyceride level, which is measured via blood analysis. Additionally, T-cells, a subset of white blood cells involved in immune defense, are linked to various conditions such as cancer, atherosclerosis, Alzheimer's disease, and mortality risk.

4.1.1 Nutritional Variables and Data Collection

Diet and nutrition play a fundamental role in health, yet assessing their impact presents challenges in both analytical and data collection contexts. Nutritional assessment involves recording dietary intake over a specified period, encoding this information, and processing it using food composition tables to estimate energy, nutrient, and other dietary intakes.

Various nutritional assessment methodologies exist, each with distinct advantages and limitations. Selecting the most appropriate method depends on the study's objectives [9]. One widely used method is the food frequency questionnaire (FFQ), often combined with single or recurring 24-h dietary recalls. Other techniques include food records (food diaries), portion size estimation (using standardized or population-average portion sizes, images, or food models), diet history, checklists, direct observation, and dietary screening [9]. In recurring 24-h recall methods, trained interviewers or automated self-administered systems record detailed dietary intake, including the type, quantity, brand, and preparation methods of foods and beverages consumed. This process is supported by structured questionnaires and memory-aiding images designed to minimize bias.

To promote balanced nutrition and prevent malnutrition, Food-Based Dietary Guidelines (FBDG) are employed as a strategic approach to encourage healthier food choices while considering environmental sustainability [3]. FBDGs aim to influence consumer behavior by educating individuals about nutrients, food, and beverage choices [11].

In this study, we utilized publicly available data from the National Health and Nutrition Examination Survey (NHANES), a large-scale program assessing the health and nutritional status of the U.S. population. Dietary data were collected using the FFQ combined with 24-h recall, ensuring a comprehensive evaluation of dietary patterns.

4.1.2 Scope of the Study

Nutritional biomarkers can serve multiple purposes: they may validate dietary interventions, act as surrogate indicators of dietary habits, or directly measure nutrient levels. When generic biomarkers are insufficient for drawing conclusions about specific food components, combining dietary intake methods with nutritional biomarkers provides more comprehensive insights [30].

This chapter focuses on identifying trends in nutrient intake and analyzing the distribution of extreme values in dietary data. We analyzed approximately 70,000 data points spanning over a decade (2000–2014). Our methodology applies extreme value theory and correlation techniques to examine dietary patterns across different weight classifications. The overarching objective is to automate pattern identification in nutrient intake and its relationship with weight, thereby enabling personalized nutrition interventions through intelligent information systems. By providing actionable insights, this research contributes to consumer education and behavior modification in dietary habits.

To clarify the examined hypotheses, we employed a machine learning framework based on the principles of what, why, and how, as described in [27].

This chapter is structured as follows:

- Section 4.2 presents an overview of nutritional biomarkers and nutritional epidemiology, along with key applications of extreme value theory in medicine.
- Section 4.3 discusses challenges and research issues relevant to this study.
- Section 4.4 details the methodology used for data analysis.
- Section 4.5 explains the implementation process and presents selected case studies to illustrate the practical usability of our approach. The section also discusses key findings.
- Section 4.6 concludes the study, summarizing insights and proposing future research directions.

4.2 Literature Review and Related Work

The Global Burden of Disease study identifies suboptimal diet as the leading risk factor for morbidity and mortality, surpassing even smoking [32].

To assess the burden of disease, time-based metrics such as Disability-Adjusted Life Years (DALYs) are used. DALYs incorporate two components:

- Years of Life Lost (YLL) due to premature mortality.
- Years Lost due to Disability (YLD), calculated based on the duration and severity of disability [38].

According to Herforth et al. [32], in 2019, dietary risk factors were responsible for approximately 8 million deaths and 180 million DALYs worldwide.

Traditional epidemiology has contributed significantly to identifying many key lifestyle and environmental risk factors for chronic disease. Traditional epidemiology has played a crucial role in identifying key lifestyle and environmental risk factors for chronic diseases. Systems epidemiology integrates standard epidemiological methodologies with modern technologies to enhance the understanding of biological and metabolic pathways.

Similarly, nutrition research benefits from combining traditional approaches with technological advancements to optimize knowledge creation and dissemination. Diet plays a pivotal role in health, and effective dietary data collection methods have significantly contributed to raising awareness [7].

Contrary to some claims, there is limited evidence suggesting that metabolic effects significantly hinder weight loss [8]. Instead, adherence to dietary prescriptions has been identified as the primary determinant of weight loss success. Studies analyzing various diet plans with differing macronutrient compositions confirm that compliance with the prescribed diet is the strongest predictor of success [8].

Preventive dietary strategies, when integrated into public health guidelines, can positively influence eating behaviors and contribute to population-wide weight management [14].

4.2.1 Research Continuity

Building on our previous research, we have achieved high precision in predicting key health states using blood-exam-based features, specifically the standard biochemistry profile.

These methodologies have been incorporated into a decision-support system that processes blood test results, evaluates health status, provides recommendations, and facilitates automated strategic health interventions [23, 25, 26].

Additionally, regulatory considerations have been explored in [27], emphasizing the importance of validation and regulatory compliance in the development of AI-driven health applications.

4.2.2 Theory and Application Highlights of Extreme Value Analysis

This section outlines the fundamental principles of Extreme Value Analysis (EVA), which is widely applied in health data analysis. EVA is primarily approached through two methodologies:

1. Block Maxima (Minima) Method

    - Extracts annual maximum (or minimum) values from time-series data, generating an Annual Maxima (Minima) Series (AMS).

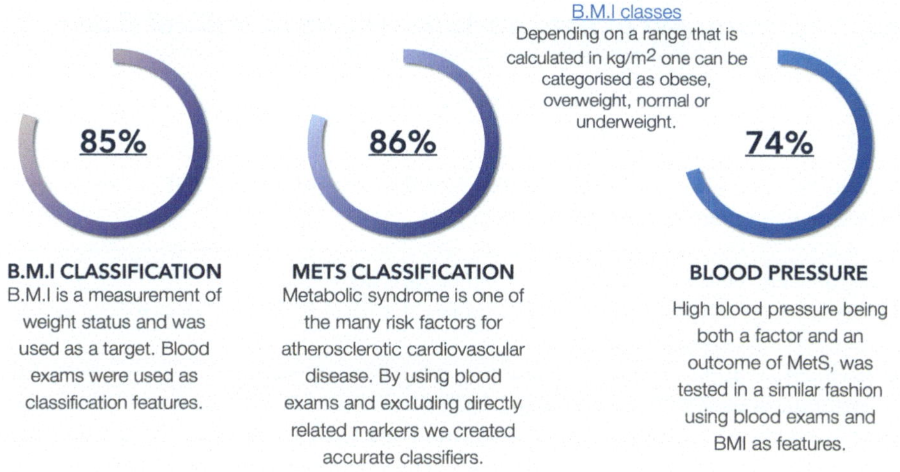

Fig. 4.1 Developed methodologies and outcomes

- AMS datasets are commonly analyzed using the Fisher-Tippett-Gnedenko theorem, which supports the fitting of the Generalized Extreme Value (GEV) distribution [6].

2. Peak-Over-Threshold (POT) Method

- Identifies peak values exceeding a predefined threshold.
- Two probability distributions are typically fitted:
  - One describing the frequency of threshold exceedances.
  - Another characterizing the size of the extreme values.
- The Pickands-Balkema-De Haan theorem states that POT extreme values asymptotically follow a Generalized Pareto Distribution (GPD), while event occurrences are modeled using a Poisson distribution [40].

The return level (R.V.) of extreme values can be approximated using the fitted distribution. This corresponds to the expected threshold exceedance once every interval T, with probability 1/T.

The Probability Density Function (PDF) of a continuous random variable represents the relative likelihood of a given observation occurring within the sample space [6].

4.2.3 Applications of Extreme Value Theory in Medical Data

Extreme Value Theory (EVT) is extensively used in medical data analysis. Some notable applications include:

- Prediction of Weekly Mortality Rates: Maud et al. applied EVT to forecast weekly deaths from pneumonia and influenza [36].
- Emergency Department Visit Patterns: EVT was used to model daily emergency department visit volumes and extreme occurrences [36].
- Mortality and Morbidity Analysis: Chiu et al. utilized EVT to assess mortality and morbidity patterns [5].
- Body Mass Index (BMI) Distribution: Flegal et al. analyzed age-specific BMI extremes using CDC growth charts [10].
- Monte Carlo Simulations for Estimators: EVT-based estimators were evaluated for faster parameter estimation in alpha-stable interference models [37].
- Risk Estimation in Finite-Time Systems: Arsenault et al. used EVT to estimate risk in systems where data collection is expensive or infeasible [1].
- Glucose Monitoring in Diabetes: EVT, specifically block maxima analysis, has been applied to track rare and extreme glucose fluctuations [35].

More EVT applications exist in the literature, but here we highlight those most relevant to our research.

4.2.4 Statistical Measures in Extreme Value Analysis

To characterize probability distributions, we compute L-moments, which are linear combinations of order statistics (L-statistics). These serve as analogs to conventional moments:

- L-Kurtosis L_4 : Measures the tail heaviness of a distribution.
  $L_4 = \frac{n * \Sigma n_i (Y_i - \tilde{Y})^4}{(\Sigma n_i (Y_i - \tilde{Y})^2)^2}$
  where Y_i represents the ith variable, $\tilde{Y}$ is the mean, and n is the sample size.
- L-Skewness $\tilde{\mu}3$: Measures asymmetry in the probability distribution.
  $\tilde{\mu}3 = \frac{\Sigma_i^N (X_i - \tilde{X})^3}{(N-1) * \sigma^3}$
  where X_i denotes random variables, $\tilde{X}$ is the mean, N is the sample size, and σ is the standard deviation.

4.3 Key Issues and Challenges

4.3.1 The Patient's Journey in the Healthcare Ecosystem

The patient's journey represents a nonlinear process that integrates various components of the healthcare ecosystem. Multiple stakeholders interact both directly and indirectly. In the micro-space, direct relationships exist between patients, physicians, and healthcare staff, including nurses and administrative personnel. Indirect relationships extend to government bodies, insurance agencies, regulatory authorities, pharmaceutical companies, and other entities beyond the scope of this study. Any systemic approach aimed at automating processes within the healthcare ecosystem constitutes an intervention in the patient's journey [27].

The journey typically begins with symptom identification, prompting a visit to a healthcare professional. As depicted in Fig. 4.2, a diagnosis is followed by laboratory examinations-both generic and specialized-leading to a prescribed treatment plan, if necessary. Follow-ups ensure the assessment of the patient's health status, determining whether further intervention or lifestyle modifications are required [25, 27]. Some cases demand continuous monitoring and specialized treatments, reinforcing the importance of patient engagement. A well-aligned healthcare ecosystem fosters patient compliance, enhancing adherence to treatment plans and encouraging proactive health management [26].

Big data and artificial intelligence (AI) play pivotal roles in decision-making, especially when integrated into intelligent knowledge systems [27]. Advanced machine learning pipelines provide new foundations for diagnostics and pathology analytics [13]. Furthermore, identifying extreme values and outliers in medical datasets can uncover novel intervention pathways, improving patient outcomes.

4.3.2 The Golden Circle of Innovation Adapted for Machine Learning

The Golden Circle of Innovation is a conceptual framework comprising three fundamental questions: what, why, and how [33]. In this study, we adapt this model to the

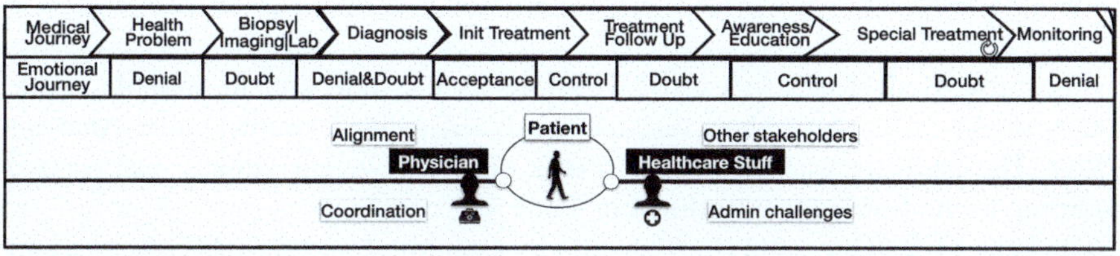

Fig. 4.2 Patient's journey

domain of machine learning, facilitating a structured approach to problem-solving and system design [27].

The What

By analyzing blood test results, we obtain a preliminary assessment of weight classification and metabolic status [20–22, 24, 28]. This analysis reveals distinct patterns across weight classes, where outliers within each weight cluster represent potential transition points from one class to another. By examining these outliers, additional dietary and metabolic patterns can be uncovered, enabling a deeper understanding of health indicators.

The Why

If patterns in dietary variables can be recognized, it becomes possible to automate recommendations and dietary strategies. This allows for the development of personalized interventions, tailored to an individual's specific health metrics and nutritional needs.

The How

Extreme values (outliers) are data points that appear infrequently within the tails of a univariate or multivariate distribution. Understanding and managing these outliers is essential for effective data analysis. Leveraging these values enables a more precise fine-tuning of dietary intake based on personalized and specific parameters. In the following sections, we detail the methodology employed to address these challenges through Extreme Value Analysis (EVA). The implementation of our approach will further illustrate the capabilities and practical utility of the proposed system.

4.4 Methodology

4.4.1 Technical Overview

In this study, Body Mass Index (BMI) serves as the independent variable, while dietary intake is the dependent variable, analyzed using Extreme Value Analysis (EVA). The analysis was conducted across different weight classes, categorized according to the BMI ranges outlined in Table 4.1.

The Peak-Over-Threshold (POT) method was applied, sorting BMI observations in ascending order. Three distinct dietary intake scenarios were examined per weight class:

Table 4.1 BMI classification

Range	Category
Less than 18	Underweight
18–25	Normal weight
25–30	Overweight
Above 30	Obese

1. Above-threshold consumption: The dietary variable exceeds a predefined threshold.
2. Sub-threshold consumption: The dietary variable is above zero but below the threshold.
3. Zero consumption: No intake of the specific dietary variable.

Dietary EVA Algorithm

Figure 4.3 presents the designed Dietary EVA Algorithm. The process follows these steps:

1. Selection of BMI class and dietary variable for analysis.
2. Descriptive statistical analysis to explore the dataset.
3. Threshold setting for identifying extreme values.
4. Extreme value plotting and L-moment computation.
5. Threshold evaluation and reiteration if needed.
6. Return value (R.V.) extraction and trend-based predictions.

The EVA methodology was implemented using Python libraries [31], with necessary modifications and adjustments to optimize performance.

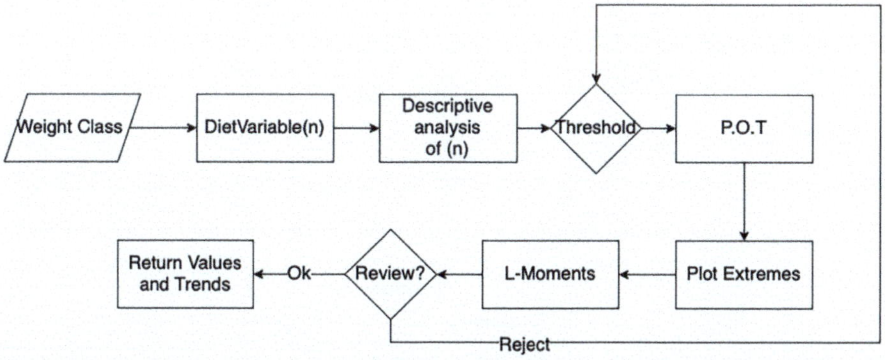

Fig. 4.3 Dietary variable analysis and intervention algorithm

4.4.2 EVA-Based Decision-Making Algorithm

Decision-making algorithms process task-related information from their environment to generate actionable insights. In healthcare, such algorithms use vital markers to assist in diagnosis and treatment planning. For instance, in [2], a food recommendation system for chronic disease patients was developed using a hybrid clustering method. Similarly, intelligent agents, as defined in [17], are software-driven entities that interact with an environment, making decisions based on observed data.

Figure 4.4 illustrates the proposed EVA-Based Decision-Making Algorithm, which integrates the previously described Dietary EVA Algorithm with clustering and correlation analysis:

1. Apply Dietary EVA Algorithm: Generate output plots and return values.
2. Extract Three Clusters:

   - (e) Above-threshold consumption.
   - (o) Below-threshold but nonzero consumption.
   - (z) Zero consumption.

3. Analyze Blood Biomarker Correlations for each cluster.
4. Filter Correlations using a predefined threshold (e.g., ±55%).
5. Compare Correlations Across Clusters:

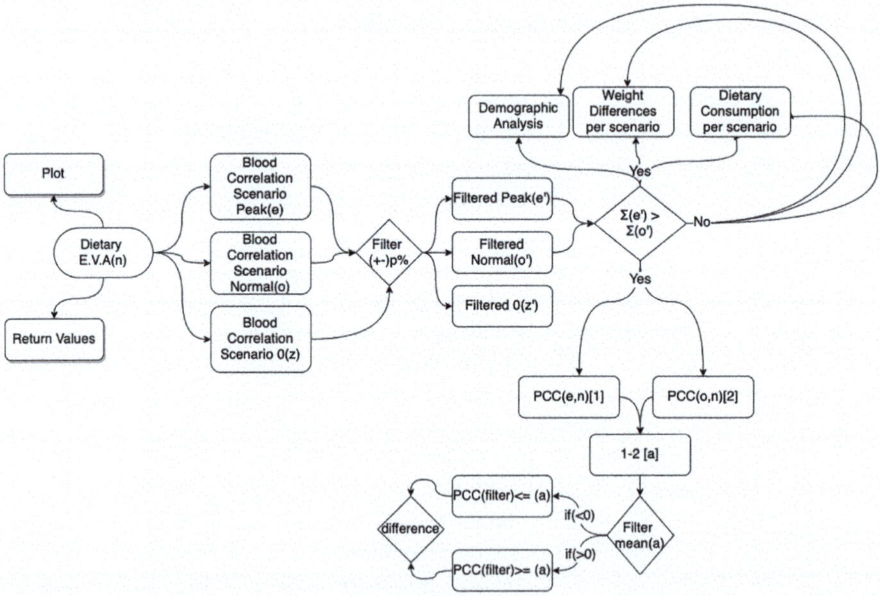

Fig. 4.4 EVA-based decision-making algorithm

- If scenario (e') shows increased correlations versus (o'), additional demographic and dietary analyses are conducted.
- If Total(e') > Total(o'), the Pearson Correlation Coefficient (PCC) between the dietary variable and blood biomarkers is extracted, determining whether the relationship is positive or negative.
- If correlations between (e') and (o') are equal or (e') < (o'), only dietary consumption per scenario is examined.

This decision-making framework predicts weight trends, determines an optimal dietary intake range, and assesses health progression based on blood biomarker correlations. Moreover, by outlining demographic patterns per scenario, the system enhances explainability, empowering users to visualize the benefits of dietary adjustments.

4.4.3 Business Issue and Systemisation

A system is a structured series of repeatable processes and operations, while systemization refers to the integration, optimization, and automation of these processes. Figure 4.5 presents the systemized EVA decision-making process, incorporating the Golden Circle of Innovation (What, Why, How) [33].

- What: The intelligent system automates dietary analysis and trend detection using EVA, providing actionable insights.
- Why: Systemization enhances health interventions, improving precision in dietary recommendations.
- How: User input (dietary intake) is analyzed using comparative EVA-based analysis, producing insights for optimized dietary strategies.

The system also includes an explainability feature, allowing users to understand how decisions are made. The user journey involves:

1. Inputting dietary intake into the system.
2. Predicting weight class using the Weight Class Predictor [27].
3. Validating weight class and proceeding to EVA-based analysis.
4. Receiving personalized dietary recommendations in a detailed report.

This approach enhances user engagement by providing clear, evidence-based recommendations tailored to individual health profiles.

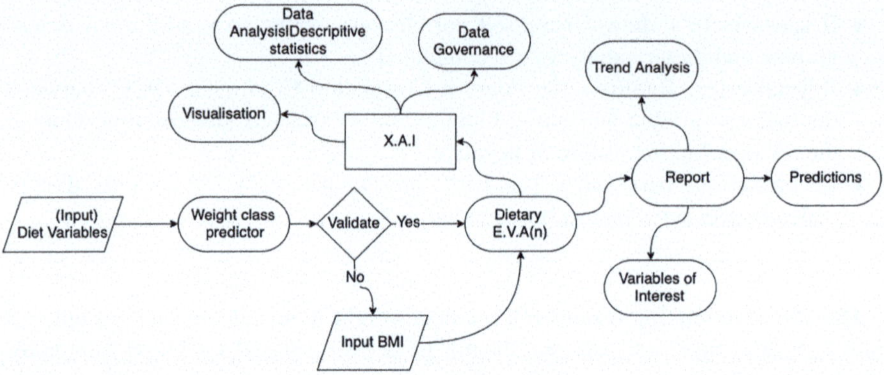

Fig. 4.5 Systemization of the EVA decision-making process

4.5 Implementation

4.5.1 Overview

The diet-related data analyzed in this study represent nutrients obtained from food, beverages, and water (tap or bottled). Nutrients from supplements or medications are not included in the dataset. Data collection was conducted in the United States using two 24-h dietary recall interviews (see Sect. 4.1.1) and was standardized according to the National Health and Nutrition Examination Survey (NHANES), operated by the Centers for Disease Control and Prevention (CDC) [15].

A comprehensive list of analyzed dietary intake variables is provided in Table 4.2, while the examined blood biomarkers, along with their aliases and respective measurement units, are detailed in Table 4.3. The results of the implemented EVA pipeline are summarized in Fig. 4.3, and the outputs of the EVA-based decision-making algorithm are examined.

The following dietary variables serve as examples to illustrate the proposed methodology: Vitamin C for the overweight category and Alcohol for the obese category.

4.5.2 Data analysis

Figure 4.6 presents selected metrics and demographics from the analyzed dataset. This section details the descriptive data analysis, outlines the Extreme Value Analysis (EVA) results, and introduces the user report generated by the systemization algorithm (Fig. 4.5).

Table 4.2 Dietary variables examined—system input data

Name	Name
Alpha-carotene (mcg)	Alcohol (gm)
Vitamin E (mg)	Vitamin B12 (mg)
Beta-carotene (mcg)	Caffeine (mg)
Calcium (mg)	Carbohydrate (gm
Total choline (mg)	Cholesterol (mg)
Copper (mg)	Beta-cryptoxanthin (mcg)
Folic acid (mcg)	Folate equivalents (mcg)
Food folate (mcg)	Dietary fiber (gm)
Total folate (mcg)	Iron (mg)
Energy (kcal)	Lycopene (mcg)
Lutein + zeaxanthin (mcg)	MFA (16–22):1 (gm)
Magnesium (mg)	Fatty acids (gm)
Moisture (gm)	Niacin (mg)
Phosphorus (mg)	Potassium (mg)
Protein (gm)	Retinol (mcg)
Selenium (mcg)	fatty acids (gm)
Sodium (mg)	Total sugars (gm)
Total fat (gm)	Theobromine (mg)
Vitamin A (mcg)	Vitamin B1 (mg)
Vitamin B12 (mcg)	Vitamin C (mg)
Vitamin D (D2 + D3) (mcg)	Vitamin K (mcg)
Selenium (mcg)	Zinc (mg)

The dataset is categorized based on BMI-defined weight classes (Table 4.1), consisting of:

- Obese: 23,000 data points
- Overweight: 21,000 data points
- Normal weight: 20,000 data points
- Underweight: 2,100 data points

Additionally, the dataset is gender-balanced with an equal distribution of males and females. Other demographic attributes include age, educational level, pregnancy status, and marital status. These variables were used either for training the EVA algorithm or for categorizing individuals based on their relation to the identified clusters. The following section applies the algorithms described in Sect. 4.4, analyzing key variables to demonstrate the functionality of the EVA-based decision-making system.

The derived results are divided into three key components:

1. Extreme Value Analysis (EVA) Outputs
   –> Identification and plotting of peaks over threshold (POT) values.

Table 4.3 Standard Biochemistry Profile

Alias	Name (Measurement)
LBXSAPSI	Alkaline phosphatase (IU/L)
LBXSASSI	Aspartate aminotransferase AST (IU/L)
LBXSATSI	Alanine aminotransferase ALT (IU/L)
LBXSC3SI	Bicarbonate (mmol/L)
LBXSCLSI	Chloride (mmol/L)
LBXSCR	Creatinine (mg/dL)
LBXSGB	Globulin (g/dL)
LBDSGLSI	Glucose, refrigerated serum (mmol/L)
LBXSGTSI	Gamma glutamyl transferase (U/L)
LBXSIR	Iron, refrigerated serum (ug/dL)
LBXSLDSI	Lactate dehydrogenase (U/L)
LBXSNASI	Sodium (mmol/L)
LBDSPHSI	Phosphorus (mmol/L)
LBXSTB	Total bilirubin (mg/dL)
LBXSTP	Total protein (g/dL)
LBXSTR	Triglycerides, refrigerated (mg/dL)
LBDSUASI	Uric acid (umol/L)

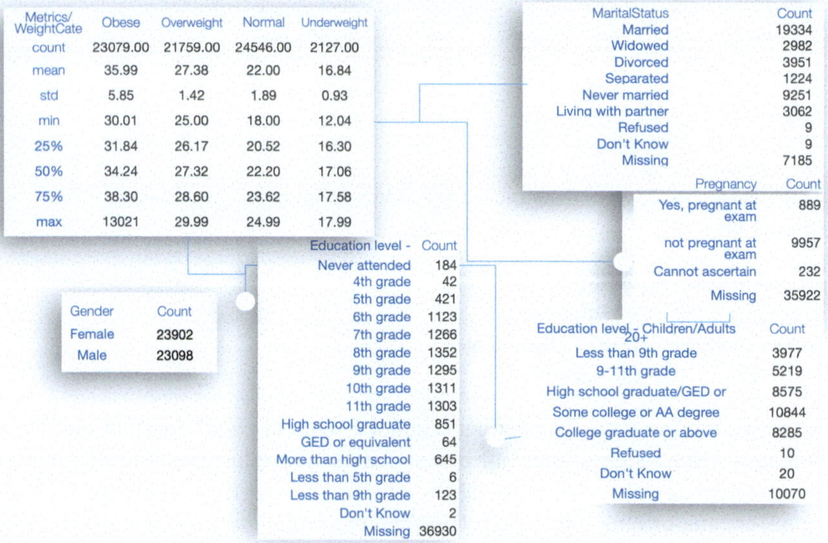

Fig. 4.6 Descriptive analysis of data and selected demographics

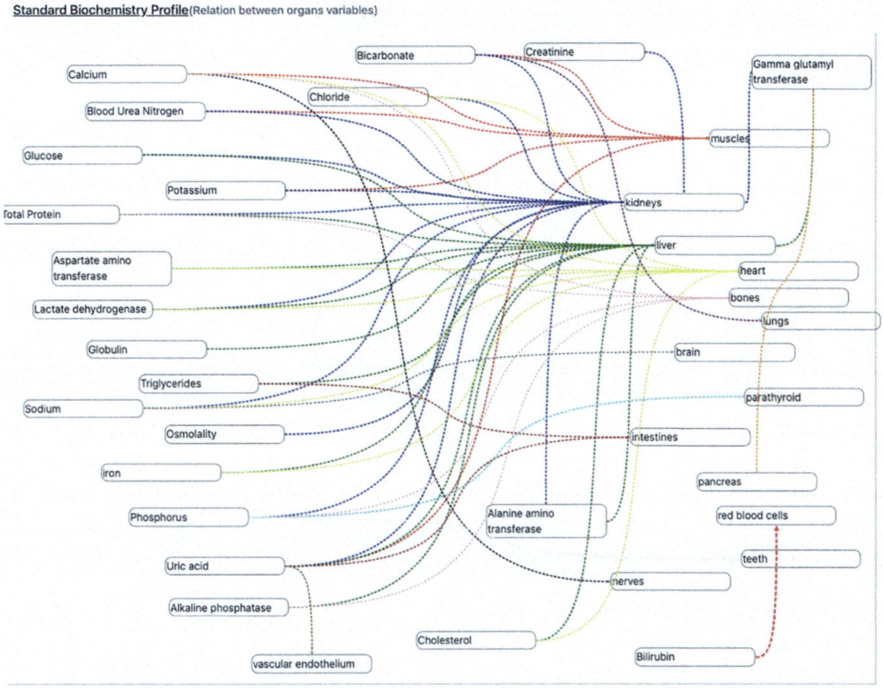

Fig. 4.7 Blood, diet, and organ functionality (screenshot from NUHealthSoft [26])

2. Return Value and Probability Density Analysis
   –> Return value plots and probability density plots based on extreme value distributions.
   –> Calculation of L-moments to define the distribution of extremes.
3. Correlation Analysis of Blood Variable
   –> Evaluation of blood biomarker correlations under different dietary scenarios.
   –> Comparative analysis based on BMI classification, dietary intake, and demographics.

The following sections provide case studies demonstrating how the analyzed data can be utilized by external users through the systemization algorithm (Fig. 4.5).

4.5.3 System Analysis and Pipeline Outputs

Vitamin C Intake in Overweight Individuals

Figure 4.8 illustrates the Vitamin C intake analysis for the overweight category using the proposed EVA-based algorithm. The threshold (t) for peak detection was set as a percentage (p) of the maximum observed intake:

$t = \max(\Sigma X_n) \times p$

For the overweight category, p = 0.34, resulting in a threshold of t = 668 mg. BMI values were sorted in ascending order, and for T = 1.6 (BMI threshold), the extreme return value for Vitamin C intake was estimated at 1017 mg, with a confidence interval of 1010–1047 mg, occurring with a probability of 1/T.

A comparison of blood biomarker correlations (Fig. 4.9) across dietary intake scenarios, following the decision-making algorithm (Fig. 4.4), indicates a higher correlation strength in the extreme intake scenario (e') compared to the normal intake scenario (o').

Figure 4.10 provides demographics and key findings, which serve as decision-making benchmarks. These include:

- Demographic attributes per scenario
- Significant blood biomarker correlations
- Average dietary intake per scenario

For the normal weight category, where Vitamin C intake is within normal limits, the average consumption was 86 mg, serving as a benchmark for dietary recommendations.

According to the EVA-based decision-making algorithm, the most prominent blood biomarker correlations in scenario (e') include:

- Aspartate aminotransferase, globulin, glucose, gamma-glutamyl transferase, iron, phosphorus, and triglycerides.

The associated organ impact multipliers include:

- Heart (2×), liver (6×), kidneys (3×), pancreas, bones, parathyroid, and intestines.

Figure 4.7 visualizes the relationship between blood biomarkers and organ functionality, derived from NUHealthSoft [26].

The impact of extreme Vitamin C consumption on BMI is minimal, with BMI values slightly higher in the normal intake range compared to the extreme range (ΔBMI = 0.074). A positive correlation was observed between Vitamin C intake and blood biomarkers, with stronger correlations in extreme consumption cases.

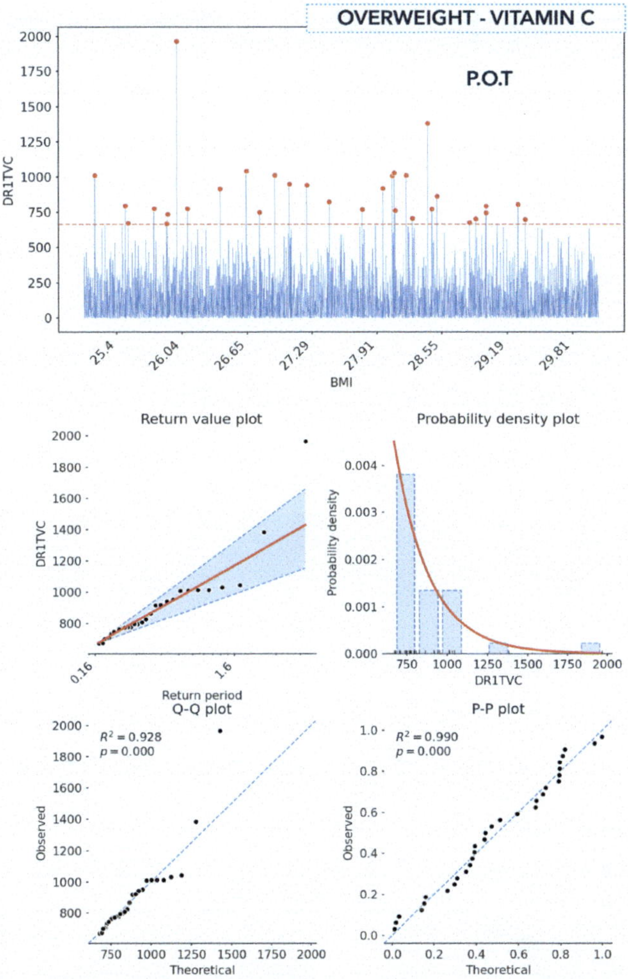

Return Intervals	return value	lower ci	upper ci
0.16	-inf	NaN	NaN
0.32	816.15	813.39	828.87
0.80	1017.27	1010.75	1047.30
1.60	1169.42	1160.06	1212.54
4.00	1370.54	1357.42	1430.98
8.00	1522.68	1506.72	1596.22
16.00	1674.83	1656.02	1761.46
40.00	1875.95	1853.39	1979.90
80.00	2028.09	2002.69	2145.14
160.00	2180.24	2151.99	2310.38

Fig. 4.8 Overweight category–EVA of vitamin C (DR1TVC)

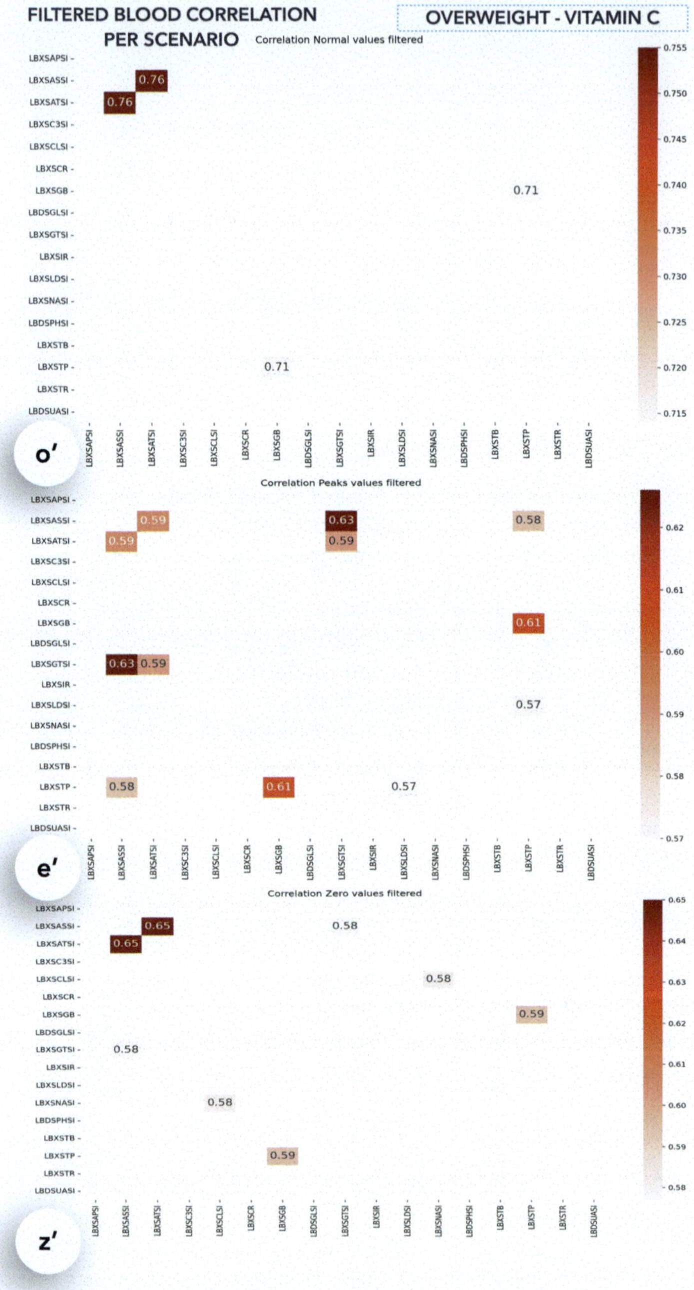

Fig. 4.9 Overweight category–vitamin C. Filtered blood variable correlation as per EVA pipeline

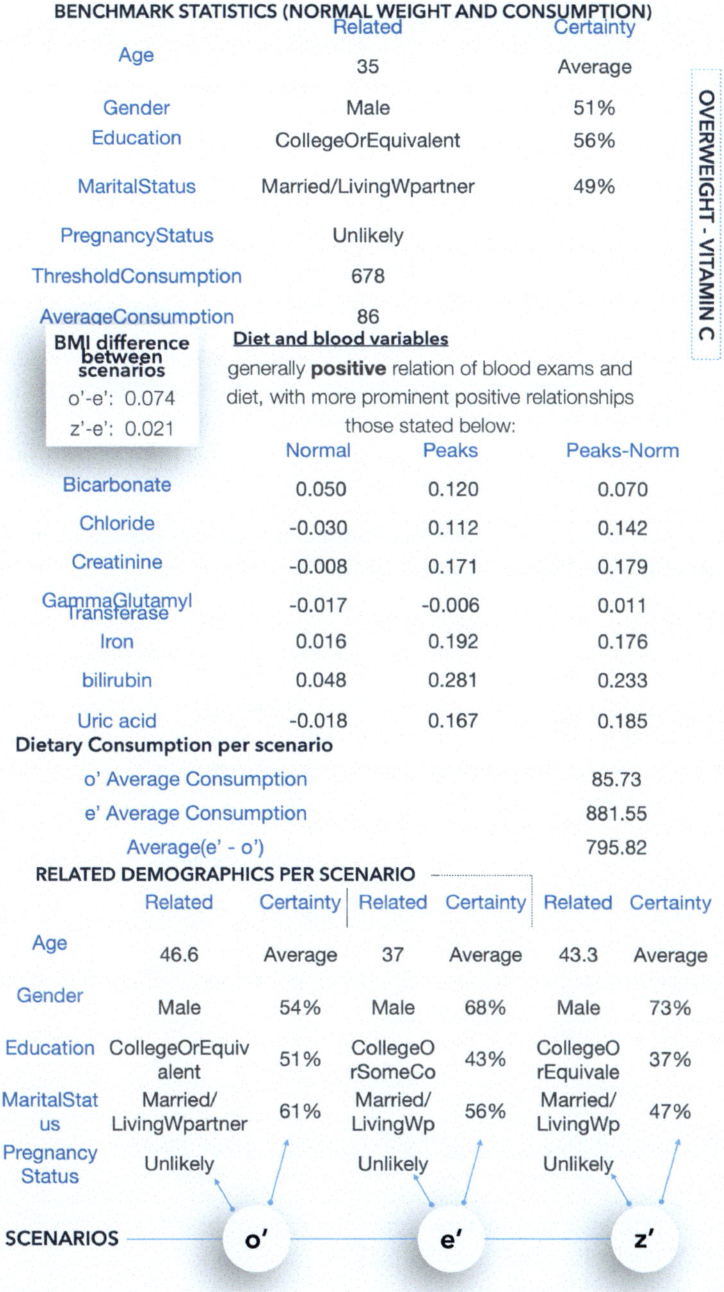

Fig. 4.10 Vitamin C, related demographics and key findings

Alcohol Intake in Obese Individuals

Figure 4.11 presents the alcohol consumption analysis for the obese category using the EVA-based algorithm. The threshold (t) was again defined as:

$t = \max(\Sigma X_n) \times p$

For obese individuals, p = 0.36, resulting in a threshold of 366.21 g. Sorting the data by BMI, the extreme return value was estimated at 366.21g (CI: 365–393g) for T = 15.8 BMI. The probability of exceeding this threshold is 1/T.

Correlation analysis (Fig. 4.12) reveals a stronger correlation in the extreme intake scenario (e') compared to normal consumption.

Key findings (Fig. 4.13):

- Prominent blood biomarker correlations: Bicarbonate, chloride, creatinine, gamma-glutamyl transferase, iron, bilirubin, and uric acid.
- Organ impact multipliers: Muscles (2×), lungs, kidneys (5×), heart (2×), liver (3×), pancreas, intestines, and vascular endothelium.

The impact of extreme alcohol consumption on BMI is negative, meaning higher alcohol intake is associated with lower BMI.

- Normal consumption BMI is higher by 0.59 compared to extreme consumption.
- Zero alcohol intake BMI is 1.7 higher than extreme consumption.

Blood biomarkers generally increase with higher alcohol intake, with greater increases in extreme consumption scenarios. The extreme average alcohol consumption is 240.62 g greater than the normal average consumption.

Demographics per scenario:

- Extreme intake (e'): Average age 37 years, lower education level than (o') and (z').
- Normal intake (o'): Average age 39 years, lower likelihood of being married or in a relationship.
- Normal-weight category: Average age 45 years, alcohol intake 38 g, serving as a benchmark for dietary recommendations.

4.5.4 Proposed Report and Recommendation Prototype

In this section, the system-generated report is outlined based on the Extreme Value Analysis (EVA) pipeline and decision-making framework.

System-Generated Report

1. Extreme consumption of dietary variable x detected based on the EVA pipeline. If dietary intake of x exceeds that of n in weight class g, a BMI increase of s may lead to a dietary intake increase of $s\prime$, and vice versa.

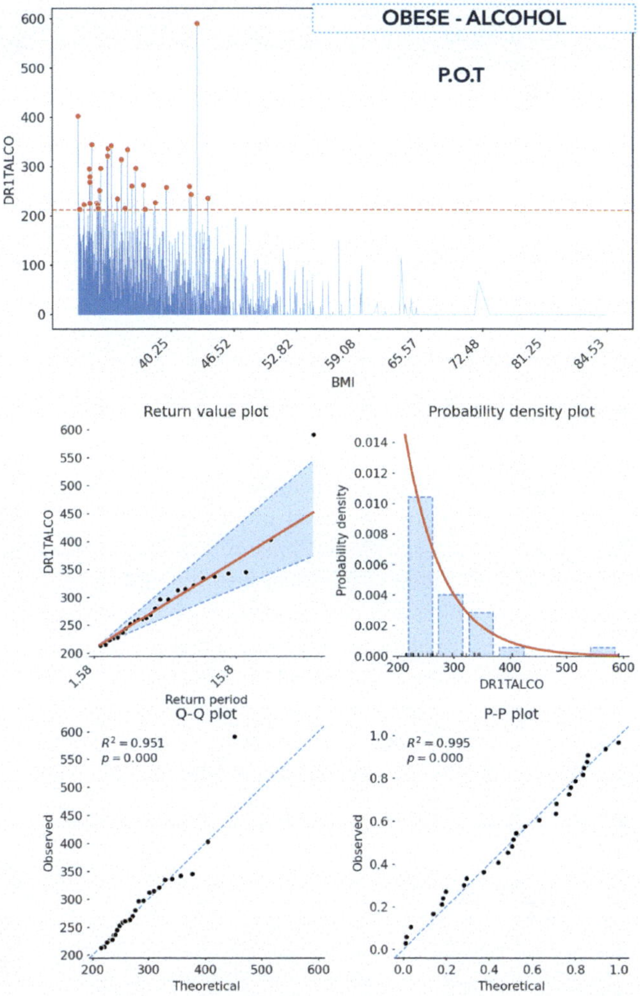

Return Intervals	return value	lower ci	upper ci
1.58	-inf	NaN	NaN
3.16	256.25	256.07	264.04
7.90	318.86	318.40	337.89
15.80	366.21	365.56	393.75
39.50	428.81	427.89	467.60
79.00	476.17	475.05	523.46
158.00	523.53	522.20	579.32
395.00	586.13	584.54	653.17
790.00	633.49	631.69	709.03
1580.00	680.84	678.85	764.90

Fig. 4.11 Obese category–EVA of alcohol (DR1TALCO)

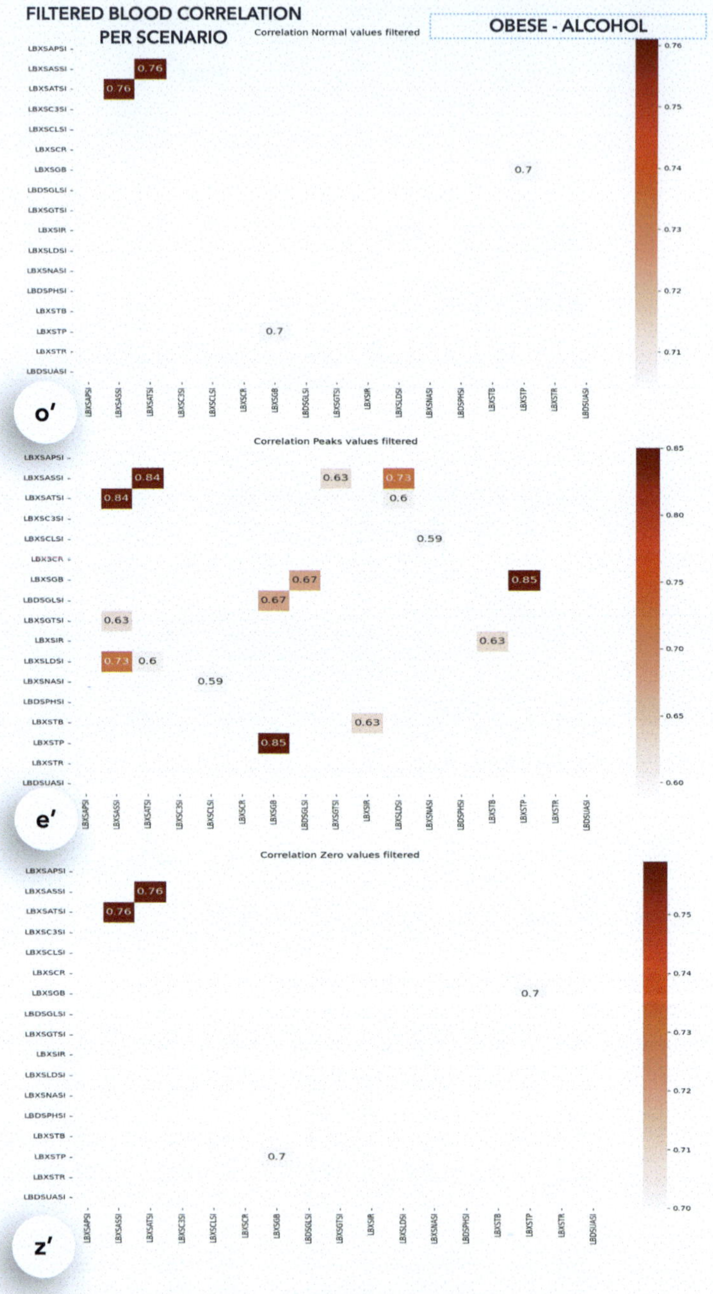

Fig. 4.12 Obese category–alcohol. Filtered blood variable correlation as per EVA pipeline

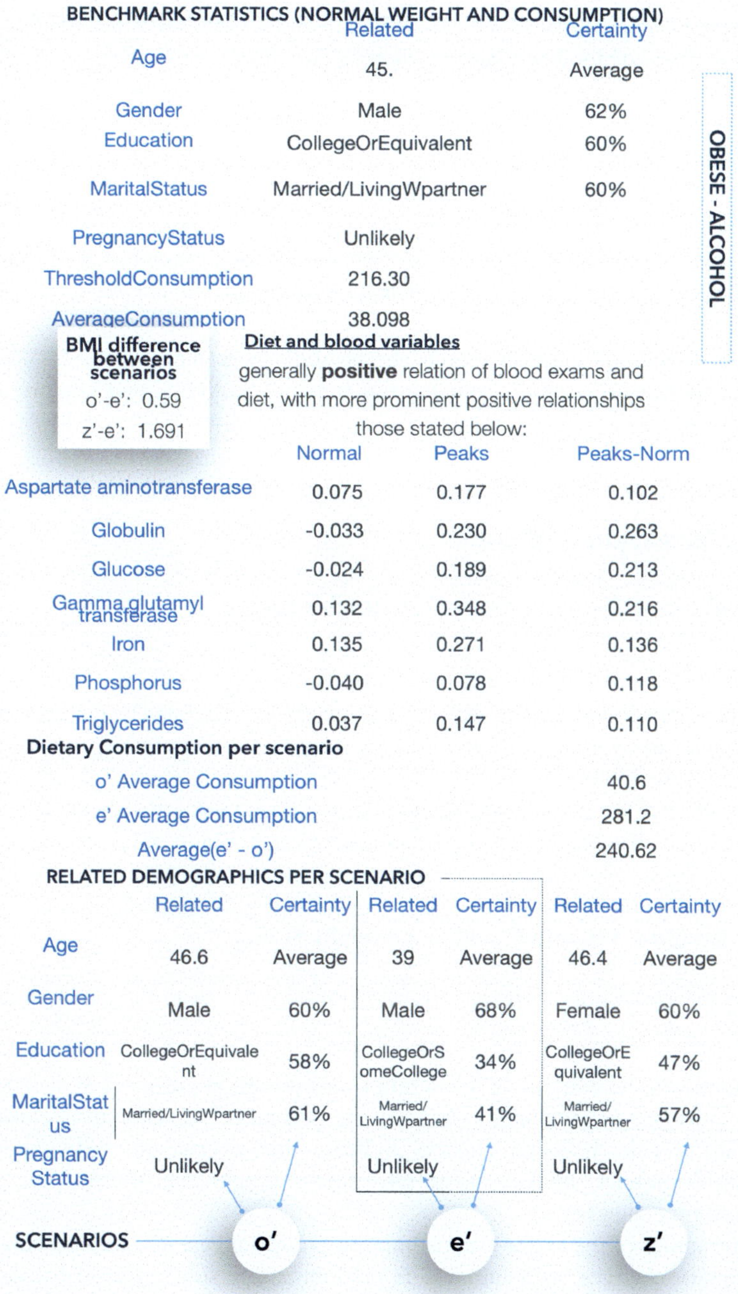

Fig. 4.13 Alcohol, related demographics and key–findings

2. Reduction in x consumption by s'' may lead to a BMI decrease of s'''.
3. Most affected blood biomarkers when dietary intake of x is extreme: B_n exhibits significant variation.
4. Stronger correlation observed between blood biomarkers B_n and $B_{n'}$ under extreme dietary intake conditions.
5. Blood biomarkers B_n are predominantly associated with the following organs: O_n (e.g., liver, kidneys, heart, etc.).
6. A general positive (or negative) relationship exists between blood biomarker B_n and dietary variable x : The strongest relationships are between x and filtered blood biomarkers $B_{\text{filtered}(n)}$.
7. User demographic details: Your age is AGE, your gender is GENDER, your education level is EDU, your marital status is STATUS, and you are a (non-)smoker.

4.5.5 *Discussion of Results*

4.5.6 *Knowledge Discovery Through Ontologies*

New knowledge can be extracted and formalized using ontologies, enabling automatic processing and the retrieval of targeted insights. As defined by Cedeno-Moreno et al. [4] and Studer et al. [34], an ontology represents an explicit specification of a shared conceptualization.

In this section, we identify and structure previously unstructured concepts related to the analyzed variables, presenting them as an extension to the proposed report.

Dietary Insights: Vitamin C and Overweight Status

Applying the analytical pipelines proposed in this study, several noteworthy findings emerge:

- Individuals in the extreme (e') Vitamin C consumption cluster (Fig. 4.10) have a lower average age compared to those in the normal (o') consumption cluster.
- BMI differences between clusters indicate that overweight individuals with extreme Vitamin C intake tend to have lower weight than those in the normal consumption category.
- Demographics of the extreme consumption group suggest a desire for weight loss and dietary changes, as Vitamin C is predominantly sourced from fruits, which are associated with healthy dietary choices.
- The high intake of Vitamin C may indicate self-initiated dietary modifications, possibly without medical guidance.

Despite the observed lower weight trends, excessive intake of any nutrient may lead to unintended health consequences. The key takeaways are the following:

- The findings indicate a willingness for lifestyle change among individuals in the extreme Vitamin C consumption cluster.
- Professional intervention is crucial to optimize dietary choices, ensuring health benefits while preventing unnecessary health risks.

Health Impacts of Excessive Vitamin C Intake

Vitamin C is essential for health, primarily sourced from fruits and vegetables such as oranges, strawberries, red peppers, and broccoli. It plays a vital role in iron absorption, growth, and development.

The recommended daily intake is:

- 75 mg/day for women
- 90 mg/day for men
- 120 mg/day during pregnancy

The upper limit for adults is 2000 mg/day, exceeding which may lead to:

- Diarrhea
- Nausea and vomiting
- Heartburn and stomach cramps
- Headaches [19]

Dietary Insights: Alcohol and Obesity

Applying our analytical pipelines to alcohol consumption patterns in obese individuals, we extract the following insights:

- BMI is lower in the extreme alcohol consumption (e') cluster compared to the normal (o') consumption cluster (Fig. 4.13).
- The average age in the extreme consumption cluster is 36.6 years, whereas in the normal consumption cluster, it is 46.6 years.
- Males are significantly more likely (68%) to belong to the extreme alcohol consumption cluster, with a higher probability of being unmarried.
- Zero alcohol consumption (z') is more prevalent among females.

Targeted Interventions

Defining specific population characteristics enables precision in public health strategies. Health awareness campaigns designed for at-risk demographics can lead to higher engagement and improved health outcomes [12].

Alcohol and Obesity

Excessive drinking is defined as consuming four or more drinks during a single occasion for women and five or more drinks for men. Heavy drinking is considered to be when 8 or more drinks are consumed per week for women and 15 or more drinks per week for men. Excessive alcohol consumption can lead to harmful health conditions and result in injuries, violence, poisoning, kidney failure, miscarriage and stillbirth and heart attacks [39].

Health Risks of Alcohol Consumption

Excessive alcohol consumption is categorized as:

- Binge drinking:
  - >=4 drinks in a single occasion (women)
  - >=5 drinks in a single occasion (men)
- Heavy drinking:
  - >=8 drinks per week (women)
  - >=15 drinks per week (men)
- Health risks include:
  - Injuries and violence
  - Liver damage and kidney failure
  - Increased risk of miscarriage and stillbirth
  - Heart disease and strokes [39]

The Impact of Diet on Blood Biomarkers

Dietary intake influences all bodily functions, with nutritional imbalances often reflected in blood biomarkers.

- Essential minerals (sodium, potassium, calcium, chloride) are critical for neurological and muscular functions.
- Macronutrients (proteins, carbohydrates) provide energy for metabolic processes.
- Endocrine and immune system functions can be compromised due to malnutrition or excessive nutrient intake.
- Inflammatory disorders often arise from prolonged dietary imbalances [18].

4.6 Conclusions

The integration of intelligent health knowledge systems with wearable tracking devices and digital platforms offers new opportunities for medical ontology development and health data representation. Such technologies facilitate:

- Self-monitoring and improved health literacy [16, 29].
- Enhanced data sharing and precision health interventions.
- Personalized recommendations that empower users to make informed health choices.

EVA By refining the usability aspects of EVA within the Golden Circle of Innovation framework (Sect. 4.3.2), we have designed a scalable algorithm for health data analysis and decision support. The proposed system will contribute to real-time health monitoring, personalized dietary management, and predictive analytics for disease prevention.

Appendix

Additional Material

Abbreviations

The following abbreviations are used in this manuscript:

BMI	Body mass index
EVA	Extreme value analysis
EVT	Extreme value theory
POT	Peaks over threshold
A.M.S.	Annual maxima series
P.D.F.	Probability density plot
R.V.	Return value-level
FFQ	Food frequency questionnaires
DALYs	Disability-adjusted life years
F.B.D.G.	Food-based dietary guidelines
Y.L.L.	Years of life lost from mortality
Y.L.D.	Years lost due to disability
CDC	Center for Disease Control and Prevention
NHANES	National Health and Nutrition Examination Survey
WHO	World Health Organisation
PCC	Pearson correlation coefficient

References

1. E. Arsenault, Y. Wang, M.P. Chapman, Towards scalable risk analysis for stochastic systems using extreme value theory, in arXiv arXiv: 2203.12689 [math.ST] (2022)
2. J.W. Baek, J.C. Kim, J. Chun, K. Chung, Hybrid clustering based health decision-making for improving dietary habits. Technol. Health Care **27**, 459–472 (2019)
3. M. Cámara, R.M. Giner, E. González-Fandos, E. López-García, J. Mañes, M.P. Portillo, M. Rafecas, L. Domínguez, J.A. Martínez, Food-based dietary guidelines around the world: a comparative analysis to update AESAN scientific committee dietary recommendations. Nutrients **13**(9), 3131 (2021)
4. D. Cedeno-Moreno, M. Vargas-Lombardo, An ontology-based knowledge methodology in the medical domain in the Latin America: the study case of republic of Panama. Acta Inform. Medica **26**, 98 (2018)
5. Y. Chiu, F. Chebana, B. Abdous, D. Bélanger, P. Gosselin, Mortality and morbidity peaks modeling: an extreme value theory approach. Stat. Methods Med. Res. **27**, 1498–1512 (2018)
6. S. Coles, J. Bawa, L. Trenner, P. Dorazio, *An Introduction to Statistical Modeling of Extreme Values*, vol. 208 (Springer, Berlin/Heidelberg, Germany, 2001)
7. M.C. Cornelis, F.B. Hu, Systems epidemiology: a new direction in nutrition and metabolic disease research. Curr. Nutr. Rep. **2**, 225–235 (2013)
8. M.L. Dansinger, J.A. Gleason, J.L. Griffith, H.P. Selker, E.J. Schaefer, Comparison of the Atkins, Ornish, Weight Watchers, and Zone diets for weight loss and heart disease risk reduction: a randomized trial. JAMA **293**, 43–53 (2005)
9. M.C. Dao, A.F. Subar, M. Warthon-Medina, J.E. Cade, T. Burrows, R.K. Golley, N.G. Forouhi, M. Pearce, B.A. Holmes, Dietary assessment toolkits: an overview. Public Health Nutr. **22**, 404–418 (2019)
10. K.M. Flegal, R. Wei, C.L. Ogden, D.S. Freedman, C.L. Johnson, L.R. Curtin, Characterizing extreme values of body mass index-for-age by using the 2000 centers for disease control and prevention growth charts. Am. J. Clin. Nutr. **90**, 1314–1320 (2009)
11. A. Herforth, M. Arimond, C. Álvarez-Sánchez, J. Coates, K. Christianson, E. Muehlhoff, A global review of food-based dietary guidelines. Adv. Nutr. **10**, 590–605 (2019)
12. S. Hoschar, L. Albarqouni, K.-H. Ladwig, A systematic review of educational interventions aiming to reduce prehospital delay in patients with acute coronary syndrome. Open Heart **7**(1), e001175 (2020)
13. R. Huss, M. Grunkin, *Artificial Intelligence Applications in Human Pathology* (WSPC, London, UK, 2022)
14. S.A. Jebb, Dietary strategies for the prevention of obesity. Proc. Nutr. Soc. **64**, 217–227 (2005)
15. C.L. Johnson, S.M. Dohrmann, V.L. Burt, L.K. Mohadjer, *National Health and Nutrition Examination Survey: Sample Design, 2011–2014* (US Department of Health and Human Services, Centers for Disease Control and Prevention, 2014)
16. R. Kent, Social media and self-tracking: representing the ‘health self’, in *Self-Tracking*. (Springer, Berlin/Heidelberg, Germany, 2018), pp. 61–76
17. M.J. Kochenderfer, T.A. Wheeler, K.H. Wray, *Algorithms for Decision Making* (MIT Press, Cambridge, MA, USA, 2022)
18. *Libre Texts Medicine—Nutrients Are Essential for Organ Function* (n.d.). https://tinyurl.com/yxsf45bs/. Accessed on 17 Nov 2022
19. *Mayo Clinic—Nutrition and Healthy Eating* (n.d.). https://www.mayoclinic.org/healthy-lifestyle/nutrition-and-healthy-eating/expert-answers/vitamin-c/faq-20058030/. Accessed on 17 Nov 2022
20. D.P. Panagoulias, D.N. Sotiropoulos, G.A. Tsihrintzis, Biomarker-based deep learning for personalized nutrition, in *2021 IEEE 33rd International Conference on Tools with Artificial Intelligence (ICTAI)*, DC, USA, Washington, Nov 2021, pp. 306–313
21. D.P. Panagoulias, D.N. Sotiropoulos, G.A. Tsihrintzis, Nutritional biomarkers and machine learning for personalized nutrition applications and health optimization, in *2021 12th International Conference on Information, Intelligence, Systems & Applications (IISA)*, Chania Crete, Greece, July 2021, pp. 1–6

22. D.P. Panagoulias, D.N. Sotiropoulos, G.A. Tsihrintzis, Nutritional biomarkers and machine learning for personalized nutrition applications and health optimization (extended journal version). Intell. Decis. Technol. **15**, 645–653 (2021)
23. D.P. Panagoulias, D.N. Sotiropoulos, G.A. Tsihrintzis, Extreme value analysis for dietary intake based on weight class, in *2022 13th International Conference on Information, Intelligence, Systems & Applications (IISA)*. Corfu, Greece, July 2022
24. D.P. Panagoulias, D.N. Sotiropoulos, G.A. Tsihrintzis, Towards personalized nutrition applications with nutritional biomarkers and machine learning, in *Advances in Assistive Technologies: Selected Papers in Honour of Professor Nikolaos G. Bourbakis - Vol. 3*, ed. by G.A. Tsihrintzis, M. Virvou, A. Esposito, L.C. Jain, vol. 28 (Springer, Cham, Switzerland, 2022), pp. 73–122
25. D.P. Panagoulias, M. Virvou, G.A. Tsihrintzis, A microservices-based iterative development approach for usable, reliable and explainable A.I.-infused medical applications using R.U.P, in *34th IEEE Conference on Tools with Artificial Intelligence (ICTAI 2022)*. Virtually, Nov 2022
26. D.P. Panagoulias, M. Virvou, G.A. Tsihrintzis, NUHEALTHSOFT: a nutritional and health data processing software tool, in *2022 16th International Conference on Signal Image Technology and Internet based Systems (SITIS)*. Dijon, France, Oct 2022
27. D.P. Panagoulias, M. Virvou, G.A. Tsihrintzis, Regulation and validation challenges in artificial intelligence-empowered healthcare applications—the case of blood-retrieved biomarkers, in *Proceedings of the 14th International Joint Conference on Knowledge-Based Software Engineering (JCKBSE 2022)*. Larnaca, Cyprus, Dec 2022
28. D.P. Panagoulias, D.N. Sotiropoulos, G.A. Tsihrintzis, SVM-based blood exam classification for predicting defining factors in metabolic syndrome diagnosis. Electronics **11**(6), 857 (2022)
29. Z. Papacharissi, *A Networked Self: Identity, Community, and Culture on Social Network Sites* (Routledge, New York, USA, 2010)
30. N. Potischman, J.L. Freudenheim, Biomarkers of nutritional exposure and nutritional status: an overview. J. Nutr. **133**(3), 873S–874S (2003)
31. *Pyextremes—Python Library* (n.d.). https://georgebv.github.io/pyextremes/. Accessed on 28 May 2022
32. J. Qiao, X. Lin, Y. Wu, X. Huang, X. Pan, J. Xu, J. Wu, Y. Ren, P.F. Shan, Global burden of non-communicable diseases attributable to dietary risks in 1990–2019. J. Hum. Nutr. Diet. **35**, 202–213 (2022)
33. J. Spruijt, T. Spanjaard, K. Demouge, The golden circle of innovation: what companies can learn from NGOs when it comes to innovation, in *Modern Marketing for Non-Profit Organizations: International Perspectives*. ed. by S. Smyczek (University of Economics in Katowice Publishing House, Katowice, Poland, 2013)
34. R. Studer, V.R. Benjamins, D. Fensel, Knowledge engineering: principles and methods. Data Knowl. Eng. **25**, 161–197 (1998)
35. M. Szigeti, T. Ferenci, L. Kovács, The use of block maxima method of extreme value statistics to characterise blood glucose curves, in *2020 IEEE 15th International Conference of System of Systems Engineering (SoSE)*. (Budapest, Hungary, 2020), pp.433–438. (June)
36. M. Thomas, M. Lemaitre, M.L. Wilson, C. Viboud, Y. Yordanov, H. Wackernagel, F. Carrat, Applications of extreme value theory in public health. PLoS ONE **11**, e0159312 (2016)
37. G.A. Tsihrintzis, C.L. Nikias, Fast estimation of the parameters of alpha-stable impulsive interference. IEEE Trans. Signal Process. **44**, 1492–1503 (1996)
38. *World health organization – Disability-adjusted life years* (n.d.). https://www.who.int/data/gho/indicator-metadata-registry/imr-details/158. Accessed: 2022-11-17
39. *World Health Organization—Alcohol Use and Your Health* (n.d.). https://www.cdc.gov/alcohol/fact-sheets/alcohol-use.htm/. Accessed on 17 Nov 2022
40. S. Xu, *Proceedings of 2013 World Agricultural Outlook Conference* (Springer, Berlin/Heidelberg, Germany, 2014)

Part II
AI-Empowered Medical Software Engineering Regulation, Validation and Explainability

Chapter 5
Iterative Microservices Approach for Explainable and Reliable AI in Medical Applications

Abstract The Rational Unified Process (RUP) is an iterative software engineering methodology that ensures structured alignment between engineers, stakeholders, and regulatory entities by breaking down development into phased, well-defined cycles. It facilitates adaptability within predefined constraints, ensuring that evolving requirements are incorporated efficiently. In distributed systems, the adoption of microservices architecture significantly enhances scalability, maintainability, and system modularity. Unlike monolithic architectures, where tightly coupled components create complexity and deployment challenges, microservices assign distinct responsibilities to independently deployable services. This modular design simplifies development, integration, and version control, enabling teams to enhance specific system functionalities without disrupting the entire platform. Furthermore, inter-service communication mechanisms, such as RESTful APIs and event-driven architectures, optimize system interoperability and real-time responsiveness. Medical applications are essential in disease prevention, treatment personalization, and long-term health management. However, their effectiveness relies on ensuring that they are adaptable to diverse patient demographics. Patients differ not only in medical conditions but also in cognitive abilities, technological familiarity, and accessibility requirements. Thus, a user-centric development approach is critical to ensuring that such systems are inclusive, intuitive, and supportive of patient needs. This chapter presents the design and implementation of an AI-powered medical application utilizing the RUP framework and microservices-based deployment model. The system development process prioritizes patient-centered requirements, ensuring that AI-driven functionalities evolve through continuous iteration, stakeholder feedback, and rigorous validation cycles. By integrating machine learning models for disease risk assessment, real-time decision support, and personalized health recommendations, the proposed system fosters adaptability, accessibility, and evidence-based healthcare optimization. The iterative nature of RUP ensures that the platform remains scalable, regulatory-compliant, and future-proof, maximizing usability and long-term impact for both patients and healthcare professionals.

Keywords Biomarkers · Nutritional biomarkers · Microservices · Rational unified process · User experience · Explainable AI

D. P. Panagoulias et al., *Artificial Intelligence-Empowered Bio-medical Applications*, Learning and Analytics in Intelligent Systems 49,
https://doi.org/10.1007/978-3-031-90174-4_5

5.1 Introduction

The advancement of precision medicine and personalized treatment heavily relies on the identification and quantification of biomarkers. These biological indicators, obtained from sources such as blood, urine, or tissue samples, play a crucial role in training machine learning models for prognostic and predictive purposes. Their integration into medical decision-making enhances diagnostic accuracy and therapeutic interventions. The development of AI-enhanced medical applications aims to bridge the gap between healthcare professionals and patients by incorporating methodologies that cater to diverse user needs.

This chapter explores the design and implementation of an AI-powered medical application built on a microservices architecture and structured following the Rational Unified Process (RUP). The approach ensures that patient-specific requirements are dynamically integrated into an adaptive software development cycle. Incorporating Artificial Intelligence (AI) into software engineering necessitates refining development methodologies to accommodate its unique challenges. Established frameworks such as the Rational Unified Process (RUP), Agile methodologies, Scrum, and microservices architecture offer structured solutions for managing the complexities inherent in AI-driven systems.

RUP, with its use-case-driven approach, iterative development cycles, and architecture-centric planning [30], provides a solid foundation for addressing AI integration challenges. Prior research has demonstrated RUP's efficacy in diverse domains, including e-commerce applications leveraging clustering for user modeling [26], and healthcare data processing, where precision and validation are paramount [22]. Additionally, [10] suggested blending RUP with Agile principles to improve adaptability-an essential requirement for evolving AI applications. Agile methodologies, particularly Scrum, emphasize iterative progress, adaptability, and continuous user feedback. Research by [36] highlighted Agile's significance in fostering human-AI collaboration, particularly in enhancing explainability and user alignment. Likewise, [23] demonstrated Agile's effectiveness in constructing interpretable AI-based decision-support systems, with Scrum's structured sprints facilitating the management of AI's dynamic requirements [29].

Microservices architecture complements these methodologies by promoting modularity, scalability, and adaptability. Studies such as [28] have examined how microservices contribute to building scalable AI platforms, while [34] illustrated their effectiveness in integrating machine learning models. Recent investigations, including JCKBSE2022's extended research [35], have further explored methodological advancements for complex AI-driven systems, particularly in trust modeling [36]. The modular nature of decision-support systems is vital, as emphasized in [33], which presented AI-powered decision-making frameworks designed for real-world applications. Integrating Agile principles and microservices architecture further optimizes system deployment and maintenance [32].

5.1.1 The Rational Unified Process (RUP)

The Rational Unified Process (RUP) is a structured software development methodology that organizes the project lifecycle into four distinct phases: Inception, Elaboration, Construction, and Transition. It encompasses six core development activities: business modeling, requirements gathering, system analysis and design, implementation, testing, and deployment. While these activities are present throughout the project, their emphasis evolves with each phase.

In the early stages (Inception and Elaboration), business modeling is critical, focusing on feasibility assessments, market analysis, and risk evaluation. The iterative nature of RUP is particularly beneficial for healthcare applications, where regulatory compliance and multi-stakeholder validation are essential before advancing to later development phases. Failure to meet these validation requirements could lead to costly redesigns and project inefficiencies.

Through RUP, this chapter illustrates how microservices architecture provides a robust structural approach for medical applications, offering operational and economic advantages. RUP's iterative approach ensures that microservices principles are refined throughout the development lifecycle, particularly during the Elaboration Phase.

5.1.2 Inception and Elaboration Phases

The Inception Phase focuses on conceptualizing the project, defining its core objectives, and evaluating feasibility. Key tasks include drafting a vision statement, conducting market research, estimating financial viability, performing risk assessments, and developing an initial business model. Deliverables from this phase include a preliminary use case, high-level project plans, and prototype designs. Success in this phase is determined by stakeholder alignment, feasibility evaluations, and requirement clarity.

The Elaboration Phase refines system requirements and solidifies architectural choices. The primary goal is to establish a stable foundation for subsequent development. Activities in this phase include expanding use case scenarios, defining architectural components, refining the development roadmap, and enhancing risk mitigation strategies. The success of this phase is measured by the reliability of system designs, the robustness of methodologies, and stakeholder agreement on technical and financial aspects [13, 27].

5.1.3 Construction and Transition Phases

During the Construction Phase, the refined system architecture is translated into functional components, with a strong focus on implementation, testing, and iterative improvements. This phase ensures the integration of all defined use cases, deployment models, and system elements.

The Transition Phase marks the release of the final product, ensuring a seamless transition to end-users. Activities in this phase include knowledge transfer, user training, post-deployment support, and continuous monitoring to facilitate adoption and usability enhancements.

5.1.4 Process Dimension—Workflows

Each RUP phase incorporates structured workflows that define roles, responsibilities, and objectives. The relationship between these workflows and the development phases is illustrated in Fig. 5.1.

- Business Modeling—Outlines core processes using structured use cases.
- Requirement Gathering—Aligns system capabilities with user expectations.

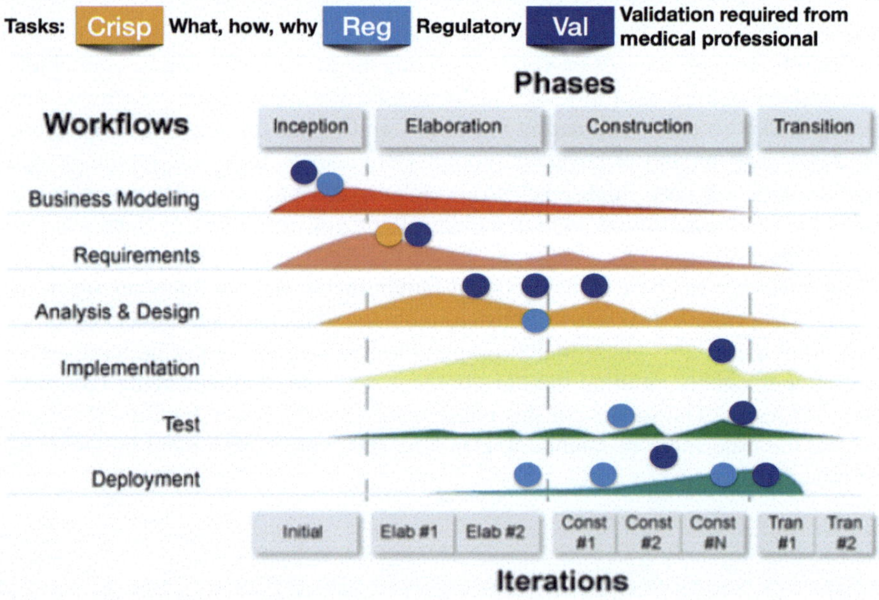

Fig. 5.1 Adapted and simplified rational unified process

- Analysis and Design—Translates requirements into structured system architecture.
- Implementation—Develops functional software components.
- Testing—Identifies and resolves defects through iterative validation.
- Deployment—Delivers the final product, ensuring smooth integration and adoption.

5.1.5 Microservices

Microservices architecture, often referred to as the microservice paradigm, structures applications as a set of loosely coupled, independently deployable services. This approach enhances scalability, maintainability, and system flexibility, making it particularly suitable for AI-driven healthcare applications.

By enabling rapid, modular development, microservices support continuous iteration and technological evolution. Their adaptability ensures that organizations can seamlessly adjust their technology stack to meet changing business requirements and emerging industry standards [37].

5.1.6 Chapter Structure

The remainder of this chapter is organized as follows:

- Section 5.2 reviews related literature.
- Section 5.3 discusses key challenges in AI-driven medical applications.
- Section 5.4 details system design using RUP methodology.
- Section 5.5 explores the functional components of the developed application.
- Section 5.6 concludes with a discussion on future research directions.

5.2 Related Work

Biomarkers serve as quantifiable biological indicators that characterize physiological and pathological states. Their applications extend to disease prediction, preventive strategies, treatment assessment, and therapeutic decision-making [17–19, 21]. Within this broad category, nutritional biomarkers provide specialized insights into dietary intake and metabolic responses. They function as objective measures of nutrient consumption, absorption, and metabolism, thereby aiding in the validation of dietary interventions and nutritional strategies. These biomarkers are particularly useful when generic biomarkers yield inconclusive results regarding specific food components. In such cases, their integration with dietary intake assessment methods enhances precision in evaluating nutritional impact and metabolic efficiency [20].

Recent studies have demonstrated that dietary patterns influence metabolic states and can be effectively analyzed using machine learning techniques. Prior research has explored the relationship between biochemical profiles and weight class prediction, leveraging machine learning models to extract insights from routine blood tests. Support Vector Machines (SVMs) and deep learning architectures have proven effective in predicting weight categories with high accuracy, based on key blood test parameters reflecting organ function and metabolic status [17–19, 21, 24].

Moreover, Extreme Value Theory (EVT) has emerged as a valuable statistical framework for analyzing dietary data and identifying outlier trends in nutrition-related health risks. EVT has been successfully applied to model extreme deviations in dietary behavior and their corresponding metabolic effects, enabling the development of targeted dietary interventions for individuals across different weight classifications [20].

The relevance of these studies is underscored by the growing recognition of obesity as a critical health determinant. Excess weight is closely linked to metabolic disorders, diabetes, and cardiovascular diseases, making early detection and intervention crucial for preventive healthcare [18]. While genetic predisposition contributes to obesity risk, environmental and behavioral factors-particularly dietary habits-play a fundamental role in weight management and long-term metabolic health [24].

The proposed system builds upon these prior findings, incorporating AI-driven methodologies to enhance biomarker analysis, refine predictive models, and optimize nutritional interventions. By leveraging microservices architecture and iterative AI methodologies, this work aims to improve the adaptability, scalability, and clinical applicability of AI-assisted medical decision-making in personalized healthcare.

5.3 Key Challenges in Implementing an Inclusive and User-Friendly Experience

Designing an AI-driven medical application that ensures inclusivity, accessibility, and user engagement poses significant challenges, particularly when considering the diverse needs of patients and healthcare professionals. While digital health solutions have advanced considerably, they often fail to accommodate all users effectively, particularly individuals with disabilities, the elderly, and those facing technological or socioeconomic barriers.

Demographic studies highlight the extent of these disparities. In the U.S. alone, approximately 60 million individuals live with some form of disability, with many experiencing significant challenges in accessing healthcare services. A survey of 714 physicians revealed that 80% believe disability negatively impacts quality of life, yet only 40% felt confident in their ability to provide equitable care. Furthermore, 18%

of physicians acknowledged systemic healthcare inequalities for disabled individuals [11]. These statistics underscore the urgent need for inclusive, human-centered design in medical applications to bridge these accessibility gaps.

5.3.1 Equal Access to Online Health Services

Despite the proliferation of digital health platforms, accessibility barriers continue to limit their effectiveness for millions of users worldwide. A wide range of neurological, cognitive, sensory, and mobility-related conditions affect how individuals interact with technology, necessitating adaptive user interfaces, multimodal access methods, and compliance with accessibility standards such as the Web Content Accessibility Guidelines (WCAG).

Some of the most prominent accessibility challenges arise from conditions affecting large segments of the global population:

- Dementia: Affects approximately 50 million people worldwide, including 10 million in Europe.
- Rheumatoid arthritis: Impacts 1% of the global population, limiting fine motor control.
- Wheelchair users: Estimated at 10 million in developed countries (1% of the population) and 120 million in developing regions, often facing navigation barriers in healthcare facilities.
- Spinal cord injuries: Around 500,000 new cases annually, primarily resulting from traffic accidents or falls.
- Dyslexia: Affects approximately 10% of the global population, making text-heavy interfaces challenging [7, 25].
- Intellectual disabilities in Europe: Affect 1% of the population, often requiring simplified information delivery methods [12].
- Visual impairments: Affect 30 million people in Europe, including 2.5 million who are blind [31].
- Hearing impairments: Between 700,000 and 900,000 individuals in Europe are profoundly deaf, with 12% of the population experiencing some level of hearing loss.
- ADHD: Impacts 4% of the adult population in Europe, requiring adaptable and structured interface designs [2, 4].

Given these statistics, ensuring equal access to health services through adaptive, inclusive design principles is imperative. Features such as screen reader compatibility, voice-assisted navigation, color contrast adjustments, text resizing, and gesture-based interfaces significantly improve usability for individuals with disabilities. The integration of AI-driven accessibility enhancements can further personalize user experiences, ensuring seamless interaction with healthcare applications across different demographics.

5.3.2 *Effective Use of Health Information Technologies*

The increasing adoption of health information technologies (HIT) introduces usability, safety, and reliability concerns, particularly in the context of electronic health records (EHRs), telemedicine, and patient management systems. Research suggests that poor interface design, inefficient workflows, and lack of user feedback mechanisms contribute to widespread dissatisfaction among both patients and healthcare professionals [3, 8].

To address these issues, the American Medical Informatics Association (AMIA) established a task force comprising academic and industry experts to improve usability standards in HIT systems, emphasizing the importance of user-centered design in enhancing patient safety and care quality [16].

Key usability challenges in health IT systems include: Many health IT systems face significant usability challenges that hinder their effectiveness and accessibility. One major issue is the complexity of user interfaces, as many systems lack intuitive navigation, leading to cognitive overload for users. Healthcare professionals often struggle with inefficient data entry processes, particularly in poorly designed electronic health records (EHRs), which increase administrative burdens and reduce the time available for patient care. Additionally, the absence of standardized design practices results in fragmentation across different platforms, making interoperability and usability inconsistent. Another critical limitation is the lack of customization options, which prevents systems from adapting to the unique needs of individual users, ultimately affecting efficiency and engagement.

To address these challenges, integrating iterative usability evaluations, real-time user feedback mechanisms, and AI-assisted automation can significantly enhance the accessibility and functionality of health IT systems. By refining interface designs based on user interactions and leveraging AI-driven tools for automated data management and workflow optimization, medical applications can reduce errors, improve efficiency, and create a more seamless experience for healthcare providers and patients alike.

5.3.3 *Aging Population and Usability in Health Applications*

The global aging population is increasing at an unprecedented rate, with projections indicating that by 2050, more than 2 billion people will be aged 60 or older. Contrary to common misconceptions, elderly individuals are not resistant to adopting digital technologies. In fact, when interfaces are intuitive, well-structured, and tailored to their needs, older adults are more likely to engage with digital platforms [6, 9].

Recent reports from Europe illustrate this trend, showing a steady rise in mobile device usage among older adults. Data indicates that 52% of individuals aged 55–74 now use mobile devices, a figure that has doubled in the past five years.

In Sweden, 72% of individuals aged 76 and older access the internet via smart devices, with 51% using them daily, underscoring their increasing reliance on technology for communication, information, and healthcare [15].

Despite this growing adoption, many health applications remain poorly optimized for elderly users, creating unnecessary barriers to accessibility and usability. Common issues include small font sizes and low-contrast displays, which make reading difficult, and complex navigation structures that contribute to cognitive overload. Furthermore, limited voice interaction features pose challenges for users with motor impairments, while the absence of multimodal accessibility options, such as touch, voice, or gesture controls, restricts interaction flexibility.

Usability studies suggest that incorporating clear visual hierarchies, simplified clinical information, intuitive icons, and structured data representations can significantly enhance engagement and improve the user experience for older adults [5]. Designing inclusive and adaptable interfaces is essential to ensuring that aging populations can fully benefit from digital healthcare innovations, enabling greater independence, improved health management, and enhanced overall well-being.

5.3.4 Access to Information, Security, and Privacy in Healthcare Systems

Ensuring respect, dignity, and equitable access to medical information is essential in patient-centered healthcare. Studies indicate that fear, anxiety, and perceived lack of control are common barriers to patient engagement. However, when healthcare professionals communicate effectively and involve patients in decision-making, satisfaction and treatment adherence improve [14].

Moreover, digital security and data privacy are pressing concerns in healthcare IT, given the industry's susceptibility to cyber threats and data breaches. The distinction between privacy and security in medical applications is crucial:

- Privacy—Focuses on safeguarding patient data through regulatory compliance, consent policies, and encryption protocols.
- Security—Involves protecting systems from cyberattacks, unauthorized access, and data leaks, ensuring integrity and confidentiality.

With the rise of big data analytics, AI-driven diagnostics, and cloud-based health platforms, implementing robust security measures such as end-to-end encryption, multi-factor authentication, and AI-powered anomaly detection is essential to safeguard patient information and ensure trust in digital healthcare solutions [1].

5.4 System Design via RUP

The Rational Unified Process (RUP) provides a structured approach for software development, ensuring scalability, adaptability, and maintainability. This chapter details the design, development, and deployment of a medical application aimed at enhancing patient care through AI-driven personalization, while adhering to security and accessibility standards.

By leveraging microservices architecture and machine learning algorithms, this system integrates predictive analytics, disease risk assessment, and automated recommendations, all within a patient-centered digital environment. RUP's iterative nature allows for continuous refinement and regulatory compliance, ensuring medical validity, transparency, and user trust.

5.4.1 Inception Phase

During the Inception Phase, the core objectives, challenges, and functional scope of the system were defined. This phase focused on concept validation, prototype development, and initial implementation, addressing critical healthcare barriers such as accessibility, AI transparency, and data security.

A key decision in this phase was to initially limit the scope to the patient's perspective, ensuring a user-centric foundation before expanding functionalities to additional stakeholders (e.g., doctors, regulators, administrators). By revamping vital aspects of the patient journey through AI, the system aims to enhance accessibility, improve disease management, and optimize preventive care strategies [22].

Core Functionalities:

- Health State Classification—Machine learning models categorize individuals based on biomarkers, metabolic health indicators, and weight class.
- Metabolic Risk Assessment—AI-driven risk models evaluate heart disease susceptibility based on clinical markers.
- Personalized Recommendations—AI suggests tailored dietary and lifestyle modifications, which adapt dynamically as more patient data is integrated.
- Scalability and Expandability—The architecture allows for new AI models and features to be integrated seamlessly.
- Regulatory and Medical Validation—Doctors can review AI-generated insights, while regulators ensure compliance with healthcare standards.
- Secure Data Exchange—Health data is encrypted, stored, and shared securely, ensuring patient confidentiality and adherence to legal frameworks.

Figure 5.2 illustrates the four main system actors:

- Patients—Input personal health data, retrieve blood test insights, and receive AI-driven health recommendations.

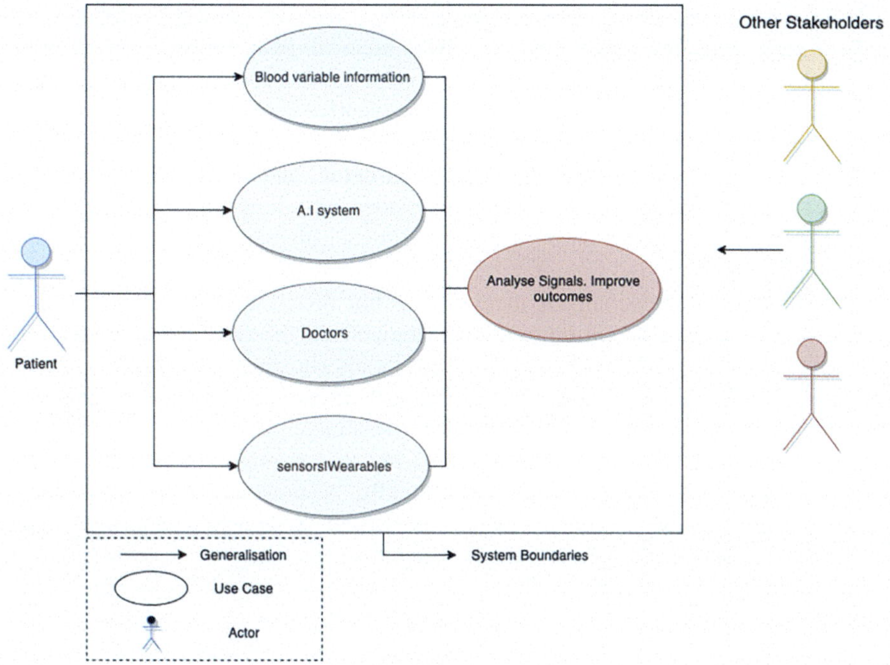

Fig. 5.2 Use case actors and basic functionality

- Doctors—Review AI-generated insights, validate predictions, and assess emerging biomarker correlations.
- Regulators—Ensure compliance with data security, methodology validation, and legal guidelines.
- Administrators—Manage system operations, oversee data integrity, and address technical or security-related concerns.

The patient journey is a highly dynamic process encompassing multiple stages, including diagnosis, treatment, behavioral adaptation, and preventive care. The integration of AI-driven interventions optimizes this experience by streamlining patient interactions, automating data analysis, enhancing accessibility, and personalizing healthcare engagement. Automated patient intake and AI-generated insights reduce administrative burdens and improve efficiency, while machine learning models analyze vast amounts of patient data to predict health risks, disease progression, and potential treatment outcomes. Accessibility is also a critical component, with AI enabling multimodal user interfaces that cater to individuals with disabilities, ensuring inclusivity in digital healthcare platforms. Furthermore, AI-driven dynamic recommendations adapt in real time based on evolving patient data, allowing for more personalized and precise healthcare management.

The patient experience can be understood through two parallel pathways: the medical path and the emotional path. The medical path follows a structured process, beginning with symptom identification, diagnosis, and treatment, ultimately leading to long-term health monitoring and preventive care. In contrast, the emotional path reflects the psychological and personal response to illness, emphasizing the importance of patient support, mental well-being, and engagement in healthcare decisions. AI-enhanced systems that consider both dimensions ensure a more holistic approach to patient care, balancing clinical efficiency with emotional and psychological support.

A key aspect of personalized medicine within this framework is the role of biomarkers in disease detection and management. AI-driven biomarker analysis facilitates early diagnosis, continuous disease monitoring, and adaptive therapeutic strategies [17]. Predictive diagnostics help identify disease predispositions, enabling early intervention and risk mitigation. Dynamic treatment plans evolve based on new biomedical and patient-generated data, ensuring that treatment strategies remain aligned with real-time health status. Additionally, AI-powered real-time decision support systems provide healthcare professionals with XAI insights, allowing them to validate recommendations and enhance diagnostic accuracy.

To ensure system inclusivity, security, and compliance, several key measures have been implemented. Adaptive user interface (UI) configurations allow diverse users to engage effectively with the system, addressing challenges related to cognitive, sensory, and motor impairments. Strict data governance protocols ensure compliance with privacy regulations and ethical guidelines, protecting patient information while allowing for secure data exchange. Moreover, role-based access controls enforce structured data permissions, safeguarding sensitive medical records and preventing unauthorized access [1].

By integrating AI into the patient journey, disease prediction, and personalized treatment approaches, this system offers an advanced, accessible, and secure medical platform that enhances both patient experiences and clinical decision-making. Through continuous improvement, machine learning adaptability, and robust security frameworks, the proposed model aims to bridge the gap between technology-driven healthcare solutions and real-world patient needs, ensuring that medical AI remains both effective and user-centric.

5.4.2 Elaboration Phase

Following the Inception Phase, the Elaboration Phase refined the technical architecture, workflows, and scalability mechanisms.

Use Cases

Building on Fig. 5.2, a more detailed use case diagram Fig. 5.3 highlights advanced functionalities:

- Patients interact with AI-driven classification, recommendations, and data visualization tools.
- Doctors validate AI predictions and adjust biomarker interpretations.
- Regulators ensure compliance with healthcare security standards.
- Users can dispute AI-generated insights, enhancing transparency.
- Accessibility tools facilitate inclusive usage.

System Components

Figure 5.4 presents the system's modular structure, detailing:

- ML Analyzer—Interprets biomedical signals for classification.
- Security Interface—Handles encryption, user authentication, and data exchange.

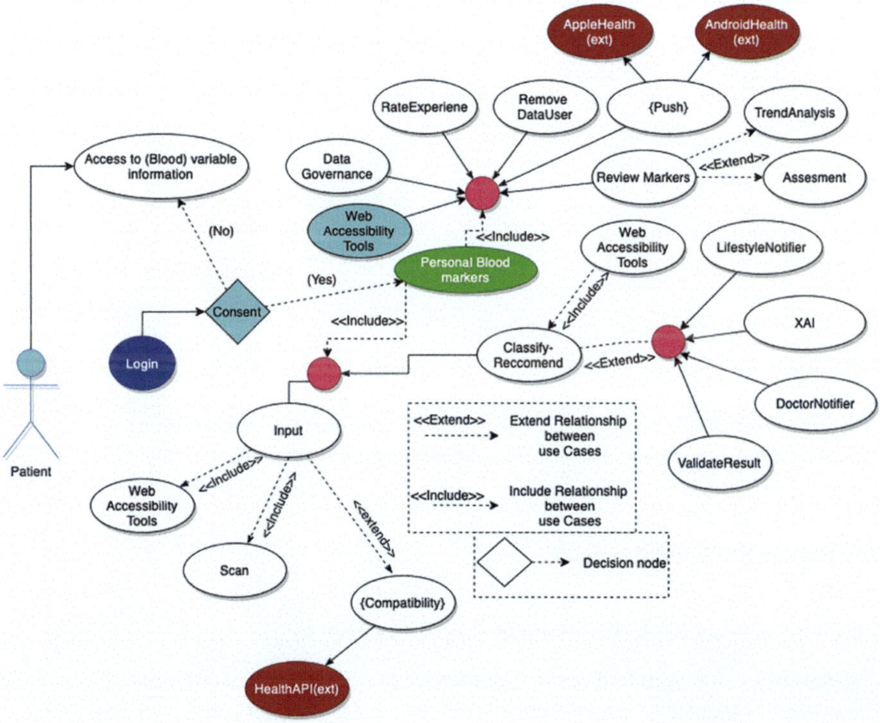

Fig. 5.3 Use case, elaboration phase

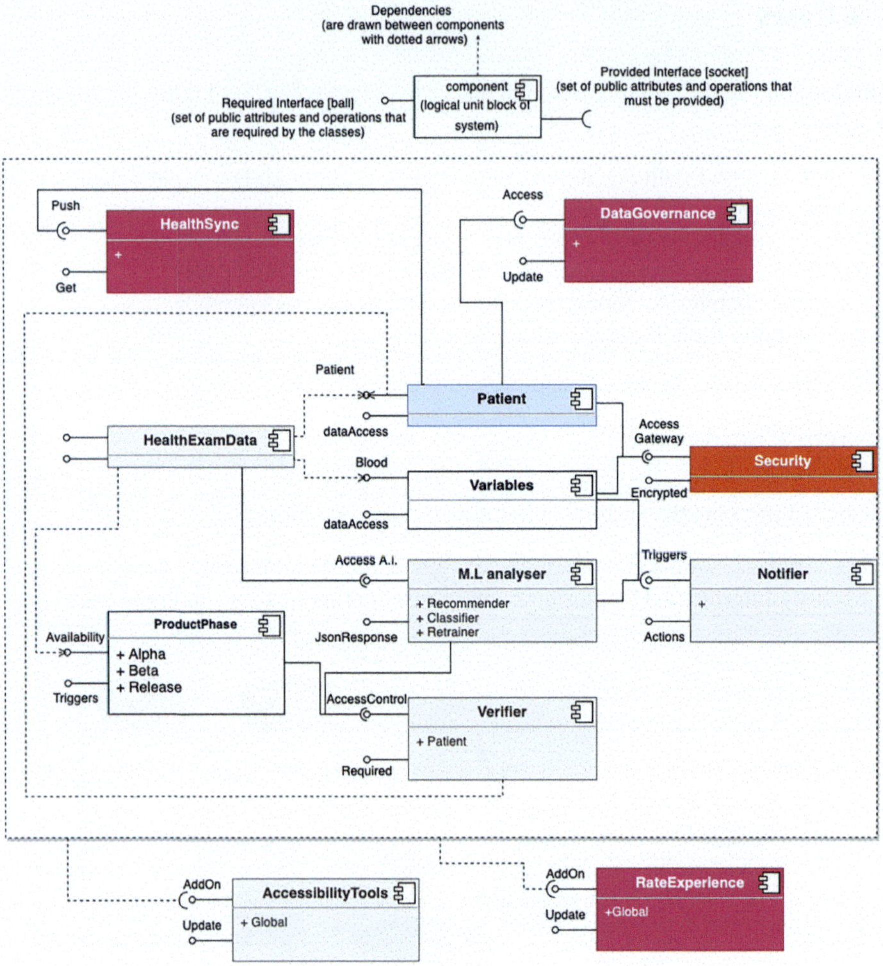

Fig. 5.4 Component diagram, elaboration phase

- Notifier—Provides alerts based on predefined health risk thresholds.
- Health Sync Interface—Integrates external medical records.

Deployment Architecture

The microservices-based deployment (Fig. 5.5) ensures:

- Scalability—Independent service deployment for new functionalities.
- Security—OpenID Connect hybrid authentication for encrypted user sessions.
- Role-Based API Access—Ensures patient data privacy.

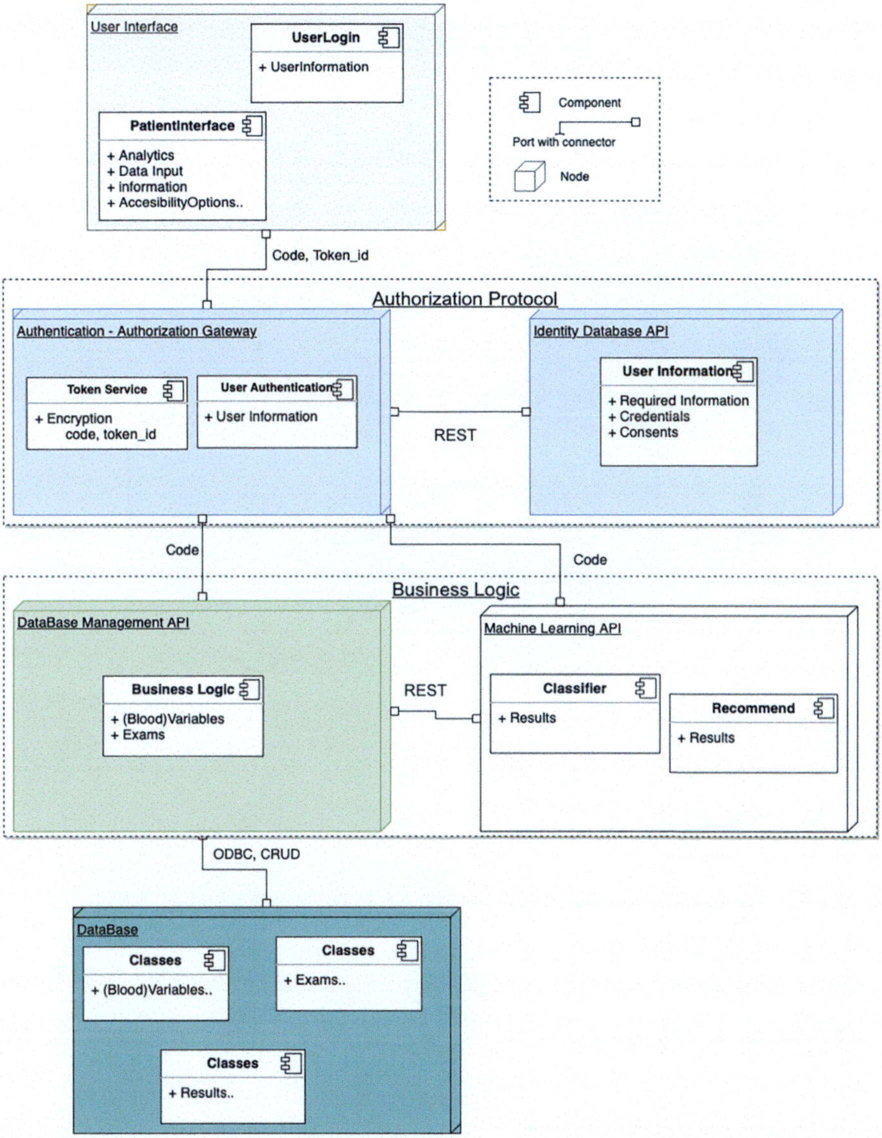

Fig. 5.5 Deployment diagram, elaboration phase

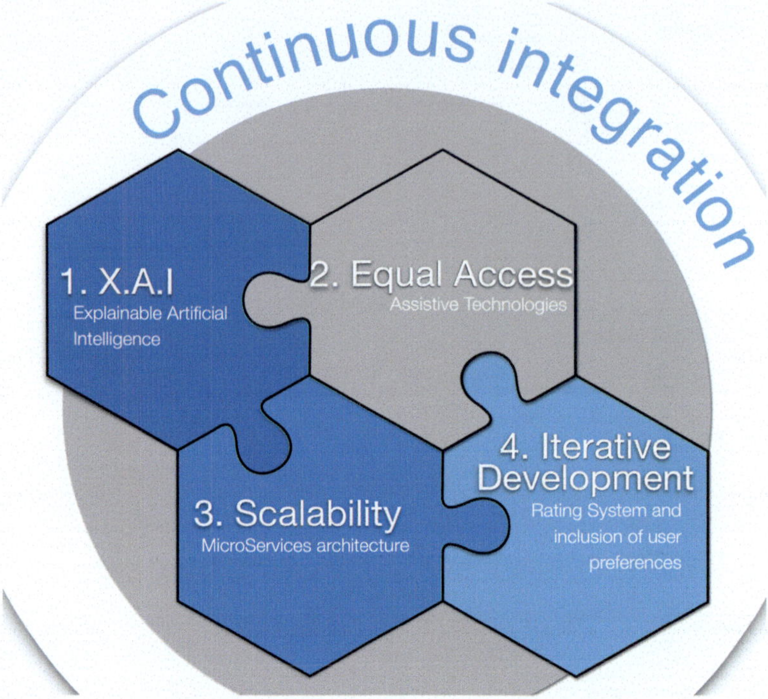

Fig. 5.6 Main Construction Pillars

5.5 Key Findings

Four core principles were extracted from the RUP methodology (Fig. 5.6):

- Explainable AI—Ensures transparency and builds trust.
- Accessibility & Inclusion—Guarantees equitable healthcare access.
- Scalability—Allows seamless expansion via microservices.
- Continuous Development—User feedback drives iterative system improvements.

5.6 Conclusions

This chapter examined the key challenges in optimizing the usability, accessibility, and reliability of an AI-powered medical application, with a particular focus on patient-centered design. To systematically address these challenges, the RUP was employed, ensuring a structured, iterative approach that accommodates evolving requirements. The adoption of microservices architecture was identified as a strategic decision, offering scalability, modularity, and flexibility-critical factors for managing the system's complex and dynamic components. Furthermore, the chapter introduced selected user interface configurations, demonstrating how the system fosters enhanced accessibility, adaptability, and patient engagement through intuitive interaction models.

While microservices offer significant benefits, their implementation introduces certain complexities. The integration of diverse technologies and independently managed services increases project complexity, necessitating specialized expertise and larger development teams. Additionally, deployment and operational costs tend to be higher compared to monolithic systems, as the distributed nature of microservices requires orchestration across multiple servers, robust inter-service communication, and enhanced security frameworks. These challenges highlight the trade-offs between scalability and resource efficiency, an area that requires further investigation. Future research will explore cost-effective deployment models, optimized resource allocation strategies, and AI-driven orchestration mechanisms to streamline microservices implementation while maintaining system performance and security.

Building on the patient-centric perspective explored in this chapter, future iterations of the system will expand the scope to incorporate additional stakeholders, including doctors, regulators, and system administrators. A comprehensive role-based management framework will be introduced to ensure seamless collaboration, secure access control, and regulatory compliance. Furthermore, a refined user feedback mechanism will be integrated, leveraging XAI principles to enhance transparency and trust. This iterative feedback loop will guide continuous system refinement, ensuring inclusivity, personalization, and engagement-driven design enhancements.

By aligning technical innovations with real-world healthcare challenges, this work aims to contribute to the evolution of AI-powered medical applications, advancing their practical usability, reliability, and accessibility in clinical and patient-centered environments.

References

1. K. Abouelmehdi, A. Beni-Hessane, H. Khaloufi, Big healthcare data: preserving security and privacy. J. Big Data **5.1**, 1–18 (2018)
2. ADHD Europe (n.d.). https://adhdeurope.eu. Accessed 5 Aug. 2022
3. C.M. Allwood, T. Kalén, User-competence and other usability aspects when introducing a patient administrative system: a case study, in *Interacting with Computers*, vol. 5.2, pp. 167–191 (1993)
4. *American academy of pediatrics* (n.d.). https://www.aap.org. Accessed 5 Aug 2022
5. M.M. Baig, H. GholamHosseini, F. Ahmad, A usability and user-experience analysis of vitalsassist: a mhealth application to monitor vital signs in acute care settings, in *2020 42nd Annual International Conference of the IEEE Engineering in Medicine and Biology Society (EMBC)*, pp. 5132–5135 (2020). https://doi.org/10.1109/EMBC44109.2020.9176467
6. G. Demiris, M.J. Rantz, M.A. Aud, K.D. Marek, H.W. Tyrer, M. Skubic, A.A. Hussam, Older adults attitudes towards and perceptions of 'smart home' technologies: a pilot study, in *Medical Informatics and the Internet in Medicine*, vol. 29.2, pp. 87–94 (2004)

7. *ELINET The European Literacy Policy Network* (n.d.). https://elinet.pro. Accessed 5 Aug. 2022
8. F. Fritz, S. Balhorn, M. Riek, B. Breil, M. Dugas, Qualitative and quantitative evaluation of EHR-integrated mobile patient questionnaires regarding usability and cost-efficiency. Int. J. Med. Inf. **81.5**, 303–313 (2012)
9. J. Goodman, J. Lundell, *HCI and the Older Population* (2005)
10. M. Hirsch, Making RUP agile, in *OOPSLA 2002 Practitioners Reports*, 1–ff (2002)
11. L.I. Iezzoni, S.R. Rao, J. Ressalam, D. Bolcic-Jankovic, N.D. Agaronnik, K. Donelan, T. Lagu, E.G. Campbell, Physicians' perceptions of people with disability and their health care: study reports the results of a survey of physicians' perceptions of people with disability, in *Health Affairs*, vol. 40.2, pp. 297–306 (2021)
12. *Inclusion Europe* (n.d.). https://www.inclusion-europe.eu. Accessed 5 Aug. 2022
13. P. Kruchten, *The Rational Unified Process: an Introduction* (Addison-Wesley Professional, 2004)
14. N. Makkar, K. Jain, V. Siddharth, S. Sarkar, Patient involvement in decision-making: an important parameter for better patient experience—an observational study (STROBE Compliant). J. Patient Exp. **6.3**, 231–237 (2019)
15. M.W. Mansson, F. Öhberg, K. Danielsson, M. Sandlund, Co-creation with older adults to improve user-experience of a smartphone self-test application to assess balance function. Int. J. Environ. Res. Publ. Health **17.11**, p. 3768
16. B. Middleton, M. Bloomrosen, M.A. Dente, B. Hashmat, R. Koppel, J.M. Overhage, T.H. Payne, S.T. Rosenbloom, C. Weaver, J. Zhang, Enhancing patient safety and quality of care by improving the usability of electronic health record systems: recommendations from AMIA. J. Am. Med. Inf. Assoc. **20.e1**, e2–e8 (2013)
17. D.P. Panagoulias, D.N. Sotiropoulos, G.A. Tsihrintzis, Biomarker-based deep learning for personalized nutrition, in *2021 IEEE 33rd International Conference on Tools with Artificial Intelligence (ICTAI)*, Washington, DC, USA, pp. 306–313, Nov. 2021
18. D.P. Panagoulias, D.N. Sotiropoulos, G.A. Tsihrintzis, Nutritional biomarkers and machine learning for personalized nutrition applications and health optimization, in *2021 12th International Conference on Information, Intelligence, Systems and Applications (IISA)*. (Chania Crete, Greece, 2021), pp.1–6. (July)
19. D.P. Panagoulias, D.N. Sotiropoulos, G.A. Tsihrintzis, Nutritional biomarkers and machine learning for personalized nutrition applications and health optimization (extended journal version). Intell. Dec. Technol. **15**, 645–653 (2021)
20. D.P. Panagoulias, D.N. Sotiropoulos, G.A. Tsihrintzis, Extreme value analysis for dietary intake based on weight class, in *2022 13th International Conference on Information, Intelligence, Systems and Applications (IISA)*, Corfu, Greece, July 2022a
21. D.P. Panagoulias, D.N. Sotiropoulos, G.A. Tsihrintzis, Towards personalized nutrition applications with nutritional biomarkers and machine learning, in *Advances in Assistive Technologies: Selected Papers in Honour of Professor Nikolaos G. Bourbakis*, ed. by G.A. Tsihrintzis, M. Virvou, A. Esposito, and L.C. Jain, vols. 3, 28 (Springer, Cham, Switzerland, 2022b), pp. 73–122
22. D.P. Panagoulias, M. Virvou, G.A. Tsihrintzis, Regulation and validation challenges in artificial intelligence-empowered healthcare applications—the case of blood-retrieved biomarkers, in *Proceedings of the 14th International Joint Conference on Knowledge-Based Software Engineering (JCKBSE 2022)*, Larnaca, Cyprus, Dec. 2022
23. D.P. Panagoulias, E. Sarmas, V. Marinakis, M. Virvou, G.A. Tsihrintzis, H. Doukas, Intelligent decision support for energy management: a methodology for tailored explainability of artificial intelligence analytics, in *Electronics*, vol. 12.21, p. 4430 (2023)
24. D.P. Panagoulias, D.N. Sotiropoulos, G.A. Tsihrintzis, SVM-Based blood exam classification for predicting defining factors in metabolic syndrome diagnosis, in *Electronics*, vol. 11.6, p. 857 (2022)
25. *PISA is the OECD's Programme for International Student Assessment* (n.d.). https://www.oecd.org/pisa/. Accessed 5 Aug 2022

26. A. Savvopoulos, M. Virvou, D.N. Sotiropoulos, G.A. Tsihrintzis, Clustering for user modeling in recommender e-commerce application: a RUP-based intelligent software life-cycle, in *Knowledge-Based Software Engineering* (IOS Press, 2008a), pp. 295–304
27. A. Savvopoulos, M. Virvou, D.N. Sotiropoulos, G.A. Tsihrintzis, Clustering for user modeling in recommender e-commerce application: a RUP-Based intelligent software life-cycle, in *Proceedings of the 2008 Conference on Knowledge-Based Software Engineering: Proceedings of the Eighth Joint Conference on Knowledge-Based Software Engineering* (NLD: IOS Press, 2008b), pp. 295–304. ISBN: 9781586039004
28. E. Schäffer, L.N. Penczek, M. Bartelt, M. Brossog, B. Kuhlenkötter, J. Franke, A microservice-And AutomationML-based reference architecture for an engineering configurator web platform. Procedia CIRP **103**, 274–279 (2021)
29. S. Shafiee, Y. Wautelet, L. Hvam, E. Sandrin, C. Forza, Scrum versus Rational Unified Process in facing the main challenges of product configuration systems development. J. Syst. Softw. **170**, 110732 (2020)
30. A.K. Shuja, J. Krebs, *IBM Rational Unified Process Reference and Certification Guide: Solution Designer (RUP)*. Pearson Education (2007)
31. *The voice of blind and partially sighted people in Europe* (n.d.). https://www.euroblind.org. Accessed 5 Aug. 2022
32. G.A. Tsihrintzis, M. Virvou, N.G. Bourbakis, L.C. Jain, Introduction to advances in information, intelligence, systems and applications, in *International Conference on Information, Intelligence, Systems, and Applications* (Springer, 2023), pp. 1–9
33. G.A. Tsihrintzis, M. Virvou, H. Doukas, L.C. Jain, Introduction to advances in artificial intelligence-empowered decision support systems, in *Advances in Artificial Intelligence-Empowered Decision Support Systems: Papers in Honour of Professor John Psarras* (Springer, 2024), pp. 1–11
34. G.A. Tsihrintzis, M. Virvou, E. Sakkopoulos, L.C. Jain, *Machine Learning Paradigms: Applications of Learning and Analytics in Intelligent Systems*, vol. 1 (Springer, 2019)
35. G.A. Tsihrintzis, M. Virvou, T. Saruwatari, *Special Collection of Extended Selected Papers on Novel Research Results Presented at the 14th International Joint Conference on Knowledge-based Software Engineering (JCKBSE2022)*, 22–24 Aug. 2022, Larnaca, Cyprus (2022). https://easyconferences.eu/jckbse2022/
36. M. Virvou, G.A. Tsihrintzis, A novel trust state-chart model for requirements engineering of trustful AI-Empowered software, in *2023 14th International Conference on Information, Intelligence, Systems and Applications (IISA)* (IEEE, 2023), pp. 1–6
37. *What are microservices?* (N.d.). https://microservices.io. Accessed 5 Aug. 2022

Chapter 6
Challenges in Regulating and Validating AI-Driven Healthcare

Abstract Biomarkers have emerged as powerful classification features in the development of neural network-based, machine learning, and AI-driven prognostic models within the field of personalized medicine. Their ability to provide quantifiable insights into a patient's health state makes them invaluable for targeted interventions, risk assessment, and precision treatment planning. Biomarkers are measurable indicators derived from blood samples, tissue analysis, or other bodily fluids, offering objective metrics that aid in medical decision-making. For example, an electrocardiogram (ECG), which records electrical signals from the heart, serves as a biomarker for cardiovascular health, enabling clinicians to evaluate heart function and detect abnormalities. By leveraging such indicators, biomarkers contribute to actionable insights, making them essential tools in patient management, disease monitoring, and treatment administration. This chapter introduces a medical application built upon a machine learning backbone that utilizes biomarkers retrieved from blood tests to define various health states, including obesity, metabolic syndrome, and systolic blood pressure anomalies. The system is designed and implemented using the Rational Unified Process (R.U.P.) and the Cross-Industry Standard Process for Data Mining (CRISP-DM), ensuring a structured and iterative development approach. By adopting and adapting industry standards, this work aims to address sector-specific requirements and challenges while designing intelligent, AI-driven solutions that add value to all healthcare stakeholders. The integration of new technologies in medicine holds the potential to redefine healthcare pathways, bridging the gap between laboratory research and real-world patient applications. However, the clinical adoption of AI-driven healthcare solutions is contingent upon strong medical validation and rigorous regulatory oversight to ensure usability, reliability, and patient safety. In the development of healthcare applications, compliance with regulatory frameworks is a key challenge, requiring continuous validation and oversight throughout the software lifecycle. A fundamental aspect of this work involves defining the conditions under which a healthcare software application qualifies as a medical device, as this distinction directly impacts development constraints, legal requirements, and compliance measures. Given the critical role of regulatory bodies in shaping healthcare technology, this chapter also explores strategies for effectively managing regulatory interactions throughout the production cycle. By incorporating validation protocols, risk

D. P. Panagoulias et al., *Artificial Intelligence-Empowered Bio-medical Applications*, Learning and Analytics in Intelligent Systems 49,
https://doi.org/10.1007/978-3-031-90174-4_6

assessment models, and transparent AI methodologies, the proposed system seeks to align technological advancements with medical standards, ensuring that AI-powered applications are clinically viable, ethically sound, and legally compliant.

Keywords Personalised medicine · Bioinformatics · Biomarkers · Pattern recognition · Microservices · Rational unified process(rup) · Innovation strategies · Golden circle of innovation · Machine learning

6.1 Introduction

Patient evaluation and personalized treatments are increasingly being refined through biomarker identification, which plays a central role in modern medical research and clinical decision-making. Biomarkers are key research variables Table 6.1 that serve as essential features in the training of prognostic and predictive machine learning models. They provide quantifiable indicators of physiological and pathological states, obtained from blood samples, tissue analysis, and other bodily fluids. Additionally, biochemical variations over time can hold significant diagnostic and prognostic value, further contributing to biomarker discovery and validation.

Over the past two decades, the medical industry has experienced a paradigm shift, moving away from the one-size-fits-all treatment model toward a personalized, patient-specific approach. The integration of biomarkers with statistical analysis and deep learning techniques has led to data-driven medical advancements, replacing outdated practices that often resulted in over- or under-treatment of patients. One notable example is the HER2 biomarker [13], a genetic and protein-based predictive marker used in clinical oncology to assess treatment response in breast cancer patients. This demonstrates how biomarker-driven AI models can optimize therapeutic interventions, ensuring precision medicine strategies that are both effective and personalized.

Table 6.1 Examples of biomarkers

Name	Description	Health results
C-reactive protein (CRP)	A protein produced by the liver in response to inflammation	Indicator of systemic inflammation, cardiovascular disease risk
Hemoglobin A1c (HbA1c)	Measures average blood sugar levels over the past three months	Diabetes diagnosis and long-term glucose control
Prostate-Specific Antigen (PSA)	A protein produced by prostate gland cells	Prostate cancer screening and monitoring
Brain Natriuretic Peptide (BNP)	A hormone secreted by the heart in response to stress	Heart failure diagnosis and severity assessment

A critical aspect of deploying AI-driven medical applications is their classification as medical devices when used for diagnostic or therapeutic purposes. As a medical device, such software is subject to regulatory oversight, necessitating compliance with local, global, or industry-specific technical standards. This regulatory framework requires supervisory oversight throughout multiple phases of the production cycle, ensuring that the software adheres to quality control measures, safety protocols, and data security standards.

For a medical application to gain acceptance within the medical community, its usability, accuracy, and clinical utility must be validated before being introduced into a clinical setting. This underscores the need for transparency and active involvement of medical professionals throughout the development, validation, and deployment stages. Medical practitioners, regulatory bodies, and other key stakeholders must be integrated into the production process to ensure the application's proper use, compliance, and patient safety. Understanding the roles of different stakeholders [11] in the patient journey is essential to defining clear contributions, regulatory checkpoints, and validation requirements. Additionally, every phase of development, from inception to deployment, must be systematically analyzed and optimized. To achieve this, the Cross Industry Standard Process for Data Mining (C.R.I.S.P-D.M.) will be employed for data-driven process evaluation, while the Rational Unified Process (R.U.P.) will serve as the structural framework for defining the building blocks of each development phase.

Given the complex interplay of regulatory, medical, and supervisory requirements, it is crucial to establish a synchronized framework that ensures quality control, safety thresholds, and transparency in AI-driven healthcare applications. By adapting established methodologies, this study proposes a structured integration of regulatory and medical supervisory bodies into the production lifecycle (Sect. 6.4). Furthermore, in Sect. 2.5, a method for enhancing neural network transparency, usability, and validation is outlined, ensuring that AI-driven medical models adhere to interpretable and clinically valid practices.

The chapter is organized as follows: Sect. 6.2 summarizes the key challenges and issues addressed in this study. Section 6.3 provides an overview of related literature, while Sect. 6.4 discusses future research directions and the tools and methodologies used for system development. Finally, Sect. 6.5 presents the conclusions and key findings drawn from this research.

6.2 Regulatory and Medical Validation in AI-Driven Healthcare Applications

Developing a healthcare device or application presents numerous regulatory and medical validation challenges, particularly when machine learning (ML) and artificial intelligence (AI) are integrated into diagnostic and therapeutic processes. AI-driven automation in medical applications must align with clinical best practices, safety standards, and regulatory frameworks to ensure that interventions remain effective, reliable, and compliant. Awareness and integration of regulatory guidelines from the early stages of product development can significantly enhance the quality, usability, and acceptance of the final product, reducing the time required for a market-ready release. This chapter focuses on the deployment of a web-based medical application designed to analyze blood test results and classify users into weight categories using deep neural networks. This classification enables a personalized recommendation system, offering nutrition-based interventions and heart health risk assessments tailored to individual health profiles.

This chapter proposes a structured methodology that simplifies the validation process from both a regulatory and medical perspective, leveraging established methodologies to enhance the development speed and efficiency of AI-driven healthcare applications. By streamlining regulatory alignment, system validation, and smart intervention deployment, AI-driven solutions can seamlessly integrate into healthcare systems, enhancing patient outcomes and medical decision-making.

In this chapter, we outline two key frameworks that support the development of medical AI systems:

- The Golden Circle of Innovation—A framework adapted for AI-driven medical applications, providing a structured "What, Why, and How" approach to designing machine learning solutions.
- The Cross-Industry Standard Process for Data Mining (CRISP-DM) and the Rational Unified Process (RUP)—These methodologies structure the AI model development process, ensuring that regulatory requirements, medical standards, and stakeholder involvement are incorporated at every stage of production.

6.2.1 Biomarkers: Key Indicators in AI-Driven Healthcare Applications

Biomarkers are critical in medical research, diagnostics, and personalized treatments, helping answer three fundamental questions: how much, what, and for whom Fig. 6.1 [17]. Many biomarkers are easily measurable and are part of routine medical exams, making them accessible for early disease detection and continuous monitoring.

Biomarkers serve as early warning indicators for disease development in population groups with specific health characteristics. Common examples include:

- Cardiovascular function—Blood pressure, heart rate, and pulse measurements.
- Metabolic health—Cholesterol and triglycerides for coronary heart disease risk.
- Obesity markers—Body weight, BMI, and waist-to-hip ratio for metabolic disorders.
- Immune system status—T-cell counts and immune response tracking.
- Stress indicators—Cortisol levels as a biomarker for chronic stress.

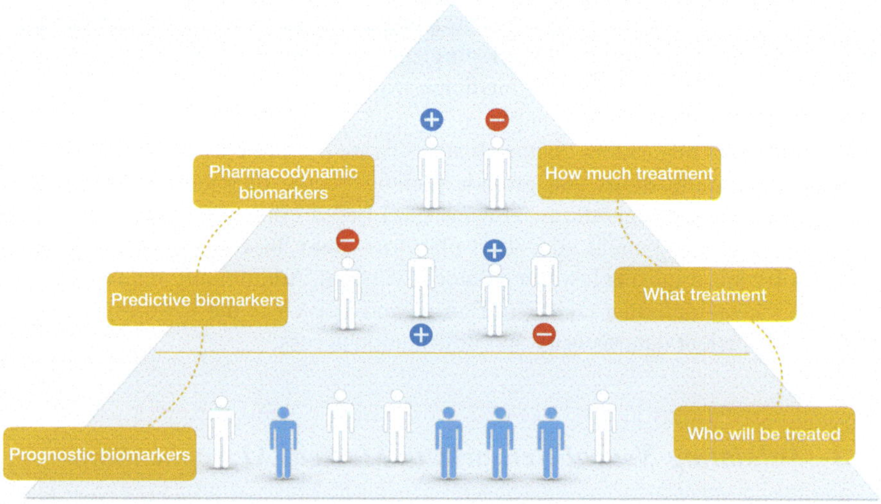

Fig. 6.1 Categories of biomarkers and clinical use cases

As illustrated in Fig. 6.1, biomarkers can be classified into:

- Pharmacodynamic biomarkers—Used to measure treatment response.
- Predictive biomarkers—Determine optimal treatment options.
- Prognostic biomarkers—Identify patients at risk for specific conditions.

An elevated biomarker beyond a specific threshold signals potential disease risk, making biomarkers crucial for preventive medicine and early intervention. Furthermore, integrating biomarker-based AI models allows for the development of novel indicators, improving risk assessment and clinical decision-making. For instance, research shows that older men are at higher risk for immune system and neuroendocrine disorders, whereas older women face higher risks associated with high systolic blood pressure and metabolic syndrome. Throughout this book, biomarkers are utilized as measurable health variables, triggering AI-driven actions, classifying patient states, and automating clinical workflows.

6.2.2 Regulation and Compliance in Medical AI Applications

Medical device regulations are governed at both regional and national levels. In Europe, the Medical Devices Directive (MDD) sets the foundation for regulatory compliance, while countries like the UK have additional enforcement agencies such as the MHRA. In the U.S., the FDA and FTC regulate diagnostic and assistive medical software, ensuring that AI-driven healthcare applications meet safety, efficacy, and ethical standards.

A healthcare application is classified as a medical device when it is used for diagnostic or therapeutic purposes. Developing such an application without regulatory approval is equivalent to practicing medicine without a license. In the U.S., the FDA ensures that AI-driven medical applications undergo pre-release validation, while the FTC monitors misleading claims about software performance [5, 9, 12].

Medical AI applications that provide patient-specific diagnoses are subject to strict medical device regulations. However, AI-based clinical support tools, such as decision aids and patient education platforms, may be categorized as general medical software, requiring lower regulatory oversight. Additionally, data protection and privacy laws-including GDPR in Europe and HIPAA in the U.S.-impose strict controls over patient data security and information access.

6.2.3 Validating Neural Networks in Medical AI Applications

Implementing artificial neural networks (ANNs) in medical applications presents unique challenges [2]. In the U.S., medical neural networks are subject to FDA approval, requiring extensive validation and explainability measures [21].

A key factor in healthcare AI usability is the measurement of its impact, ease of use, and clinical relevance [26]. This study evaluates AI usability through: • Patient accessibility—The extent to which AI improves patient decision-making. • Healthcare provider efficiency—Reduction in administrative workload and enhanced diagnostic accuracy. • Regulatory compliance—Alignment with industry standards to ensure ethical AI deployment.

AI continuous integration is achieved by incorporating doctors, regulators, and patients into the feedback loop, ensuring that new AI models and system updates are validated before release. A built-in review system measures ease of use, perceived usability, and impact, ensuring that AI interventions remain patient-centered, safe, and transparent.

6.2.4 The Role of Machine Learning in Health Care

To structure and optimize machine learning (ML) applications in healthcare, we propose an adaptation of the Golden Circle of Innovation [24]. Originally developed as a framework for organizational strategy and problem-solving [22], the Golden Circle model consists of three core elements-What, Why, and How-which we apply to the design, deployment, and validation of machine learning pipelines in healthcare (Fig. 6.2). By adopting this structured methodology, we can align stakeholder needs, define clear development goals, and explain the role of machine learning interventions to medical professionals, regulators, and end users.

The Golden Circle approach, as outlined in [24], systematically examines the necessity, implementation, and impact of innovation:

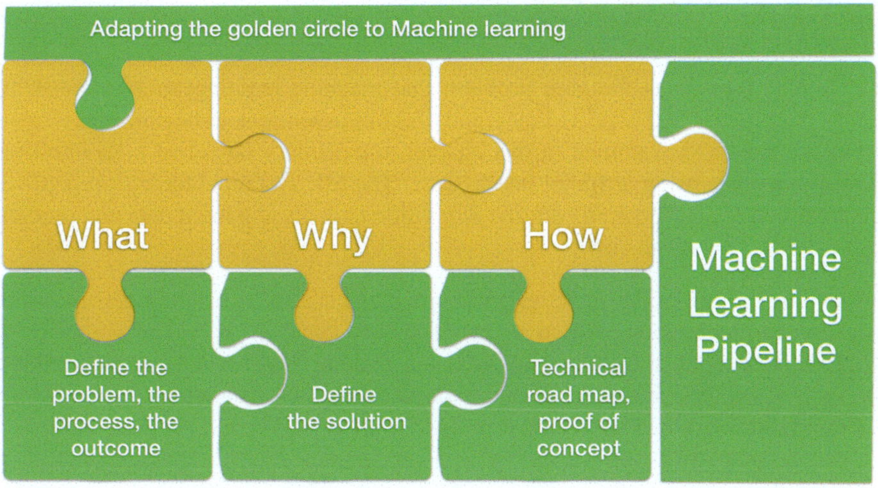

Fig. 6.2 The (adapted) golden circle of innovation

- Why innovation is necessary—Understanding the fundamental need for change and the motivation behind implementing new solutions.
- How innovation is achieved—Developing structured mechanisms to foster innovation through effective managerial and technological frameworks.
- What innovation produces—Defining both innovation as a process and innovation as an outcome, ensuring practical implementation and measurable benefits.

In this framework, the first dimension of "What" explores the motivation for innovation, addressing key drivers, stakeholder involvement, and implementation direction. It evaluates whether innovation emerges bottom-up (data-driven needs from clinical practice) or top-down (policy and regulatory-driven requirements), and whether its source is internal (developed within an institution) or external (collaborative efforts across organizations) [20]. The second dimension of "What" focuses on the form of innovation, categorizing it as a product, service, or systemic improvement [3]. Additionally, this dimension considers the magnitude of impact, distinguishing between technical innovation (algorithmic advancements) and administrative innovation (workflow optimization in healthcare delivery).

By integrating these innovation principles into machine learning-driven healthcare applications, we define a structured approach that ensures stakeholder alignment, regulatory compliance, and AI explainability. Below, we outline the adaptation of this model to machine learning in healthcare.

The What in Machine Learning

The "What" represents the core problem that machine learning aims to solve. It defines the process and expected outcome of AI interventions in healthcare.

For the use case examined in this chapter, the primary objective is to improve clinical outcomes using artificial intelligence. The ML system automates repetitive clinical processes, leveraging pattern recognition to analyze patient data and predict health states. This results in:

- Reducing the emotional and cognitive burden on patients by offering personalized health information and predictive analytics.
- Enhancing patient engagement and decision-making by personalizing recommendations and interventions based on biomarker-driven classifications.
- Minimizing workload for medical professionals by automating routine administrative tasks, allowing physicians to focus on complex decision-making and direct patient care.

By structuring ML-based healthcare applications around a clear problem statement, we ensure that AI integration is purposeful, efficient, and patient-centered.

The Why in Machine Learning

The "Why" defines the rationale behind adopting machine learning solutions in healthcare. It emphasizes the transformative potential of AI in delivering more accurate, efficient, and personalized healthcare services.

In this chapter, the motivation for AI implementation is driven by the need to define and classify health states, enabling automated intervention triggers. By implementing machine learning, we:

- Enhance diagnostic accuracy—AI models analyze biomarkers, medical records, and real-time patient data to classify health states with high precision.
- Enable early intervention—Automated risk prediction and personalized recommendations allow for preventive healthcare strategies rather than reactive treatments.
- Streamline patient monitoring—AI assists in tracking disease progression, treatment efficacy, and adherence to medical protocols, ensuring continuous patient care optimization.
- Support regulatory compliance—AI models can be designed with explainable AI (XAI) principles, making them transparent, interpretable, and aligned with medical validation requirements.

By answering "Why" machine learning is necessary, we validate its role in enhancing medical workflows, improving patient outcomes, and optimizing healthcare delivery.

The How in Machine Learning

The "How" details the technical roadmap, specifying tools, methodologies, and architectural considerations that drive AI implementation in healthcare. This includes:

- Machine learning models and algorithms—Selection of supervised learning, deep neural networks, and reinforcement learning for predictive analysis.
- Neural network architecture—Design of AI models to process biomarker data, electronic health records, and patient history.
- Integration with clinical workflows—Embedding AI tools within electronic health record (EHR) systems, decision-support platforms, and remote monitoring tools.
- Validation and regulatory alignment—Implementing AI fairness metrics, validation datasets, and performance benchmarking to ensure clinical reliability.
- CRISP-DM methodology—Applying Cross-Industry Standard Process for Data Mining (CRISP-DM) to structure data collection, processing, modeling, evaluation, and deployment.

The accuracy of machine learning models serves as a proof of concept, demonstrating the feasibility and effectiveness of AI-driven healthcare applications. The proposed adaptation of the Golden Circle of Innovation ensures that the development and deployment of AI models remain clinically relevant, technically sound, and ethically robust.

6.3 Related Work

The importance of data protection and its associated challenges in the distributed computing era is a well-documented area of study. Data security concerns are particularly relevant in the healthcare sector, where confidentiality, integrity, and accessibility of medical records must be ensured. In [19], issues related to data privacy and secure data sharing are explored, addressing emerging threats in decentralized systems. Additionally, in [10], alternative learning methodologies, such as Synchronous and Asynchronous Learning, are examined in the context of General Data Protection Regulation (GDPR) compliance. These studies highlight the complexity of integrating digital health platforms with evolving regulatory frameworks while ensuring personalized yet privacy-compliant data handling approaches.

The role of intelligent systems in personalized software applications is analyzed in [23], where adaptive AI-driven solutions are investigated to optimize user-specific interactions. Furthermore, machine learning techniques inspired by artificial immune systems (AIS) are explored in [23], presenting a biologically inspired computational model for enhancing pattern recognition, anomaly detection, and security in intelligent systems. These findings underscore the potential of AI in personalized healthcare applications, where adaptive learning algorithms can enhance medical decision-making, patient engagement, and system reliability.

A strong correlation between blood biomarkers, nutrition, and weight classification has been established through machine learning models, particularly neural networks. In [14] and [15], AI-based nutritional biomarker analysis demonstrated the capability of predicting weight categories based on standard blood test variables. Further classification methodologies aimed at identifying weight-related patterns in blood tests-specifically biochemical profiles-were examined in [16] and [17], where statistical modeling and AI-driven classification techniques were tested on clinical datasets.

A more refined classification approach was proposed in [18], where support vector machines (SVMs) were used to enhance predictive accuracy in BMI and metabolic syndrome classification. The study achieved an 84% accuracy in correctly identifying body mass index (BMI) and metabolic syndrome and a 74% accuracy in predicting systolic blood pressure levels. These results emphasize the efficacy of AI-based models in automated metabolic health assessments, validating the use of machine learning-driven diagnostic tools in clinical practice.

Building on these previous findings, this research integrates machine learning models into a healthcare application, leveraging biomarker-based AI analysis to classify health states and generate personalized recommendations. The objective is to utilize the existing benchmark models as a foundation for a scalable, regulatory-compliant AI-driven healthcare platform. By implementing this intelligent system within a structured development framework, the proposed model seeks to accelerate deployment timelines, streamline compliance processes, and enhance usability for both medical professionals and patients. This approach enables faster adoption of AI-driven clinical tools, ensuring that healthcare applications remain efficient, transparent, and aligned with real-world medical needs.

This study contributes to existing research by extending machine learning methodologies into a practical, deployable framework, demonstrating how AI-powered tools can be effectively integrated into medical workflows while ensuring compliance with regulatory standards. Future work will focus on optimizing model accuracy, improving explainability in AI decision-making, and enhancing real-time adaptation to patient-specific health trends.

6.4 Building a Blood Exam-Based Personalised Recommender System

The proposed web-based medical application classifies patients using machine learning (ML) algorithms, extracting clinically relevant insights from blood exam data. The methodology for classification and prediction was validated in [14–18], where biomarker-based ML models demonstrated high accuracy in predicting weight categories, metabolic syndrome, and systolic blood pressure risks.

This application applies ML-driven classification to determine weight status and metabolic state based on established metabolic syndrome guidelines. The system then assesses the risk of high blood pressure and provides personalized dietary and lifestyle recommendations, leveraging patterns identified in healthy individuals. The approach integrates clinical parameters and lifestyle factors, ensuring that recommendations remain evidence-based and adaptive.

Stakeholders play a critical role in the system's deployment:

1. Healthcare professionals monitor AI-driven classifications, validate results, and intervene when necessary to ensure clinical accuracy.
2. Regulatory authorities ensure that the system complies with legal, ethical, and safety standards for AI in healthcare.
3. Patients interact with the system, receiving real-time insights and recommendations based on their blood test results.

To structure the machine learning pipeline and ensure its robust implementation, we have adopted two industry-standard methodologies:

1. The Cross-Industry Standard Process for Data Mining (C.R.I.S.P-D.M.)—Used to define data-related processes, feature extraction, and predictive modeling.
2. The Rational Unified Process (R.U.P.)—Applied to analyze the different phases of software implementation, regulatory compliance, and production cycles.

By combining these two methodologies, we aim to optimize AI integration in healthcare applications, ensuring scalability, regulatory transparency, and enhanced patient outcomes.

6.4.1 Development Methodologies

Cross Industry Standard Process for Data Mining

The C.R.I.S.P-D.M. framework is an industry-standard methodology for structuring data analysis and machine learning workflows. Given that machine learning relies heavily on data preprocessing and feature engineering, C.R.I.S.P-D.M. is a vital tool for ensuring quality control, repeatability, and model transparency.

C.R.I.S.P-D.M. was developed in 1996 by SPSS as a structured approach for handling large-scale data mining projects. It consists of six iterative phases (Fig. 6.3), each focusing on different aspects of data-driven problem solving:

- Business Understanding—Defines the key problem and intended outcome.
- Data Understanding—Explores dataset characteristics, structure, and statistical distributions.
- Data Preparation—Involves feature selection, preprocessing, and transformations.
- Modeling—Applies machine learning models to identify patterns and predictive insights.
- Evaluation—Assesses model performance, explainability, and clinical validity.

Fig. 6.3 Elements involved in C.R.I.S.P

- Deployment—Integrates the trained model into the final AI-driven system for real-world application.

Various iterations of C.R.I.S.P-D.M. have been developed, including quality assurance models [25] and optimization techniques for machine learning applications [4]. In the context of this chapter, C.R.I.S.P-D.M. is used to refine the patient journey by leveraging pattern recognition. By integrating AI-driven automation into healthcare workflows, unnecessary steps in the patient journey can be minimized, allowing for more efficient diagnostics and treatment pathways. Machine learning models can analyze patient data in real-time, providing immediate assessments that reduce the need for redundant medical visits and manual evaluations. This automation enhances early disease detection by identifying biomarker deviations before symptoms become apparent, enabling proactive interventions and personalized preventive strategies. Additionally, AI-generated insights help alleviate the workload on healthcare professionals by streamlining routine decision-making processes, allowing physicians to focus on more complex cases while ensuring that patients receive timely and accurate recommendations. Furthermore, by addressing administrative and educational barriers, AI-powered systems improve communication between patients and medical professionals, making health information more accessible and understandable. This ensures that patients remain well-informed about their conditions, treatment options, and lifestyle modifications, ultimately fostering better engagement, adherence to medical advice, and overall healthcare outcomes.

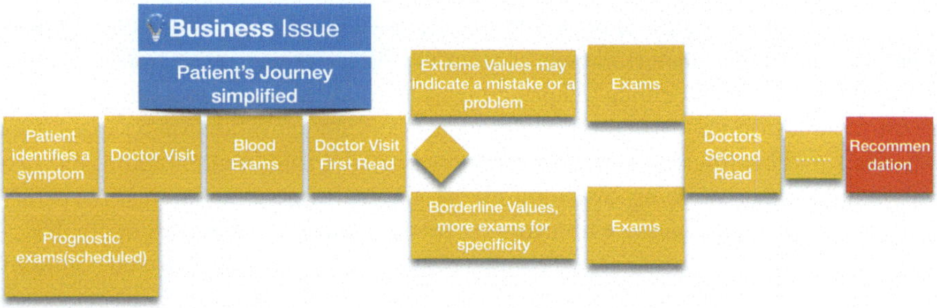

Fig. 6.4 Simplified patient's journey

A simplified version of the patient journey is shown in Fig. 6.4, illustrating the business issue of C.R.I.S.P-D.M. in a healthcare setting. The journey begins with symptom recognition, followed by a clinical visit, medical examinations, and physician evaluations. Traditionally, this process requires multiple patient visits and manual interpretation of blood test values before a diagnosis is made.

By integrating machine learning into the patient journey, we can automate routine classification tasks, improve efficiency, and enhance diagnostic accuracy. This methodology lays the foundation for a structured AI-driven intervention system.

RUP

RUP is an agile software development methodology designed to structure complex software engineering projects. It divides the project lifecycle into four iterative phases:

1. Inception—Defines business needs and feasibility.
2. Elaboration—Refines system architecture and risk assessment.
3. Construction—Implements software components.
4. Transition—Deploys and validates the final product.

Each phase incorporates six core software development disciplines. Namely, Business modeling, Requirements gathering, Analysis and design, Implementation, Testing and Deployment.

RUP's structured, iterative approach ensures that stakeholders-including regulatory authorities-remain engaged at every stage. In healthcare applications, this is crucial, as regulatory validation and compliance are required before moving forward with development. Failure to secure medical approval can lead to costly redesigns and regulatory setbacks [1].

By integrating C.R.I.S.P-D.M. into R.U.P., we define a development framework where data science, regulatory compliance, and software engineering are seamlessly connected. AI-driven healthcare applications must adhere to stringent legal and ethical standards to ensure patient safety, data privacy, and regulatory compliance. By integrating governance frameworks from the outset, these applications can align with healthcare policies, ethical AI principles, and medical device regulations, ensuring trust and transparency in their deployment. Continuous validation by regulatory authorities and healthcare professionals plays a crucial role in maintaining the system's reliability, as ongoing oversight allows for real-time assessments, necessary refinements, and compliance with evolving medical guidelines. Additionally, leveraging an iterative development cycle enhances the efficiency of AI-driven solutions by identifying and resolving potential errors early in the process. This approach not only improves system accuracy and performance but also accelerates innovation while mitigating risks, ultimately leading to more effective, safe, and scalable AI applications in healthcare.

The C.R.I.S.P-D.M. methodology (Crisp), regulatory oversight (Reg), and validation processes (Val) are embedded within each RUP phase, ensuring full compliance and transparency.

Logical View of Proposed Application

As shown in Fig. 6.6, the system operates through two tiers:

1. Basic System—Patients receive automated classifications based on blood test values. AI-driven recommendations are generated based on weight, metabolic risk, and heart disease probability.
2. Advanced System—Patients contribute health data to retrain and refine ML models. Regulators and physicians validate new AI models before deployment, ensuring that ML updates meet clinical standards.

If AI confidence in a classification falls below a predefined threshold, additional human review is required. Physicians validate uncertain classifications, and further medical tests are requested if necessary. The system also includes:

- Role-based UI designs for patients, doctors, and regulators.
- Real-time alerts and system transparency mechanisms to improve trust in AI-driven decision-making.

By integrating structured AI workflows, this system enhances efficiency, compliance, and scalability, paving the way for broader adoption of ML in clinical decision-making.

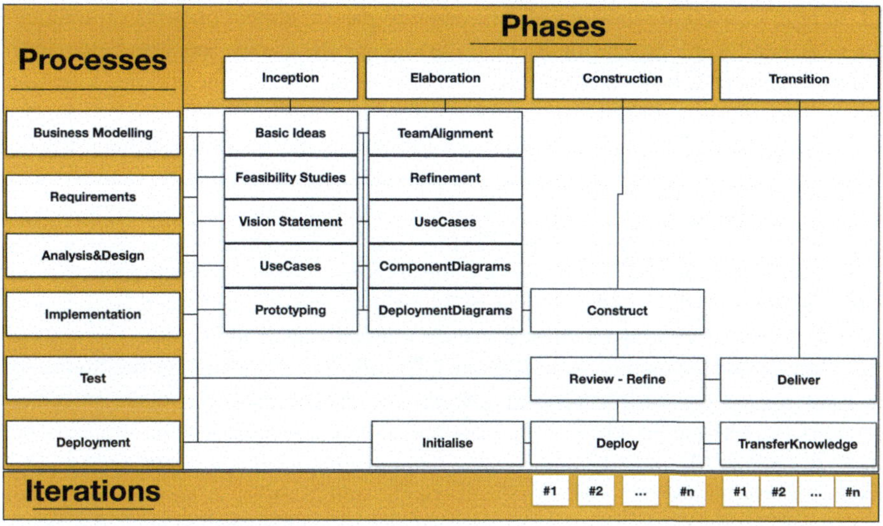

Fig. 6.5 RUP methodology with implementation phases

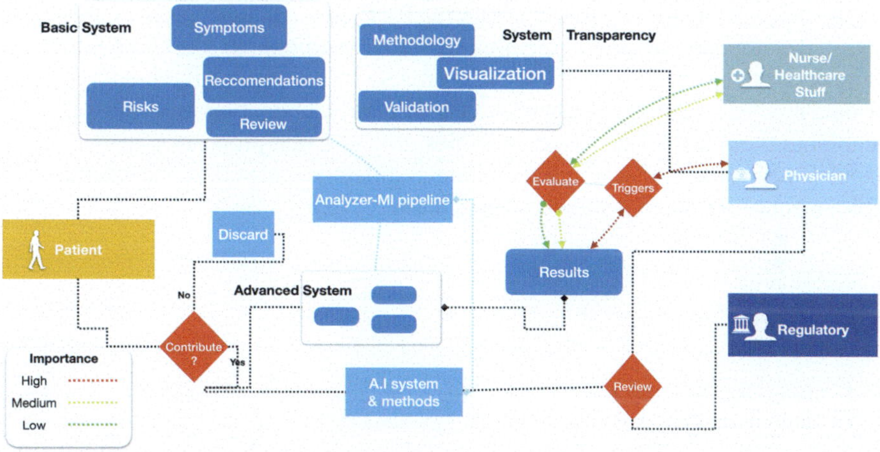

Fig. 6.6 Analysis and design, the logical view case model

6.5 Conclusion and Research Key Findings

When modeling a medical application, a solid starting point is understanding the patient's journey. Smart technologies have the potential to alleviate many of the challenges physicians face [8], minimizing the time gaps between different stages of the patient's experience. By integrating customized tools into the healthcare process, patients can benefit from more streamlined interactions, improved accessibility, and more effective care delivery [7]. Enhancing patient education and communication with healthcare professionals helps bridge existing gaps, ensuring that individuals are better informed about their conditions and treatment options. Creating greater awareness and facilitating better symptom administration can reduce the necessity for specialized and expensive treatments, accelerate decision-making processes, decrease the severity of conditions, and ultimately lower fatality rates [6]. The web-based application developed in this study, which leverages blood biomarkers to guide personalized interventions, will be presented at the conference to demonstrate its potential outcomes and the methodologies employed in its development.

In this chapter, the Golden Circle of Innovation has been adapted as a framework to define the role of machine learning in healthcare and provide a structured validation model for medical professionals. By offering a transparent methodology for assessing AI-driven interventions, healthcare practitioners can make informed decisions on whether to adopt or dismiss specific AI-generated recommendations. The C.R.I.S.P-D.M. and RUP methodologies have also been incorporated to align stakeholder requirements, regulatory constraints, and usability concerns. Given that medical validation and regulatory compliance remain critical challenges in the development of AI-powered medical applications, these elements have been embedded directly into the production cycle roadmap. As shown in Fig. 6.5, validation and regulatory supervision are integrated within the RUP methodology, ensuring that each stage of development adheres to required standards. Additionally, the C.R.I.S.P-D.M. framework is embedded within the RUP process, providing a structured approach for data analysis, feature selection, model training, and continuous evaluation. By combining these established methodologies, this research proposes a development model that enhances transparency, facilitates regulatory approval, and accelerates the deployment of AI-based healthcare applications.

References

1. S. Ahangama, D.C. Choon Poo, Unified structured process for health analytics. Int. J. Med. Health Biomed. Bioeng. Pharm. Eng. **8.11**, 798–806 (2014)
2. M.A. Ahmad, C. Eckert, A. Teredesai, Interpretable machine learning in healthcare, *Proceedings of the 2018 ACM International Conference on Bioinformatics, Computational Biology, and Health Informatics*, pp. 559–560 (2018)
3. H.W. Chesbrough, *Open Innovation: The New Imperative for Creating and Profiting from Technology* (Harvard Business Press, 2003)
4. W. Duch, R. Adamczak, K. Grabczewski, A new methodology of extraction, optimization and application of crisp and fuzzy logical rules, in *IEEE Transactions on Neural Networks*, vol. 12.2, pp. 277–306
5. K.M. Fargen, D. Frei, D. Fiorella, C.G. McDougall, P.M. Myers, J.A. Hirsch, J. Mocco, The FDA approval process for medical devices: an inherently flawed system or a valuable pathway for innovation? J. Neurointerventional Surg. **5.4**, 269–275 (2013)

6. J.P. Gregg, L. Tianhong, K.Y. Yoneda, Molecular testing strategies in non-small cell lung cancer: optimizing the diagnostic journey. Trans. Lung Cancer Res. **8.3**, 286 (2019)
7. M.L. Jacobs, J. Clawson, E.D. Mynatt, My journey compass: a preliminary investigation of a mobile tool for cancer patients, in *Proceedings of the SIGCHI Conference on Human Factors in Computing Systems*, pp. 663–672 (2014)
8. D. Kauw, P.R. Huisma, S.K. Medlock, M.A.C. Koole, E. Wierda, A. Abu-Hanna, M.P. Schijven, B.J.M. Mulder, B.J. Bouma, M.M. Winter, et al., Mobile health in cardiac patients: an overview on experiences and challenges of stakeholders involved in daily use and development, in *BMJ Innovations*, vol. 6.4 (2020)
9. D.E. McNiel, R. Binder, Current regulation of mobile mental health applications. J. Am. Acad. Psychiatry Law **46**, 204–11 (2018)
10. E. Mougiakou, S. Papadimitriou, M. Virvou, Synchronous and asynchronous learning methods under the light of general data protection regulation, in *2020 11th International Conference on Information, Intelligence, Systems and Applications (IISA* (IEEE, 2020), pp. 1–7
11. N.S. Nipa, M. Alam, M.S. Haque, Identifying relevant stakeholders in digital healthcare, in *Proceedings of the Applied Intelligence and Informatics: First International Conference, AII 2021, Nottingham, UK*, 30–31 July 2021, vol. 1 (Springer, 2021), pp. 349–357
12. I. Nygaard, "What Does" FDA Approved "Mean for Medical Devices?" in *Obstetrics and Gynecology-New York*, vol. 111.1, p. 4 (2008)
13. N. Omar, B. Yan, M. Salto-Tellez, HER2: an emerging biomarker in non-breast and non-gastric cancers", in *Pathogenesis* vol. 2.3, pp. 1–9 (2015)
14. D.P. Panagoulias, D.N. Sotiropoulos, G.A. Tsihrintzis, Biomarker-based deep learning for personalized nutrition, in *Proceedings of the 33rd IEEE International Conference on Tools with Artificial Intelligence (ICTAI)*, Virtually, pp. 73–122, Nov. 2021a
15. D.P. Panagoulias, D.N. Sotiropoulos, G.A. Tsihrintzis, Nutritional biomarkers and machine learning for personalized nutrition applications and health optimization, in *Proceedings of the Twelfth IEEE International Conference on Information, Intelligence, Systems and Applications*, Chania, Greece, pp. 731–733, July 2021b
16. D.P. Panagoulias, D.N. Sotiropoulos, G.A. Tsihrintzis, Nutritional biomarkers and machine learning for personalized nutrition applications and health optimization. Intell. Decis. Technol. **15**, 645–653 (2021)
17. D.P. Panagoulias, D.N. Sotiropoulos, G.A. Tsihrintzis, Towards personalized nutrition applications with nutritional biomarkers and machine learning, in *Advances in Assistive Technologies: Selected Papers in Honour of Professor Nikolaos G. Bourbakis*, ed. by G.A. Tsihrintzis, M. Virvou, A. Esposito, L.C. Jain. vols. 3, 28 (Springer, Cham, Switzerland, 2022), pp. 73–122
18. D.P. Panagoulias, N. Dionisios, N. Sotiropoulos, G.A. Tsihrintzis, SVM-Based blood exam classification for predicting defining factors in metabolic syndrome diagnosis, in *Electronics*, vol. 11.6, p. 857
19. E. Politou, E. Alepis, M. Virvou, C. Patsakis, *Privacy and Data Protection Challenges in the Distributed Era*, vol. 26 (Springer, 2022)
20. S.E. Reid, U. De Brentani, The fuzzy front end of new product development for discontinuous innovations: a theoretical model. J. Product Innov. Manag. **21.3**, 170–184 (2004)
21. R. Schnall, H. Cho, J. Liu, et al., Health information technology usability evaluation scale (Health-ITUES) for usability assessment of mobile health technology: validation study. JMIR mHealth uHealth **6.1**, e8851 (2018)
22. A. Sharma, B.J. Singh, Understanding lean six sigma 4.0 through golden circle model, in *EasyChair Preprint*, vol. 3314 (2020)
23. D. Sotiropoulos, G.A. Tsihrintzis, *Machine learning paradigms: artificial immune systems and their applications in software personalization* (2016)

24. J. Spruijt, T. Spanjaard, K. Demouge, The golden circle of innovation: what companies can learn from ngos when it comes to innovation, in *Modern Marketing for Non-Profit Organizations: International Perspectives*. ed. by S. Smyczek (University of Economics in Katowice Publishing House, Katowice, Poland, 2013)
25. R. Studer, V.R. Benjamins, D. Fensel, Knowledge engineering: principles and methods. Data Knowl. Eng. **25**, 161–197 (1998)
26. L. Zhou, J. Bao, I.M.A. Setiawan, A. Saptono, B. Parmanto, et al., The mHealth app usability questionnaire (MAUQ): development and validation study. JMIR mHealth uHealth **7.4**, e11500 (2019)

Chapter 7
Framework for AI Explainability Leveraging User Acceptance and Health Literacy Models

Abstract Recent advances in AI-enabled systems and machine learning applications across multiple sectors have highlighted the critical need for robust, tailored frameworks to enhance system usability and trustworthiness. In the healthcare domain, where verifying methods and outcomes is paramount, establishing explainability requires a comprehensive, systemic approach that balances depth with appropriate abstraction levels. The qualities of explainability and interpretability are central to ensuring that AI technologies are both user-friendly and reliable, thereby playing a key role in technology acceptance. In this chapter, we introduce a novel framework for defining explainability requirements in AI-driven systems by integrating the principles of the Technology Acceptance Model (TAM). Our framework leverages targeted machine learning techniques-including hierarchical clustering and k-means, among others-to develop a user model that supports multi-layered, personalized explainability. Furthermore, it incorporates a rule-based mechanism that adjusts the level of trustworthiness based on user perceptions and their proficiency with AI. We demonstrate the application of this framework in the context of AI-powered medical systems, with the objectives to (1) evaluate and quantify physicians' familiarity with technology and AI, (2) create tailored layers of explainability that correspond to individual user needs and trust levels, and (3) foster an environment that promotes transparency and validation. To assess doctors' technological proficiency, we employ the Rapid Estimate of Adult Literacy in Medicine (REALM), a tool widely used in healthcare to improve communication between clinicians and patients.

Keywords AI-empowered software engineering · Biomarkers · Explainability · Diffusion of innovation theory · Technology acceptance model

7.1 Introduction

AI serves as a pivotal force in bridging user needs with advanced tools, leveraging health-related data and physiological measurements to enhance medical pathways and optimize treatments. However, challenges related to relevance and potential misuse persist. The validation of ML-based medical systems remains a pressing

D. P. Panagoulias et al., *Artificial Intelligence-Empowered Bio-medical Applications*, Learning and Analytics in Intelligent Systems 49,
https://doi.org/10.1007/978-3-031-90174-4_7

concern, with biases arising from data inconsistencies or stakeholder influence in the evaluation process. To ensure patient safety and data integrity, robust frameworks must be established to guarantee transparency and interpretability. The complexity of AI-assisted medical recommendations presents inherent risks, necessitating stringent validation measures to mitigate potential threats to patient well-being.

Explainable AI (XAI) has emerged as a crucial field of research XAI [8, 20, 53, 54] provides the necessary structure to ensure AI-powered medical systems operate in a manner that is safe, transparent, and interpretable for both healthcare professionals and patients. AI systems are often regarded as intelligent due to their ability to simulate human reasoning and generate contextually relevant recommendations. Their effectiveness is measured based on factors such as communication capability, knowledge representation, intentionality, and adaptability [48]. With the rise of ML-powered language models, such as large language models (LLMs), AI systems are now evaluated on additional dimensions, including reasoning efficiency, processing speed, and conciseness [35].

This chapter introduces a novel framework for personalized XAI, combining machine learning techniques with innovation theories. Specifically, the proposed system integrates clustering methodologies with a rule-based decision model to enhance AI transparency. The rule-based approach employs the Technology Acceptance Model (TAM) to assess user behavior concerning AI adoption, leveraging clustering methods to quantify user perceptions of system usefulness and ease of use. Originally introduced by Fred Davis, TAM is a well-established framework for evaluating the factors that influence technology adoption. It posits that perceived ease of use and perceived usefulness are key mediators in determining whether users accept or reject new technologies.

Within this context, hierarchical clustering and k-means are applied to construct user models that enable personalized, multi-layered explainability. Two key external variables are considered: AI literacy and the level of abstraction in explanations. Our proposed framework integrates machine learning with a rule-based model to determine the degree of trustworthiness needed, adapting transparency levels according to the user's perception of usefulness and system usability, as outlined by TAM. This tailored approach ensures that AI explanations align with individual user requirements, ultimately fostering greater trust in AI applications.

To validate this methodology, we conducted a case study within the medical domain. Our approach focuses on three key objectives in the integration of AI in medical decision-making. First, we assess and quantify physicians' perspectives on AI adoption, their level of AI literacy, and their expectations regarding different abstraction levels of explainability. Understanding these factors is crucial for aligning AI tools with clinical workflows and practitioner needs. Second, we develop personalized explainability layers tailored to users' capabilities and trust requirements. This ensures that AI-generated explanations are both comprehensible and relevant, fostering confidence in AI-assisted decision-making. Lastly, we establish a transparent framework that enhances the interpretability and validation of AI-driven medical decisions. By improving transparency and reliability, this framework aims to increase trust and adoption of AI-powered healthcare systems.

In assessing comprehension-literacy levels, we considered tools like the Rapid Estimate of Adult Literacy in Medicine (REALM), commonly used in healthcare to bridge communication gaps between clinicians and patients. The validation of this methodology is examined within the healthcare sector. We present and analyse the findings from a survey investigating the extent of AI adoption among medical professionals.

The structure of this chapter is as follows: Sect. 7.2 provides an overview of foundational human behavioral theories and relevant prior research. Section 7.3 details the development of our innovative framework, which defines AI literacy requirements and the necessary levels of explainability abstraction. We introduce this framework as PINXEL (Personalised INterface for eXplainability basEd on Levels) Fig. 7.2, a model designed to enhance transparency in AI-driven applications. This methodology integrates multiple layers of explainability using the clustering approach illustrated in Fig. 7.6 and discussed in Sect. 7.4, where we employ PINXEL to analyze survey responses and define personalization layers through TAM-based clustering, as depicted in Figs. 7.2, 7.6, and 7.4. The objective is to improve AI adoption rates by incorporating explainability and transparency as fundamental design principles. Section 7.5 implements these explainability layers within NuhealthSoft, using it as a testbed for the proposed framework. Finally, Sect. 7.6 summarizes the chapter and outlines future research directions.

7.2 Related Work

Several theories have been widely utilized to analyze, predict, and explain human behavior concerning technology adoption and the integration of innovative solutions. These models offer distinct perspectives on how users interact with new technologies, enabling a structured evaluation of acceptance patterns. The Diffusion of Innovation (DOI) Theory, introduced by Rogers, explains that innovation adoption is driven by how a population perceives its value. If a new idea or product is seen as novel and beneficial, its diffusion is more likely, accelerating its adoption [47]. Closely related is the Bass Diffusion Model, which mathematically categorizes adopters into innovators and imitators, illustrating the adoption lifecycle of emerging technologies [25, 34].

Another key perspective is the Unified Theory of Acceptance and Use of Technology (UTAUT), which integrates elements from eight models, including the Technology Acceptance Model (TAM). It highlights key adoption drivers such as performance expectancy, effort expectancy, social influence, and facilitating conditions [22, 52]. Complementing this, the Theory of Planned Behavior (TPB) asserts that behavioral intentions are shaped by attitudes, subjective norms, and perceived control over the behavior [1, 56].

Adoption patterns are also influenced by Critical Mass Theory, which suggests that once a technology gains enough adopters, its uptake accelerates and becomes self-sustaining [14, 28]. Additionally, Social Cognitive Theory emphasizes the impact of social influence and observational learning in decision-making, particularly how individuals adapt behaviors based on external interactions [4, 7]. Finally, the Consumer Decision Process Model outlines the structured stages consumers go through before adopting a new product or technology, from awareness and evaluation to the final decision [5, 13].

For this study, we selected the Technology Acceptance Model (TAM) due to its clarity, extensive acceptance, and adaptability to digital systems where automated decision-making is essential. Additionally, TAM incorporates external factors that influence adoption, making it suitable for assessing AI-driven medical applications.

Informational Communication Gap in Medicine

The disparity in communication between healthcare professionals and patients presents a critical challenge for inclusive and responsible healthcare. Addressing this gap is essential to ensure that patients fully comprehend medical information, as differences in literacy and familiarity with medical terminology can lead to misunderstandings [15, 46].

Several standardized assessments are used to evaluate health literacy, ensuring patients can comprehend and engage with medical information effectively. The Rapid Estimate of Adult Literacy in Medicine (REALM) measures a patient's ability to read and pronounce common medical terms related to anatomy and diseases, providing insight into their basic medical vocabulary [32]. The Test of Functional Health Literacy in Adults (TOFHLA) assesses a patient's ability to interpret written health-related instructions, such as prescription labels and appointment slips, helping determine their capacity to navigate healthcare information [45]. The Newest Vital Sign (NVS) evaluates both literacy and numeracy skills by analyzing a patient's ability to interpret a nutrition label, reflecting their proficiency in understanding health-related quantitative data [55].

7.2.1 Technology and Applications

Our research focuses on developing NuhealthSoft, an AI-powered medical application built on a microservices architecture [40]. This system utilizes machine learning models to analyze health data, generating classifications, predictions, and personalized recommendations. By extracting key features from blood test results, we achieved high precision in predicting critical health indicators. Specifically, weight category prediction reached 85% accuracy [36–38, 43], metabolic syndrome detection achieved 86% accuracy, and high systolic blood pressure prediction attained 74% accuracy.

These methodologies have been integrated into a digital platform that processes blood test data to evaluate health conditions, provide evidence-based guidance, and

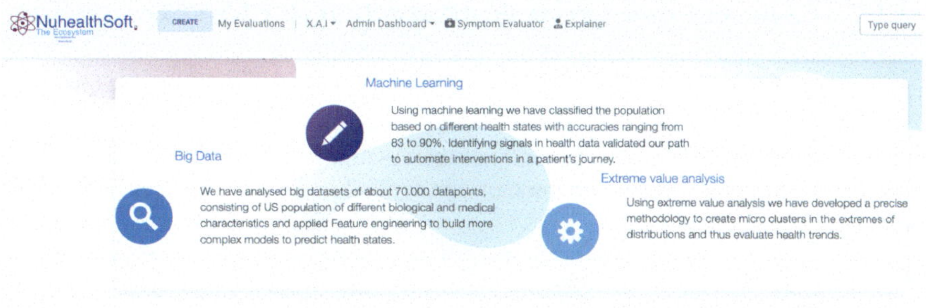

Fig. 7.1 Screenshot from NuhealthSoft

automate personalized interventions [41]. Ensuring regulatory compliance was a key consideration, addressing healthcare standards and validation requirements [42]. Additionally, extreme value analysis was applied to dietary intake assessments, enhancing the system's ability to derive meaningful health insights [39]. In this chapter, we used NuhealthSoft as a testbed to assess PINXEL (Fig. 7.1).

Given the significance of medical validation, the role of healthcare professionals and regulatory bodies is crucial in the development of NuhealthSoft. This necessity motivated our research to ensure that doctors' requirements are adequately addressed while providing a transparent, user-centric framework for system validation. NuhealthSoft employs machine learning models alongside nutritional and biochemical data to classify users based on health concerns such as blood pressure, weight, and metabolic syndrome [36–38, 42, 43]. Additionally, it generates personalized nutritional recommendations based on an individual's laboratory results and medical profile.

7.3 The PINXEL Framework

PINXEL—an acronym for Personalised INterface for eXplainability basEd on Levels is our proposed framework, that establishes a structured approach to personalized AI explainability, leveraging different levels of AI literacy and explanation abstraction. PINXEL integrates the Technology Acceptance Model (TAM) to assess user characteristics and competencies, defining clusters based on these attributes. Each cluster receives a tailored explainability layer, ensuring an adaptive and user-specific approach.

7.3.1 AI Literacy and Levels of Abstraction in PINXEL

Technological literacy is the ability to comprehend, use, and critically evaluate technology within various contexts. It also includes the capacity to make informed decisions regarding technological applications. As digital innovation advances, technological literacy becomes a fundamental aspect of education and professional development.

AI literacy, a subset of technological literacy, pertains to an individual's ability to differentiate AI-powered technologies from conventional ones, understand AI models' operational principles, and evaluate their implications. Research indicates that AI literacy can be categorized into multiple levels [33]. Novice users may only recognize AI-based applications, while more proficient users can analyze datasets, understand data collection mechanisms, and assess AI models' interactions with their environment.

The level of abstraction in explanations refers to the granularity of AI system details presented to users. In machine learning, higher abstraction levels offer simplified representations, whereas lower abstraction levels incorporate technical details, making them suitable for expert audiences [16, 29, 30].

Since TAM defines perceived usefulness and perceived ease of use as key determinants of technology acceptance, these two parameters are associated with AI literacy and level of abstraction in PINXEL. Users with higher self-efficacy-their confidence in their ability to use AI-tend to perceive AI as more useful, leading to higher adoption rates [21, 26, 49, 51]. Similarly, system usability and clarity, facilitated by an appropriate abstraction level, contribute to perceived ease of use [10, 24].

AI Literacy Levels in PINXEL

PINXEL [44] structures AI literacy into three progressive levels-basic, average, and expert-tailoring explanations to users' knowledge and ensuring transparency in AI systems. This stratified approach aligns with educational methodologies, progressively introducing complexity as users advance [12, 23]. Similar methods, such as REALM, are used in medical literacy assessments.

At the basic level, AI concepts are presented in a user-friendly narrative format, simplifying core ideas such as machine learning, neural networks, and data-driven decision-making. Semantic grouping organizes related concepts, making them more intuitive. Users gain foundational knowledge of Natural Language Processing (NLP), Machine Learning (ML), and Large Language Models (LLMs) [11].

The average level introduces technical breakdowns, covering model architecture, training processes, and key AI components. Topics include neural network layers [17], activation functions [19], weight adjustments [9], loss functions [31], and fairness considerations [6]. Evaluation metrics such as confusion matrices and classification reports help users assess model performance [2, 18].

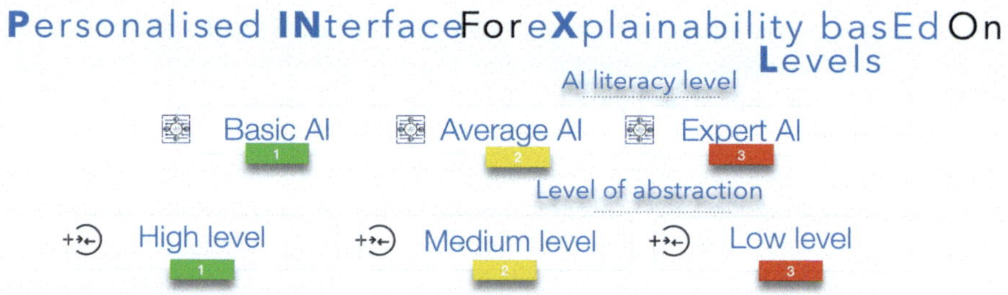

Fig. 7.2 PINXEL framework

At the expert level, explanations delve into AI decision-making using global and local explainability techniques [3]. Global approaches, including feature importance ranking, Partial Dependence Plots (PDP), Accumulated Local Effects (ALE), and dimensionality reduction, provide insights into overall model behavior. Local interpretability tools, such as LIME, SHAP, and Anchor Explanations, focus on individual predictions, offering scenario-based insights into AI decisions.

This multi-tiered approach ensures that AI transparency is adaptable to diverse user expertise, bridging the gap between technical complexity and real-world applications.

Levels of Abstraction in PINXEL

In computing, abstraction simplifies complexity by concealing low-level details, allowing users to interact with high-level representations. PINXEL organizes abstraction into three layers, depicted in Fig. 7.3. AI explanations in PINXEL are structured into three abstraction levels, adapting to user expertise and interpretability needs.

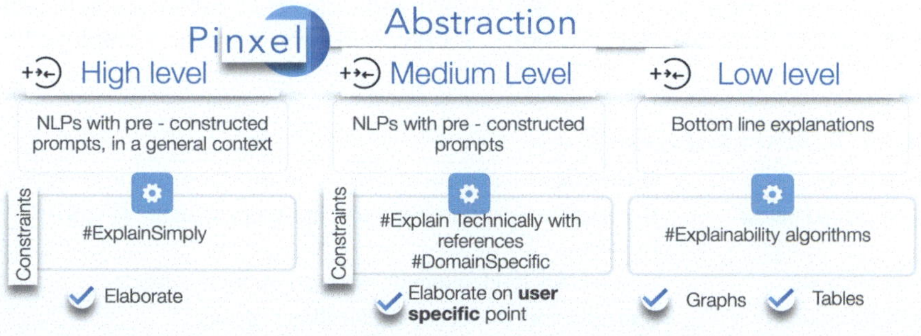

Fig. 7.3 PINXEL, levels of abstraction

At the high abstraction level, Natural Language Processing (NLP) generates simplified explanations using structured prompts, such as “Explain simply...,” ensuring alignment with the user’s AI literacy level. The medium abstraction level refines these explanations by incorporating technical justifications and domain-specific references. Users gain more control over how they interact with AI outputs while maintaining structured guidance. For users requiring full transparency, the low abstraction level presents raw data, performance metrics, and graphical analyses. Standard Explainable AI (XAI) visualization techniques, including feature importance plots and model evaluation graphs, offer in-depth insights into AI decision-making.

7.3.2 *Personalization Through Clusters and Depth of Explainability Based on REALM*

The association of perceived ease of use and perceived usefulness with external variables-AI literacy and level of abstraction-enables the definition of explainability requirements and transparency layers. However, balancing personalization with system efficiency is crucial. Over-personalization may introduce excessive complexity, making the system difficult to navigate, while an optimal number of clusters ensures that the framework remains both effective and computationally efficient.

The primary objective is to enhance user perception by improving both usefulness and ease of use, while maintaining an understandable and manageable personalization strategy. A large number of clusters can lead to over-personalization, which may overwhelm users and obscure key information [50]. Secondly, system performance and efficiency must be considered-restricting the number of clusters helps optimize computational resources and enhance processing efficiency [18].

In medical applications, as highlighted in Sect. 7.2, the Rapid Estimate of Adult Literacy in Medicine (REALM) suggests an optimal number of four (4) clusters [32], as detailed in Table 7.1.

For the PINXEL framework, three levels of AI literacy and three levels of abstraction have been defined. As a result, we establish three clusters corresponding to each dimension (AI literacy and abstraction). A structured questionnaire is employed to

Table 7.1 REALM-Based literacy clusters

Raw score	Grade range
3rd Grade and below	Best understands content through illustrations, audio, or video formats
4th to 6th grade	Low literacy level; difficulty reading prescription labels
7th to 8th grade	Struggles with most patient education materials
High School	Capable of reading most patient education materials

Table 7.2 Clustering of AI Literacy and abstraction levels

Cluster	Likert range	Abstraction level	AI Literacy level	Explainability depth
1	(0)–2.6	High Level Abstraction (HLA)(1)	Basic AI Literacy (BAIL)(*1)	1
2	2.6–3.8	Medium Level Abstraction (MLA)(2)	Average AI Literacy (AAIL)(*2)	4
3	3.8–5	Low Level Abstraction (LLA) (3)	Expert AI Literacy (EAIL)(3)	9

categorize users based on Likert-scale responses, forming the basis for clustering. Table 7.2 outlines the Likert-based clustering approach.

Mathematical Formulation for Clustering

The clustering process is generalized through the following formulation:

1. Define the Likert scale L as the set of possible responses: $L = \{0, 1, 2, 3, 4, 5\}$
2. Define the set of questionnaire items: $Q = \{q_1, q_2, q_3, ..., q_n\}$
3. Each response is represented as an n-dimensional vector R in the Likert space: $R = (r_1, r_2, r_3, ..., r_n), r_i \in L$
4. Define the number of clusters K , where: $\{C_1, C_2, ..., C_K\}$ and the maximum number of clusters is constrained to: $K_{max} = 3$
5. Assign vector R to the most representative cluster based on its similarity to cluster centroids, using a clustering algorithm (e.g., K-Means, Hierarchical Clustering, DBSCAN):
   $C_k = \{R \mid \text{Similarity}(R, C_k) > \text{Similarity}(R, C_j), \forall j \neq k\}$

7.3.3 Rule-Based Personalization

PINXEL establishes a rule-based system that aligns AI literacy levels with perceived usefulness and abstraction levels with perceived ease of use.

As illustrated in Fig. 7.4, explainability depth is determined by the predefined levels of AI literacy and abstraction. The following rules govern the explainability assignment process:

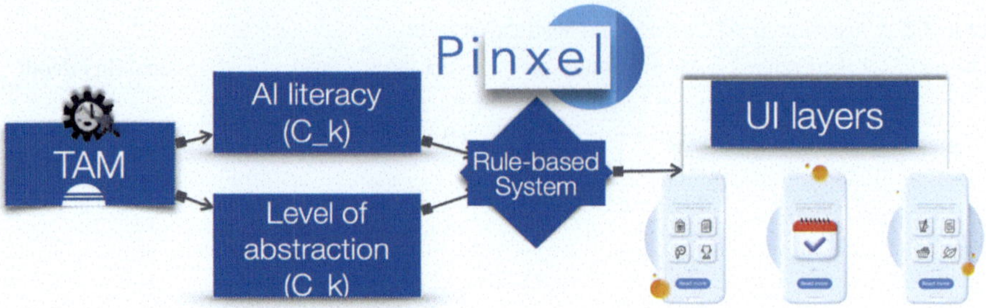

Fig. 7.4 Determining explainability depth and transparency layers

$$'Basic\ AI\ literacy'(BAIL)(1) \quad 'Average\ AI\ literacy'(AAIL)(2)$$

$$'Expert\ AI\ literacy'(EAIL)(3) \quad 'High\ Level\ Abstraction'(HLA)(1)$$

$$'Medium\ Level\ Abstraction'(MLA)(2) \quad 'Low\ Level\ Abstraction'(LLA)(3)$$

Rule-Based Decision Logic

The rule-based system assigns the depth of explainability as follows:

- IF BAIL (1) AND HLA (1) THEN Explainability Depth = 3
- IF EAIL (3) AND LLA (3) THEN Explainability Depth = 9
- IF EAIL (3) AND HLA (1) THEN Explainability Depth = 3.

For example, a user with Basic AI Literacy and a High Level of Abstraction will be provided an explainability layer that corresponds to the themes illustrated in Fig. 7.5, using simplified AI explanations. Similarly, an expert user requiring detailed transparency will receive highly granular explanations. Each rule leads to a unique

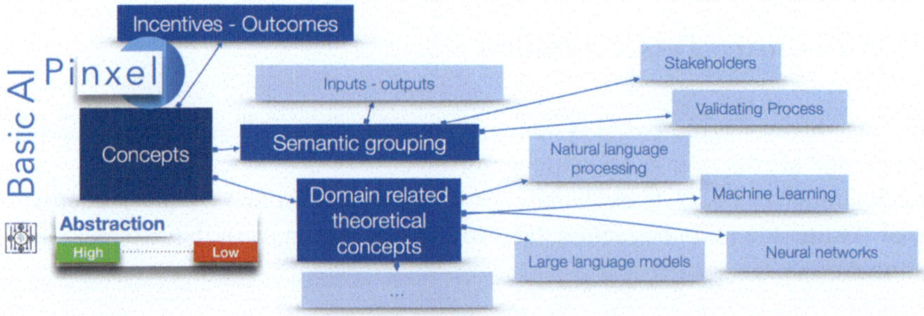

Fig. 7.5 Basic AI

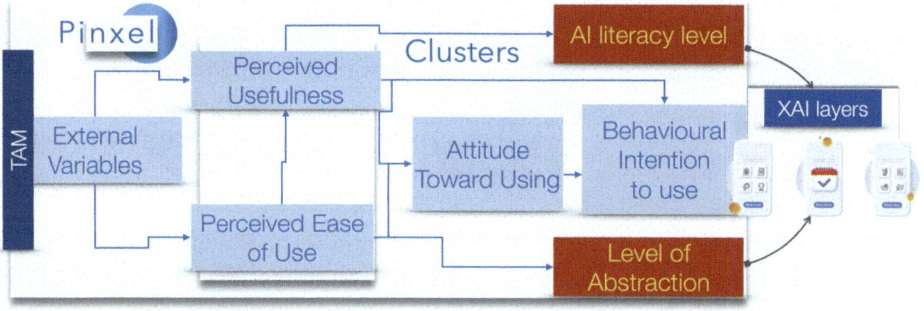

Fig. 7.6 Adapting TAM for PINXEL

explainability output, ensuring that the maximum number of rules aligns with the number of clusters. This structured approach allows efficient customization, improving AI transparency without overwhelming system complexity.

7.4 PINXEL—Personalization Analysis for NuhealthSoft

To determine the appropriate levels of Explainable AI (XAI) and define the corresponding depth of explainability using Technology Acceptance Model (TAM), a structured questionnaire was developed, distributed, and analyzed.

The survey targeted medical students and practicing doctors, collecting responses from 40 participants, of which 66.7% were men and 33.3% were women, representing diverse specializations, professional experience, and employment sectors. Among the 40 respondents, 39 completed the questionnaire, while one participant opted out. The questionnaire was electronically filled out anonymously, with no personal data collected. To ensure reliability and validity, over 70% of participants completed the form under supervision, with the support of a researcher involved in the study.

7.4.1 Demographics

Figure 7.7 presents key demographic insights from the first section of the questionnaire. The majority of participants were over 40 years old, while 10 respondents were undergraduate students under 30 years old. The employment sector distribution showed a nearly even split between public and private hospital clinicians.

Respondents specialized in various medical fields, as depicted in Fig. 7.7. Their work environments also varied: 14 respondents worked in private practices with at least two employees, while 6 respondents worked in institutions with over 10 employees. Additionally, 5 respondents held positions as hospital unit directors, and 2 were university academics.

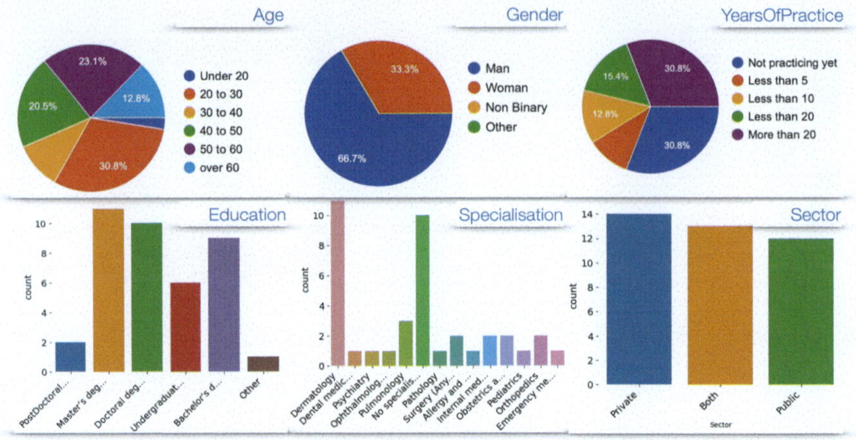

Fig. 7.7 Demographic distribution of survey participants

Regarding education, 2 respondents had completed postdoctoral training, while 10 held doctoral degrees, and 11 possessed master's degrees. The participants' English proficiency was evaluated as above average. Their professional experience ranged widely, with 30% of participants having over 20 years of experience in the medical field.

7.4.2 PINXEL—Perceived Ease of Use (TAM) and Levels of Abstraction

The survey questions assessing perceived ease of use Table 7.3 were designed to measure abstraction levels and the participants' attitude toward AI-driven innovations in medicine. Perceived ease of use refers to the extent to which individuals believe that using a given technology will be straightforward and free of major difficulties.

Question Structure: • Q1 and Q4 examine participants' attitudes toward new medical therapies and pharmaceutical innovations. • Q2 to Q6 investigate technological adoption trends in healthcare, including telehealth, medical automation, and digital symptom research. • Findings: Understanding these attitudes allows for tailored AI interfaces that enhance usability and trust by aligning with individual user preferences.

Clustering Analysis: Hierarchical clustering was applied to responses from Table 7.3, using silhouette coefficients to determine the optimal number of clusters (Fig. 7.8). Initially, six clusters were identified, but one participant selected "0" for all questions, leading to a final five-cluster model.

Table 7.3 Perceived ease of use and corresponding abstraction level survey questions

Question	Perceived ease of use assessment
Q1	Likelihood of recommending new therapies or medicines early in their introduction phase
Q2	Usage of medical devices with automation for treatment suggestions or outcome predictions
Q3	Perception of telehealth's effectiveness in comparison to in-person consultations
Q4	Opinion on whether clinical trials for new medical technologies should prioritize speed or safety
Q5	Attitude toward patients conducting online symptom research
Q6	Importance of human interaction in healthcare, especially for severe conditions

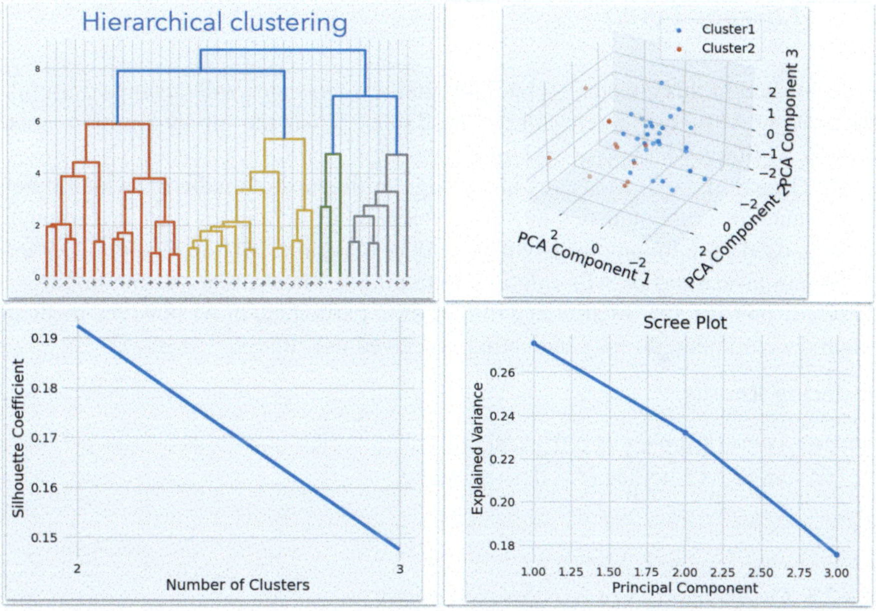

Fig. 7.8 Clusters based on perceived ease of use and abstraction levels

Key Observations: • Lowest-rated question: Q5, indicating concerns over patients self-diagnosing based on online searches. • Highest-rated question: Q4, suggesting a strong preference for prioritizing safety over rapid technological adoption. • Q2 Scores: Indicate that most respondents had not used AI-powered medical devices for treatment recommendations. • Clustering Results: Identified two primary clusters, both requiring a Medium Level of Abstraction (MLA) (score: 2) (Table 7.4).

Table 7.4 Perceived ease of use clusters—abstraction levels

Question	Cluster 1 (Mean)	Cluster 2 (Mean)	Overall average
Q1	3.48	3.5	3.49
Q2	2.24	1.4	1.82
Q3	3.31	3.0	3.16
Q4	4.48	2.8	3.64
Q5	1.72	2.9	2.31
Q6	2.24	3.2	2.72
Total count	29	10	
Average score	2.93	2.8	

7.4.3 *PINXEL—Perceived Usefulness (TAM) and AI Literacy Levels*

As established in this chapter, technology literacy correlates with perceived usefulness, making AI literacy a relevant metric. Table 7.5 lists the survey questions used to assess AI literacy and perceived usefulness.

Key Questions: Q1: Measures knowledge in diagnostic medicine, which relies on data analytics and statistics-key AI concepts. Q2: Evaluates trust in computer vision as used in autonomous driving-a widely recognized AI application. Q3 and Q4: Assess awareness of AI's role in automating medical tasks and the impact of AI literacy on trust. Q5: Measures willingness to participate in AI-powered medical system development, directly indicating perceived usefulness.

Clustering Results:

- C1: Expert AI Literacy (EAIL)—Score: 3
- C2: Average AI Literacy (AAIL)—Score: 2

Table 7.5 Clusters of perceived usefulness—AI literacy

Question	Cluster 1 (Mean)	Cluster 2 (Mean)	Overall average
Q1	3.40	3.0	3.20
Q2	3.33	3.25	3.29
Q3	3.96	2.91	3.70
Q4	4.11	3.25	3.44
Q5	4.55	2.75	3.65
Total count	27	12	
Average score	3.87	3.03	

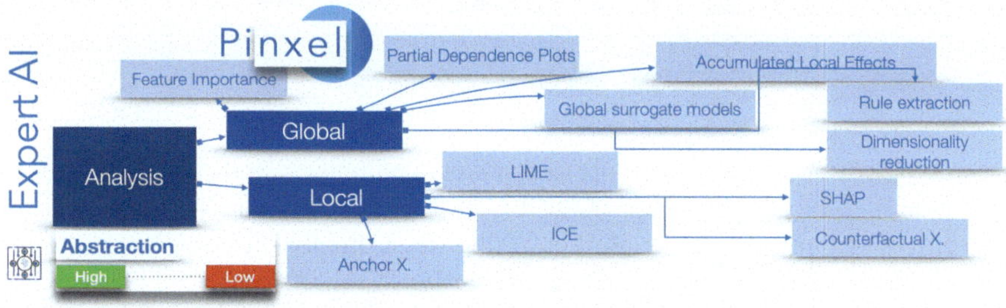

Fig. 7.9 Expert AI

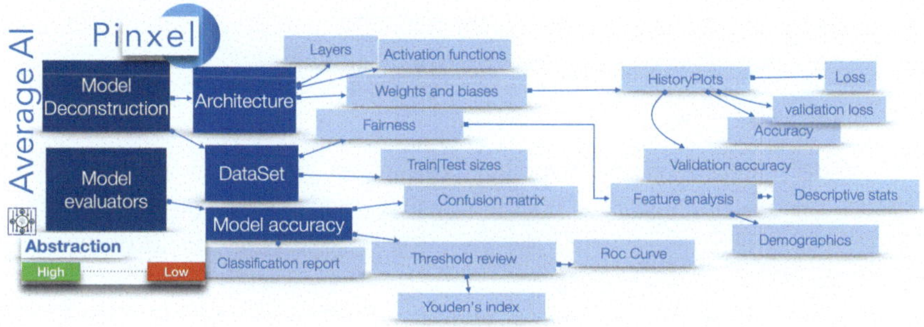

Fig. 7.10 Average AI

Final Implementation Strategy:

- (C1) IF EAIL(3) AND MLA(2) THEN Explainability Depth = 6
- (C2) IF AAIL(2) AND MLA(2) THEN Explainability Depth = 4

Conclusion: The explainability structure follows the methodology depicted in Figs. 7.9 and 7.10, employing a Medium Level of Abstraction (MLA) to ensure clear, AI-powered medical explanations (Fig. 7.3) (Fig. 7.11).

7.5 Implementation Examples in NuhealthSoft

Figure 7.12 presents an overview of the system architecture, illustrating its layered structure. The foundational first layer consists of the database (D.b) controllers, which manage the core business logic, machine learning classification algorithms, and recommendation mechanisms.

In the second layer, the system organizes semantic structures, which include input variables, output results, stakeholder interactions, and relevant contextual information. These are derived from the database controllers and structured into explainable AI (XAI) objects. Additionally, JSON-formatted summaries of machine learning

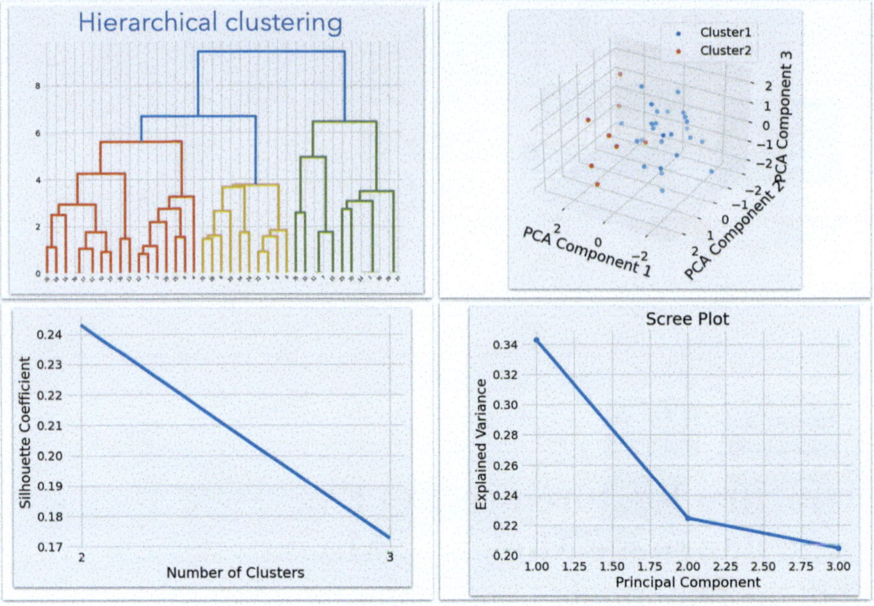

Fig. 7.11 AI literacy—perceived usefulness

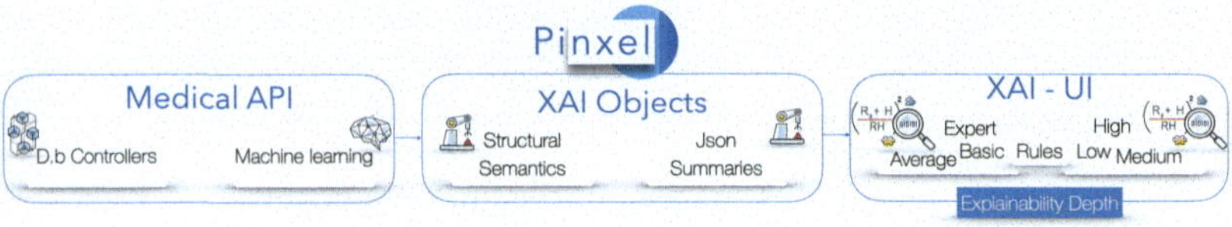

Fig. 7.12 XAI as a feature in NuhealthSoft

parameters are stored at this level, allowing for customized presentation based on AI literacy levels and abstraction depth, as discussed in earlier sections.

The final (third) layer comprises the user interfaces where the PINXEL framework is implemented. These interfaces adapt dynamically to predefined explainability levels (Basic, Average, Expert) to ensure transparency and user comprehension.

7.5.1 Addressing Key Challenges in Responsible AI Development

The concepts detailed in this chapter contribute to a framework for responsible AI, ensuring transparency, interpretability, and inclusivity. The following subsections illustrate real-world implementations of PINXEL within NuhealthSoft, demonstrating how different AI literacy levels and abstraction depths affect system usability.

PINXEL—Average AI Literacy with Low Level of Abstraction

Figure 7.13 provides an example of an Average AI Literacy user interacting with the system at a Low Level of Abstraction. The architecture section in the figure delineates the layers of the neural network model.

- The input layer consists of 18 neurons, where raw feature vectors are processed.
- The dropout layer (second layer) prevents overfitting by randomly omitting 50% of neurons during training.
- The final dense layer generates a binary classification output.

The parameters tab in the figure displays training variables, including weights and biases, which influence the learning process. These parameters evolve dynamically to optimize predictions through iterative loss function calculations and parameter adjustments.

The total number of parameters in each layer follows the formula:

Parameters = (Input Neurons + 1) × Output Neurons where the +1 accounts for bias terms.

Other elements in Fig. 7.13 include:

- Weight and Bias Distribution: Visualization of how model parameters evolve during training.

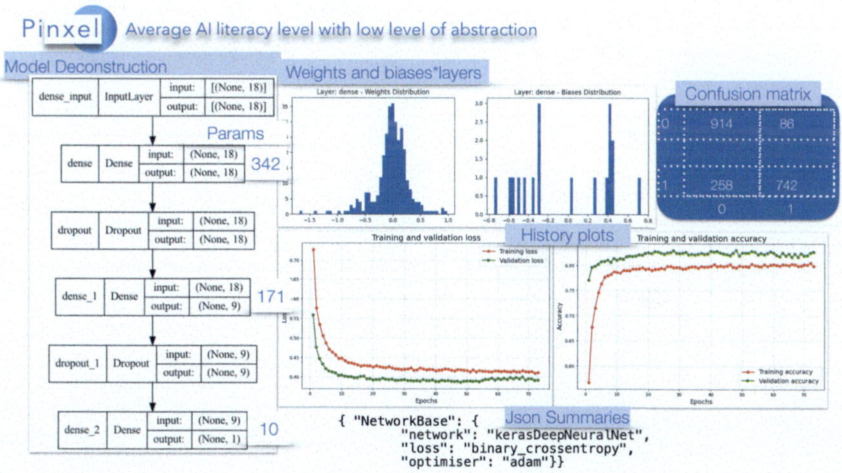

Fig. 7.13 Model deconstruction and evaluation metrics

- Confusion Matrix: Depicts model accuracy based on test data.
- Training History: Plots illustrate validation loss and accuracy trends over multiple epochs.

PINXEL—High AI Literacy with Low Level of Abstraction

Figure 7.14 demonstrates a High AI Literacy implementation with a Low Level of Abstraction, leveraging Shapley values for interpretability.

- The system analyzes Datapoint 60 (out of 10,000 samples) to evaluate feature impact on a model prediction.
- A prediction of “1” indicates overweight/obese classification, while “0” corresponds to underweight/normal.
- Force Plots reveal the influence of individual blood variables on the final classification.
- Waterfall Plots break down the step-by-step contribution of each feature toward the output.
- Decision Plots illustrate cumulative feature effects, clarifying how individual variables shift predictions.

For global model explanations, Fig. 7.15 presents a feature importance ranking using Shapley values [27].

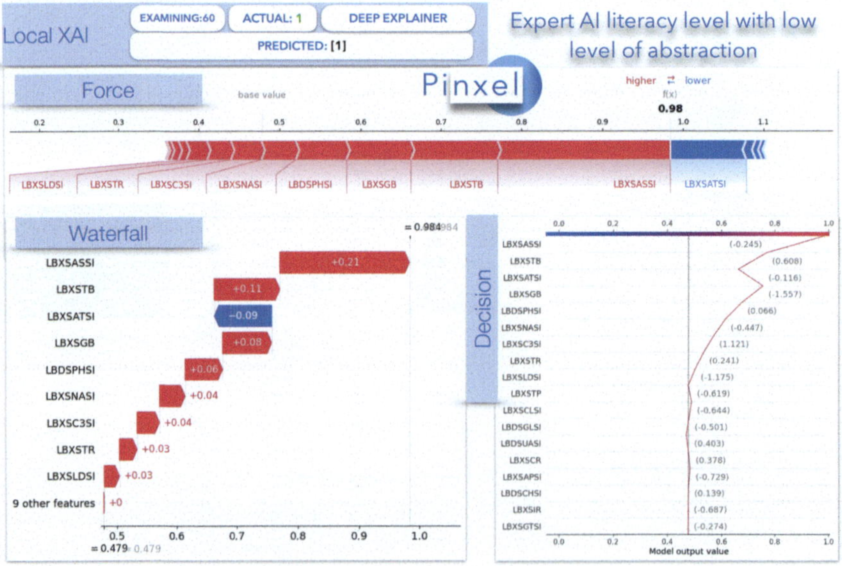

Fig. 7.14 Local XAI—weight classification based on blood tests

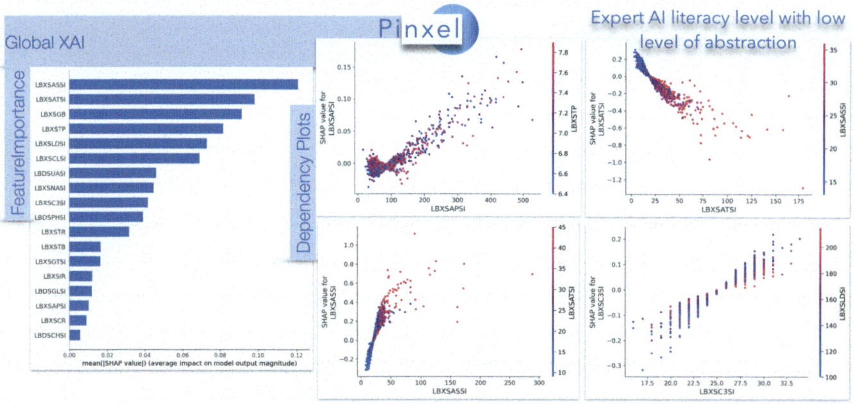

Fig. 7.15 Global XAI—feature importance in weight classification

- The feature importance plot calculates the average contribution of each variable in model predictions.
- Partial Dependence Plots (PDPs) illustrate how a single feature value (x-axis) influences model output (y-axis).
- Feature interactions are color-coded, showing how secondary variables impact prediction probabilities.

PINXEL—Basic AI Literacy with Low Level of Abstraction

Figure 7.16 showcases user interface elements designed for Basic AI Literacy users, emphasizing semantic grouping.

- In this example, Chloride levels are examined.
- The first screen displays coded names, measurement units, and contextual information.
- The second screen elaborates on how chloride levels correlate with organ functions (e.g., kidneys, heart).
- The third screen visualizes connections between chloride and other biochemistry variables.

This graphical representation helps users understand ML decision-making, making AI more accessible and intuitive.

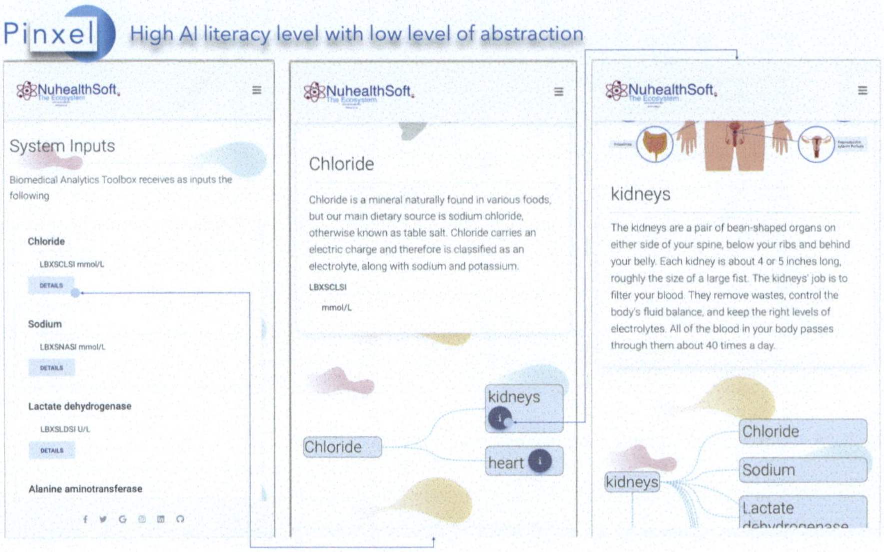

Fig. 7.16 Semantic groupings in NuhealthSoft

PINXEL Low AI Literacy with High Level of Abstraction

Figure 7.17 illustrates a chatbot-driven explainability interface, tailored for Low AI Literacy users who require high abstraction.

- The chatbot provides simplified explanations using predefined prompts.
- The structured prompts guide users through basic ML concepts before introducing more technical details.
- The interface enables progressive knowledge acquisition, ensuring that users gradually build AI understanding.

This layered approach ensures that users with limited AI knowledge can still interact meaningfully with the system.

7.6 Summary of Results and Future Works

7.6.1 Summary of Findings and Future Directions

This chapter introduced PINXEL, a novel framework designed to enhance personalization within Explainable Artificial Intelligence (XAI). PINXEL tailors explanation depth by factoring in AI literacy levels and required abstraction levels, ensuring that

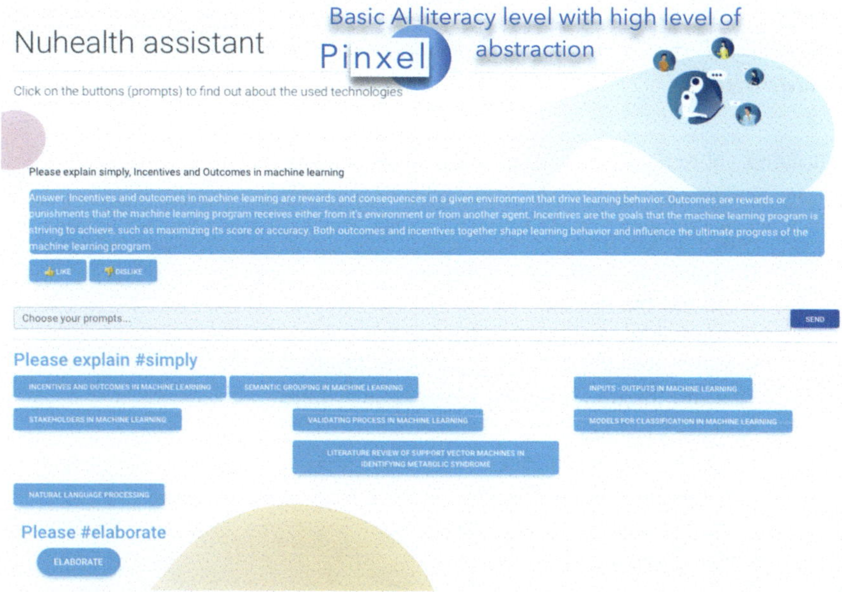

Fig. 7.17 AI chatbot for low AI literacy users

AI-driven systems provide interpretable insights suited to diverse users. The framework relies on clustering methods applied to questionnaire responses to assess users' knowledge and familiarity with AI concepts. By integrating multi-layered user interface (UI) elements, particularly in sensitive applications such as healthcare, PINXEL enables users to interpret, validate, and trust AI-powered systems.

The framework leverages a rule-based approach to dynamically adjust XAI layers based on Technology Acceptance Model (TAM) principles. The explainability components are categorized based on thematic complexity (AI literacy), while abstraction levels are addressed using large language models (LLMs), visual representations, and structured data formats (tables and graphs).

To assess the effectiveness of PINXEL, we analyzed responses from targeted questionnaires and conducted a real-world implementation study within NuhealthSoft, an AI-based medical application. This provided practical insights into how varying XAI layers influence user comprehension and system usability.

Appendix

Abbreviations

The following abbreviations are used in this manuscript:

ML	Machine Learning
RAI	Responsible Artificial Intelligenc
AI	Artificial Intelligenc
XAI	Explainable Artificial Intelligenc
REALM	Rapid Estimate of Adult Literacy in Medic
TOFHLA	Test of Functional Health Literacy in Adults
NVS	Newest Vital Sig
PDP	Partial Dependence Plots
ALE	Accumulated Local Effect
NLP	Natural Language Processing
LLM	Large language Mod
TAM	Technology Acceptance Mod
LIME	Local Interpretable Model-Agnostic Explanations
SHAP	SHapley Additive exPlanations
HLA	High level of Abstractio
MLA	Medium level of Abstractio
LLA	Low level of abstractio
ICE	Individual Conditional Expectatio
BAIL	Basic AI Literacy
AAIL	Average AI Literacy
EAIL	Expert AI Literacy

References

1. I. Ajzen, The theory of planned behavior, in *Organizational Behavior and Human Decision Processes*, vol. 50.2, pp. 179–211 (1991)
2. E. Alpaydin, *Introduction to Machine Learning*, 4th ed. (MIT Press, 2020)
3. A.B. Arrieta, N. Díaz-Rodríguez, J. Del Ser, A. Bennetot, S. Tabik, A. Barbado, S. García, S. Gil-López, D. Molina, R. Benjamins et al., Explainable Artificial Intelligence (XAI): concepts, taxonomies, opportunities and challenges toward responsible AI. Inf. Fusion **58**, 82–115 (2020)
4. A. Bandura, Social cognitive theory of personality. Handb. Pers. **2**, 154–96 (1999)
5. A.B. Bangsa, B.S. Bodo, Linking sustainable product attributes and consumer decision-making: insights from a systematic review. J. Clean. Prod. **245**, 118902 (2020)
6. S. Barocas, M. Hardt, A. Narayanan, Fairness and machine learning: limitations and opportunities. Fairness Mach. Learn. (2018). https://fairmlbook.org
7. M.R. Beauchamp, K.L. Crawford, B. Jackson, Social cognitive theory and physical activity: mechanisms of behavior change, critique, and legacy. Psychol. Sport. Exerc. **42**, 110–117 (2019)

8. P. Biecek, T. Burzykowski, *Explanatory Model Analysis: Explore, Explain and Examine Predictive Models* (Chapman and Hall/CRC, 2021)
9. C.M. Bishop, *Pattern Recognition and Machine Learning* (Springer, 2006)
10. A.J. Cañas, J.D. Novak, Concept mapping using CmapTools to enhance meaningful learning. Knowl. Manag. E-Learn. **4.4**, 341–354 (2012)
11. X. Chen, B. Liu, Y. Goyal, L. Wang, A. Srinivas, H. Wu, Y. Grandvalet, A. Joulin, Rethinking few-shot image classification: a good embedding is all you need? (2020). arXiv:2009.11929
12. A. Collins, R. Halverson, *Rethinking Education in the Age of Technology: The Digital Revolution and Schooling in America* (Teachers College Press, 2018)
13. D.F. Cox, S.U. Rich, Perceived risk and consumer decision-making—the case of telephone shopping. J. Mark. Res. **1.4**, 32–39 (1964)
14. D. Dobija, A. Hryckiewicz, M. Zaman, K. Puławska, Critical mass and voice: board gender diversity and financial reporting quality. Eur. Manag. J. **40.1**, 29–44 (2022)
15. K.M. Gallagher, J.A. Updegraff, Health message framing effects on attitudes, intentions, and behavior: a meta-analytic review. Ann. Behav. Med. **43.1**, 101–116 (2012)
16. S.L. Garfinkel, *Simson Garfinkel's Picturing Programs: An Introduction to Computer programming* (MIT Press, 2019)
17. I. Goodfellow, Y. Bengio, A. Courville, *Deep Learning* (MIT Press, 2016). http://www.deeplearningbook.org
18. T. Hastie, R. Tibshirani, J. Friedman, *The Elements of Statistical Learning: Data Mining, Inference, and Prediction*, 2nd ed. (Springer, 2009). https://web.stanford.edu/~hastie/ElemStatLearn/
19. S.S. Haykin, *Neural Networks and Learning Machines*, 3rd ed. (Pearson, 2009)
20. A. Holzinger, R. Goebel, R. Fong, T. Moon, K.-R. Müller, W. Samek, xxAI-beyond explainable artificial intelligence, in *xxAI-Beyond Explainable AI: International Workshop, Held in Conjunction with ICML 2020, 7–18, 2020*. Revised and Extended Papers (Springer, Vienna, Austria, 2022), pp. 3–10
21. C.-L. Hsu, H.-P. Lu, Why do people play on-line games? An extended TAM with social influences and flow experience, in *Information and Management*, vol. 41.7, pp. 853–868 (2004). https://doi.org/10.1016/j.im.2003.08.014
22. F. Jacob, E.H. Grosse, S. Morana, C.J. König, Picking with a robot colleague: a systematic literature review and evaluation of technology acceptance in human–robot collaborative warehouses, in *Computers and Industrial Engineering*, p. 109262 (2023)
23. P. Jarvis, *Adult Education and Lifelong Learning: Theory and Practice*. (Routledge, 2004)
24. F. Karray, M. Alemzadeh, J.A. Saleh, M. Nours A. Arab, Human-computer interaction: Overview on state of the art. Int. J. Smart Sens. Intell. Syst. **1.1**, 137–159 (2008)
25. R.R. Kumar, P. Guha, A. Chakraborty, Comparative assessment and selection of electric vehicle diffusion models: a global outlook, in *Energy*, vol. 238, p. 121932 (2022)
26. T.-C. Lin, C.-C. Tsai, Identifying Taiwanese university students' physics learning profiles and their role in physics learning self-efficacy, in *Research in Science and Technological Education* vol. 34.3, pp. 319–337 (2016). https://doi.org/10.1080/02635143.2016.1157160
27. S.M. Lundberg, S.-I. Lee, A unified approach to interpreting model predictions (2017). arXiv:1705.07874
28. M.L. Markus, Toward a "critical mass" theory of interactive media: universal access, interdependence and diffusion, in *Communication Research*, vol. 14.5, pp. 491–511 (1987)
29. D. Marr, *Vision: A Computational Investigation into the Human Representation and Processing of Visual Information* (MIT Press, 1982)
30. T.M. Mitchell, *Machine Learning* (McGraw Hill, 1997)
31. K.P. Murphy, *Machine Learning: A Probabilistic Perspective* (MIT Press, 2012)
32. P.W. Murphy, T.C. Davis, S.W. Long, R.H. Jackson, B.C. Decker, Rapid estimate of adult literacy in medicine (REALM): a quick reading test for patients. J. Read. **37.2**, 124–130 (1993)
33. D.T.K. Ng, J.K. Lok Leung, S.K. Wah Chu, M.S. Qiao, Conceptualizing AI literacy: an exploratory review, in *Computers and Education: Artificial Intelligence*, vol. 2, p. 100041 (2021)

34. J.A. Norton, F.M. Bass, A diffusion theory model of adoption and substitution for successive generations of high-technology products. Manag. Sci. **33.9**, 1069–1086 (1987)
35. OpenAI. GPT-4 Technical Report (2023). arXiv:2303.08774. https://api.semanticscholar.org/CorpusID:257532815
36. D.P. Panagoulias, D.N. Sotiropoulos, G.A. Tsihrintzis, Biomarker-based deep learning for personalized nutrition, in *2021 IEEE 33rd International Conference on Tools with Artificial Intelligence (ICTAI)*, Washington, DC, USA, Nov. 2021, pp. 306–313
37. D.P. Panagoulias, D.N. Sotiropoulos, G.A. Tsihrintzis, Nutritional biomarkers and machine learning for personalized nutrition applications and health optimization, in *2021 12th International Conference on Information, Intelligence, Systems and Applications (IISA)* (Chania Crete, Greece, July 2021), pp. 1–6
38. D.P. Panagoulias, D.N. Sotiropoulos, G.A. Tsihrintzis, Nutritional biomarkers and machine learning for personalized nutrition applications and health optimization (extended journal version). Intell. Decis. Technol. **15**, 645–653 (2021)
39. D.P. Panagoulias, D.N. Sotiropoulos, G.A. Tsihrintzis, Extreme value analysis for dietary intake based on weight class, in *2022 13th International Conference on Information, Intelligence, Systems and Applications (IISA)*, Corfu, Greece, July 2022
40. D.P. Panagoulias, M. Virvou, G.A. Tsihrintzis, A microservices-based iterative development approach for usable, reliable and explainable A.I.-infused medical applications using R.U.P., in *34th IEEE Conference on Tools with Artificial Intelligence (ICTAI 2022)*, Virtually, Nov. 2022a
41. D.P. Panagoulias, M. Virvou, G.A. Tsihrintzis, NUHEALTHSOFT: a nutritional and health data processing software tool, in *2022 16th International Conference on Signal Image Technology and Internet based Systems (SITIS)*, Dijon, France, Oct. 2022b
42. D.P. Panagoulias, M. Virvou, G.A. Tsihrintzis, Regulation and validation challenges in artificial intelligence-empowered healthcare applications—the case of blood-retrieved biomarkers, in *Proceedings of the 14th International Joint Conference on Knowledge-Based Software Engineering (JCKBSE 2022)*, Larnaca, Cyprus, Dec. 2022c
43. D.P. Panagoulias, N.S. Dionisios, G.A. Tsihrintzis, SVM-Based blood exam classification for predicting defining factors in metabolic syndrome diagnosis. Electronics **11.6**, 857 (2022)
44. D.P. Panagoulias, M. Virvou, G.A. Tsihrintzis, A novel framework for artificial intelligence explainability via the technology acceptance model and rapid estimate of adult literacy in medicine using machine learning, in *Expert Systems with Applications*, vol. 248, p. 123375 (2024). ISSN: 0957-4174. https://doi.org/10.1016/j.eswa.2024.123375. https://www.sciencedirect.com/science/article/pii/S0957417424002409
45. R.M. Parker, D.W. Baker, M.V. Williams, J.R. Nurss, The test of functional health literacy in adults: a new instrument for measuring patients' literacy skills. J. Gen. Int. Med. **10**, 537–541 (1995)
46. R.M. Parker, C.R. Scott, N. Lurie, Health literacy: a policy challenge for advancing high-quality health care, in *Health Affairs*, vol. 22.4, pp. 147–153 (2003)
47. E.M. Rogers, *Diffusion of Innovations* (Simon and Schuster, 2010)
48. R.C. Schank, What Is AI, Anyway? in *AI Magazine*, vol. 8.4, p. 59, Dec. 1987. https://doi.org/10.1609/aimag.v8i4.623. https://ojs.aaai.org/aimagazine/index.php/aimagazine/article/view/623
49. R. Scherer, F. Siddiq, J. Tondeur, The technology acceptance model (TAM): A meta-analytic structural equation modeling approach to explaining teachers' adoption of digital technology in education, in *Computers and Education*, vol. 128, pp. 13–35 (2019). https://doi.org/10.1016/j.compedu.2018.09.009
50. N. Tang, Y. Du, X. Wang, X. Wang, The impact of personalization on search behavior and self-efficacy, in *Information Processing I and Management*, vol. 52.2, pp. 373–383 (2016)
51. T. Teo, Factors influencing teachers' intention to use technology: Model development and test, in *Computers and Education*, vol. 57.4, pp. 2432–2440 (2011). https://doi.org/10.1016/j.compedu.2011.06.008

52. V. Venkatesh, M.G. Morris, G.B. Davis, F.D. Davis, User acceptance of information technology: toward a unified view, in *MIS Quarterly*, pp. 425–478 (2003)
53. G. Vilone, L. Longo, Explainable artificial intelligence: a systematic review (2020). arXiv:2006.00093
54. M. Virvou, The emerging Era of human-AI interaction: Keynote address, in *2022 13th International Conference on Information, Intelligence, Systems I and Applications (IISA)*, pp. 1–10 (2022). https://doi.org/10.1109/IISA56318.2022.9904422
55. B.D. Weiss, M.Z. Mays, W. Martz, K.M. Castro, D.A. DeWalt, M.P. Pignone, J. Mockbee, F.A. Hale, Quick assessment of literacy in primary care: the newest vital sign, in *The Annals of Family Medicine*, vol. 3.6, pp. 514–522 (2005)
56. A. Yuriev, M. Dahmen, P. Paillé, O. Boiral, L. Guillaumie, Pro-environmental behaviors through the lens of the theory of planned behavior: a scoping review, in *Resources, Conservation and Recycling* **155**, p. 104660 (2020)

Part III
Empowering Medical Software with Generative Artificial Intelligence

Chapter 8
Enhancing Domain-Specific Interactions in Large Language Models

Abstract This chapter presents an advanced AI-driven framework designed to enhance Large Language Models (LLMs) and Machine Learning methodologies through rule-based augmentation, optimizing diagnostic decision-making in Primary Care. The proposed approach introduces a structured methodology, depicted via a process diagram, which defines the integration of generative AI functions within a rule-enriched paradigm. To achieve this, the framework decomposes generative AI workflows into modular, functional segments, supporting the development of an implementation-oriented data flow model. A pivotal element of this system is the interaction model, which establishes the theoretical foundation for user engagement within an AI-assisted medical application. Named "Med|Primary AI Assistant," this tool aids users by interpreting symptoms and generating diagnostic guidance. The framework leverages cutting-edge Natural Language Processing (NLP) advancements to create a domain-specific knowledge structure, coupled with a systematic mechanism for generating contextually relevant medical insights. Beyond structuring AI-driven interactions, this methodology integrates a dynamic content adaptation mechanism, ensuring personalized and context-aware engagement tailored to user needs. Additionally, a rule-based evaluation strategy-grounded in contextual analysis and dialogue theory-provides an algorithmic framework for assessing content precision and response quality within the AI-powered diagnostic assistant.

Keywords AI-empowered software engineering · Generative AI · Dialogue theory · Large language models · Natural language processing · Rule-augmented systems · Medical diagnosis · Evaluation

8.1 Introduction

Healthcare processes are inherently multifaceted, encompassing intricate doctor-patient interactions, diverse diagnostic and therapeutic approaches, adherence to lifestyle modifications, and ongoing preventive interventions. The patient's healthcare journey is nonlinear, comprising an interconnected sequence of events and decision points [13, 19, 27, 40]. Diagnostic workflows, for instance, integrate multiple

D. P. Panagoulias et al., *Artificial Intelligence-Empowered Bio-medical Applications*, Learning and Analytics in Intelligent Systems 49,
https://doi.org/10.1007/978-3-031-90174-4_8

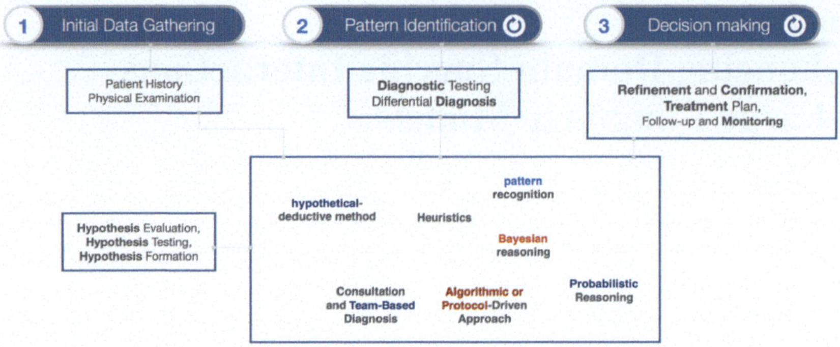

Fig. 8.1 Medical diagnosis pathways

methodologies influenced by patient symptoms, clinical expertise, and the availability of diagnostic technologies [1]. As shown in Fig. 8.1, this process typically begins with data acquisition, including medical history and potentially a physical examination. These inputs are analyzed to identify patterns guiding clinical judgment. This iterative approach refines initial assessments, shaping treatment strategies, monitoring patients, and tracking disease progression.

One of the most transformative advancements in this domain is the integration of Large Language Models (LLMs). These models excel in processing medical literature, recognizing patterns in symptomatology, and supporting clinical reasoning. Prominent examples include OpenAI's GPT, the foundation of ChatGPT, which demonstrates high proficiency in generating human-like, contextually relevant responses [22]. Other key LLMs include Google's BERT [5], Meta's LLaMA [39], and Stanford's Alpaca, a fine-tuned adaptation of LLaMA [36]. Despite their strengths, these models exhibit limitations, necessitating a multi-faceted integration strategy to balance efficiency with cost-effectiveness.

Technological innovations continue to reshape healthcare by digitizing critical processes, reducing the administrative and financial burden on primary care, and enhancing patient experiences. The adoption of Natural Language Processing (NLP), Big Data analytics, and Machine Learning (ML) techniques [7, 34, 51] has led to process optimization, automation, and workflow acceleration. These advancements facilitate administrative efficiency, minimize manual intervention, and improve medical decision-making [26, 38].

As a transformer-based model, GPT has demonstrated versatility across multiple disciplines, including education, decision support, cultural heritage, and healthcare. By leveraging vast datasets, it generates domain-specific insights applicable to a broad range of scenarios. However, domain constraints necessitate careful refinement and systematic evaluation to ensure optimal performance [42].

Beyond medical applications, ChatGPT has been incorporated into AI-assisted decision-making frameworks. The LYRICEL system, for example, combines ChatGPT with rule-based methodologies to produce customized content for cultural e-learning [25]. Research on AI trust dynamics underscores the importance of designing ChatGPT-integrated systems with transparency and reliability in mind [44]. Additionally, structured evaluations have explored its role in supporting AI-powered decision-making processes [41].

In educational contexts, ChatGPT has been utilized in intelligent tutoring systems, fostering personalized learning and cultural content creation, including poetry and lyric generation [25, 47]. Studies have assessed its pedagogical effectiveness, including frameworks such as FEPER, which evaluate its alignment with educational principles and ethical considerations [46]. Further analysis has examined AI-assisted learning and the evolving nature of human-AI collaboration [45, 47].

In the medical field, multimodal LLMs have demonstrated promise in diagnostic reasoning, treatment planning, and workflow optimization. Research on ChatGPT's application in lung cancer diagnosis and staging suggests its potential for supporting complex medical decisions [23, 24]. Enhancing vision-based LLMs with domain-specific rules has improved their interpretative accuracy in medical imaging and symptom analysis [28, 31]. The Dermacen Analytica methodology extends these advancements by incorporating multimodal ML and transformer-based LLMs to enhance tele-dermatology applications [31]. Additionally, benchmarking tools such as COGNET-MD systematically assess LLM performance in clinical environments, ensuring compliance with medical standards [29]. Further research highlights GPT's capability to analyze multimodal medical data, demonstrating adaptability in addressing complex healthcare challenges [33].

Overall, while LLMs such as GPT have demonstrated significant potential across various domains, their implementation requires domain-specific adaptation. Whether applied in education, cultural heritage, decision support, or healthcare, structured evaluation methodologies remain essential to refining performance. Within the medical field, where precision and reliability are paramount, systematic assessments are critical to ensuring AI-driven solutions effectively contribute to clinical decision-making and patient care.

8.1.1 Problem Statement

Earlier investigations introduced a structured three-phase framework for evaluating LLM applications—particularly ChatGPT—in medical diagnosis and treatment [23]. Studies assessing ChatGPT's communication effectiveness in radiology [11] and oncology [8] indicated that its performance varies based on contextual factors. Notably, findings suggest that ChatGPT and similar LLMs yield better results when supervised by medical professionals who critically assess their outputs.

Table 8.1 Dialogue rule augmentation

Question (q)	Domain (d)	Medical (m)	Answer (a)	AnswerContent (c)	Use case examples
Informational	Yes	Yes	Direct or elaborated	DefineAbility (Ab), reflect (Re)	Figure 8.10
Open ended	Yes	Yes	Direct or elaborated	DefineAbility (Ab), reflect (Re)	Figure 8.9
Any	No	No	(Deferred) dialogue reset	DefineAbility (Ab), reflect (Re)	Figure 8.12
Any	No	Yes	(Deferred) refer to specialist	DefineAbility(Ab), Reflect(Re)	Figure 8.11

Building upon these findings, this study introduces an advanced AI-powered system that combines *a rule-based decision mechanism with an LLM engine, supplemented by external machine learning and analytical APIs*. The key contributions of this framework include:

- Implementation of NLP algorithms to construct a structured domain-specific knowledge Blueprint, ensuring responses are tailored to specialized medical contexts and improving user interaction.
- Development of an AI-user interaction framework based on predefined dialogue rules, detailed in Table 8.1. These rules establish a theoretical foundation for evaluating LLM performance, ensuring compliance with domain constraints that mimic real-world medical consultations.
- Evaluation of the system's performance using GPT-4 through structured multiple-choice assessments to test its capability in symptomatology analysis across general pathology.
- Introduction of a structured methodology, depicted via a process diagram, that defines generative AI workflows and functions within a rule-enhanced paradigm. This enables the design of AI-driven systems that adhere to domain-specific constraints.

The proposed framework integrates machine learning tools that replicate diagnostic reasoning strategies commonly employed by general practitioners during initial consultations. These enhancements allow the system to emulate cognitive processes, clinical decision-making patterns, and diagnostic workflows, providing users with structured medical guidance and preliminary diagnostic recommendations. The AI-assisted functionalities outlined above have been fully integrated into a rule-augmented system and are further elaborated upon in subsequent sections of this chapter.

More specifically, the chapter is structured as follows: Sect. 8.2 provides an overview of the foundational theories and contextual background, covering both

Large Language Models (LLMs) and the current landscape of Primary Care on a global scale. Section 8.3 introduces the developed rule-augmented AI-empowered system, outlining its core components and functionality. Section 8.4 delves into the system architecture, examining its design from both micro and macro perspectives. Section 8.5 presents a comprehensive evaluation of the system's performance, detailing the assessment methodology and key findings. Finally, Sect. 8.6 concludes the chapter by summarizing the main contributions, highlighting key insights, and discussing potential directions for future research.

8.2 Background Theories and Context

This section provides a comprehensive foundation for the methodologies employed in our study. We examine key aspects of Natural Language Processing (NLP) and its extensive applications within the medical domain. The integration of NLP and Large Language Models (LLMs) in Healthcare has been transformative [10], with these technologies being utilized for extracting critical information from Electronic Health Records (EHRs), supporting clinical decision-making, and analyzing patient feedback and sentiments [16, 17].

8.2.1 Natural Language Processing (NLP)

NLP, a core subfield of AI, enables computers to process and interpret human language, facilitating tasks such as translation, sentiment analysis, and conversational interfaces. NLP has evolved from early rule-based approaches to sophisticated machine learning models, leading to the development of advanced frameworks such as GPT and BERT [6, 35, 43]. Fundamental NLP processes include tokenization, part-of-speech tagging, named entity recognition, and syntactic parsing. The primary objective is to achieve an advanced understanding of human language, enabling information extraction and reasoning simulation for task completion. NLP methodologies range from pattern-matching techniques to complex ML-based models. Some notable approaches include:

- **Pattern Matching and Rule-Based Systems**: Early NLP systems relied on predefined lexical rules. A well-known example is the ELIZA chatbot [49], one of the first attempts at simulating human-like conversations.
- **Machine Learning Models**: Approaches such as Naïve Bayes, Support Vector Machines (SVMs), and Decision Trees have been widely used for text classification and sentiment analysis.
- **Neural Networks**: Deep learning architectures, including Recurrent Neural Networks (RNNs) and Convolutional Neural Networks (CNNs), have been employed to capture sequential language patterns and contextual dependencies.

- **Embedding Models**: Techniques such as Word2Vec, GloVe, and FastText generate dense vector representations of words, capturing their semantic relationships [48].
- **Sequence-to-Sequence Models**: Encoder-decoder architectures with attention mechanisms have been extensively used for tasks such as machine translation and text summarization [4].
- **Large Language Models (LLMs)**: LLMs, designed to handle diverse NLP tasks without task-specific training data, are explored in detail in the next section.

8.2.2 *Large Language Models (LLMs)*

BERT employs bidirectional processing, analyzing words based on both preceding and succeeding context, enhancing its comprehension of sentence structures. Both models are built upon the Transformer architecture introduced in [43], utilizing an attention mechanism to prioritize significant words in a sequence. The self-attention mechanism enables models to capture long-range dependencies, substantially improving language comprehension. Unlike GPT's unidirectional approach, BERT's bidirectional training method improves contextual understanding.

The GPT series, including GPT-3 and GPT-4, are autoregressive models known for their capability to generate coherent and contextually appropriate text. These models interpret input sequences, leveraging learned linguistic structures to generate relevant responses. GPT models are particularly effective in content generation, conversational AI, and text-based reasoning tasks.Before neural networks and Transformer architectures became dominant in NLP, statistical models played a key role, including Hidden Markov Models, Markov Models, n-gram Models, Conditional Random Fields, Latent Dirichlet Allocation.

Despite significant advancements, NLP and LLM applications face several limitations, particularly in medical contexts. AI-driven models, particularly Large Language Models (LLMs), come with several inherent limitations that can impact their reliability and applicability in various domains, including healthcare. One significant issue is hallucination, where AI-generated responses, although appearing credible, may contain fabricated or incorrect information [3, 21]. This can be particularly problematic in medical contexts, where misinformation could lead to serious consequences.

Another concern is bias, as LLMs learn from vast datasets that often reflect societal prejudices. This can result in disparities in responses, reinforcing existing biases rather than providing objective analysis [2]. Additionally, these models suffer from a lack of explainability, making it difficult to understand how specific outputs are generated. The absence of clear, interpretable reasoning behind AI decisions poses a challenge in domains that require transparency, such as healthcare and legal applications [14, 15].

Moreover, real-time validation remains a challenge, as LLM-generated content is based on pre-trained knowledge and may not incorporate the most recent medical

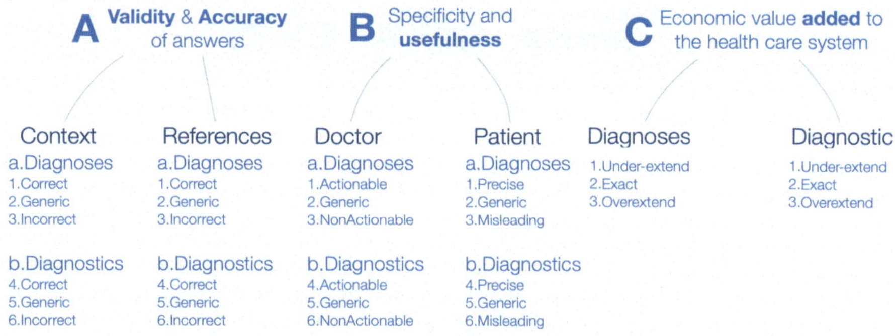

Fig. 8.2 Methodology for evaluating domain-specific proficiency of LLMs

advancements or evolving clinical guidelines [3]. This limitation underscores the need for continuous updates or external validation mechanisms.

In mathematical reasoning, LLMs demonstrate limited computational accuracy, struggling with complex equations and requiring integration with specialized mathematical engines to perform precise calculations. Additionally, token limitations impose constraints on input size, which can affect the coherence and context retention of long-form responses. While increasing token limits can alleviate this issue, it comes at the cost of higher computational demands.

These limitations highlight the necessity of refining AI systems through rule-based augmentation, external validation sources, and enhanced interpretability mechanisms to ensure their safe and effective deployment in critical domains. Prior research [23, 24] introduced a methodology for evaluating LLM accuracy and reliability in medical contexts. This evaluation framework employs a three-tiered scoring scale (1–3) to assess Actionability, Correctness, Precision and Content Scope. The evaluation process, conducted under medical supervision, assesses (A) response validity and contextual accuracy, (B) response relevance for both physicians and patients, and (C) the potential economic impact of AI integration. The assessment pipeline is illustrated in Fig. 8.2.

8.2.3 *Transformers, Attention Mechanism and State of the Art*

The Transformer model, central to state-of-the-art LLMs, is powered by a self-attention mechanism that allows it to capture relationships between words, regardless of their positions within a sequence. Figure 8.3 presents a structured overview of Transformer architecture, detailing its key components.

The **Transformer Model** is built upon *self-attention*, enabling it to capture relationships between words regardless of their position within a sequence.

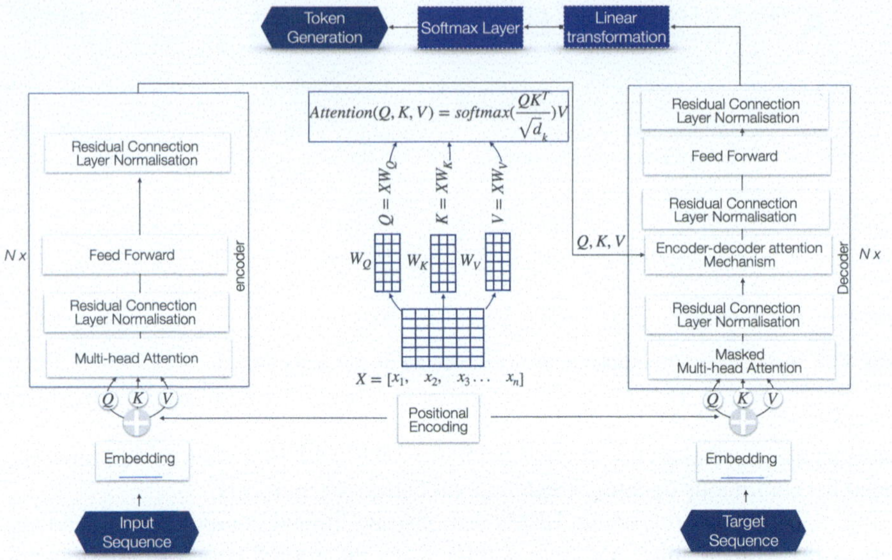

Fig. 8.3 Transformer data flow architecture

Encoder

The encoder processes input data through multiple layers of *multi-head attention* and *feed-forward networks*, followed by residual connections and normalization.

- **Embeddings**: Converts words or tokens into numerical vectors that encode semantic and syntactic information.
  - *Input Sequence Embedding*: Transforms input tokens into fixed-dimension vectors.
  - *Positional Encoding*: Adds information about token order.
- **Multi-Head Attention**: Enables the model to focus on different aspects of the data.
  - *Q, K, V Vectors*: Queries, Keys, and Values generated via learned linear projections.
  - *Scaled Dot-Product Attention*: Computes attention scores to weigh information from different positions:

$$\text{Attention}(Q, K, V) = \text{softmax}\left(\frac{QK^T}{\sqrt{d_k}}\right)V.$$

  - *Concatenation and Linear Transformation*: Merges attention outputs from multiple heads.

- **Feed-Forward Network**: Applies transformations independently at each position.
- **Residual Connection and Normalization**: Stabilizes learning by maintaining gradients.

Decoder

The decoder generates outputs by attending to both the input sequence and previously generated tokens.

- **Target Sequence Embedding**: Converts target tokens into vectors.
- **Masked Multi-Head Attention**: Ensures predictions consider only past tokens during training.
- **Encoder-Decoder Attention**: Aligns decoder input with encoder output.
- **Feed-Forward Network and Normalization**: Enhances feature representation.
- **Final Processing**:
  - *Linear Transformation and Softmax*: Maps output to vocabulary size and generates probability distributions.
  - *Token Selection and Generation*: Selects the most likely token using decoding strategies such as greedy search or beam search.
  - *Sequence Construction and Post-Processing*: Assembles tokens into structured output while ensuring syntactical and grammatical coherence.

This architecture allows the Transformer to efficiently model contextual dependencies, enabling precise and coherent sequence generation.

8.2.4 Med\Primary AI Assistant: An Extension of NuhealthSoft

Building upon the NuhealthSoft platform [26], **Med|Primary AI Assistant** is designed to provide AI-driven medical guidance, mirroring general practitioners' diagnostic methodologies. The system integrates rule-based augmentation to enhance AI-generated responses, ensuring alignment with domain-specific requirements.

8.3 Med|Primary AI Assistant

Figure 8.4 outlines the objectives, tasks, and available tools that define the operational framework of Med|Primary AI Assistant. The system is designed to execute these objectives and tasks by leveraging available tools, forming a structured *Blueprint of domain-specific knowledge*. To address the inherent limitations of Large Language Models (LLMs), the system incorporates rule-based enhancements, leading to a *rule-augmented application*.

The rule-based augmentation serves three primary functions:

- Engineering prompts based on domain-specific knowledge.
- Extracting semantically significant terms and linking them to external classifiers, services, and health sensors.
- Establishing an evaluation framework to ensure alignment with medical standards and domain constraints based on dialogue theory.

8.3.1 System Description

Med|Primary AI Assistant, integrated into the NuhealthSoft suite, utilizes LLMs, specifically GPT-4, for processing user queries and generating medical insights.

GPT-4 is a state-of-the-art multimodal model capable of processing text-based inputs to generate comprehensive text responses and assist in medical consultations. It surpasses earlier models by exhibiting enhanced reasoning abilities and deeper contextual understanding. While primarily optimized for conversational AI, it remains proficient in performing structured completion tasks.

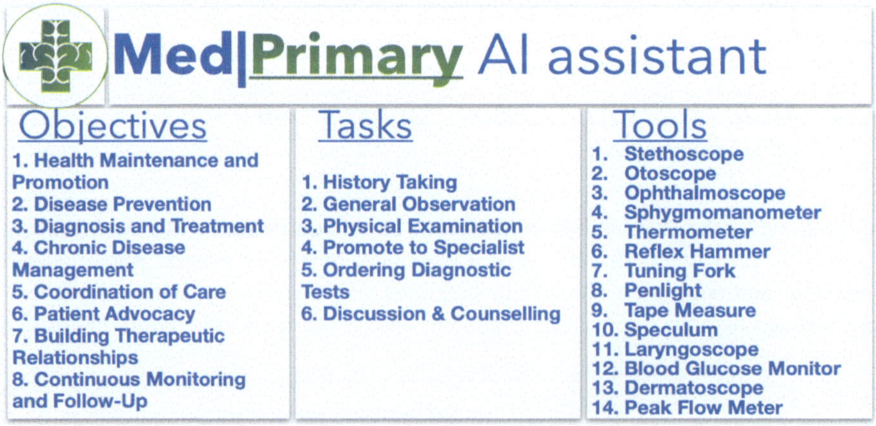

Fig. 8.4 General practitioner (assistant)—system objectives, tasks, and tools

In addition to LLM-based interactions, the system incorporates analytical services and machine learning (ML) models for extracting key insights from health data. The inclusion of domain constraints ensures that only relevant information is processed and conveyed, optimizing token usage and maintaining computational efficiency. This approach enhances the validity and precision of responses, ensuring that AI-generated advice aligns with clinical best practices.

8.3.2 *Use Cases*

Users interact with Med|Primary AI Assistant in two primary ways:

- **Unconstrained Interaction**: Users freely describe their symptoms and current health conditions, receiving potential diagnoses, recommended diagnostic tests, or referrals to medical specialists. While this approach allows unrestricted input, the system utilizes structured knowledge inputs to guide responses, ensuring medical relevance and response accuracy.
- **Guided Interaction**: Users follow a structured step-by-step process, enabling the system to gather comprehensive symptomatology data, patient history, and age-related considerations. This approach ensures a more refined diagnostic assessment. In both cases, external health sensors can provide supplementary physiological data, which is analyzed by integrated ML and analytical services [32].

Figure 8.5 illustrates the two primary actors in the system:

- **Patient**: Provides symptom descriptions either directly through free text input or via structured forms. Health sensors can supply additional physiological data. The patient also reviews the AI-generated responses for accuracy and relevance.

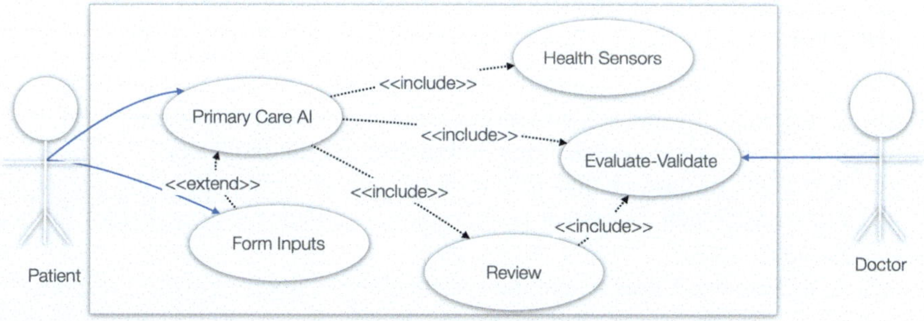

Fig. 8.5 Med|Primary AI assistant—use case interaction model

- **Doctor**: Evaluates and validates AI-generated interactions, including both patient inputs and system outputs. The physician also considers patient feedback on the AI's recommendations. This validation cycle is crucial for refining the system, improving diagnostic accuracy, and addressing AI limitations in Primary Care.

8.4 System Architecture Analysis

This section details the structural components, data flow, and organization of the Med|Primary AI Assistant. As illustrated in Fig. 8.4, the system's specific objectives, tasks, and tools serve as fundamental building blocks. For instance, any service facilitating diagnostic analysis must fall within the competencies of a general practitioner, thereby forming the *Blueprint of domain-specific knowledge* for the system.

8.4.1 Modelling the Domain Space

The system primarily manages inputs and outputs within patient-practitioner interactions, focusing on structured question-and-answer exchanges [9, 12, 50].

In any domain, questions and answers can be classified into six (6) primary types:

1. **Questions**:
   (a) *Informational*: Queries seeking specific factual information.
   (b) *Instructional*: Task-related queries, often resembling commands.
   (c) *Reflective*: Clarification or confirmation of previous statements.
   (d) *Rhetorical*: Questions meant for emphasis rather than answers.
   (e) *Open-ended*: Designed to encourage discussion or detailed responses.
   (f) *Closed-ended*: Can be answered with a simple "yes" or "no."
2. **Answers**:
   (a) *Direct*: Concise and straightforward responses.
   (b) *Elaborated*: Responses providing additional context or insights.
   (c) *Clarifying*: Aimed at resolving ambiguities within the query.
   (d) *Reflected*: Responses that align with the tone or sentiment of the question.
   (e) *Deferred*: When an answer is unavailable, guidance is provided for further inquiry.
   (f) *Non-Answers*: Cases where the system chooses not to provide a response.

8.4.2 Domain Settings

Questions and answers form the foundation of medical conversations, as depicted in Fig. 8.6. Within a healthcare setting, most interactions involve *informational* and *open-ended* questions followed by *direct* or *elaborated* responses. The system (Table 8.1) also supports *deferred answers* when a query necessitates specialist evaluation or falls outside the AI assistant's domain (*Dialogue Reset*).

Dialogue Rules ensure:

- A structured framework for performance evaluation.
- The model's ability to operate within constraints designed to mimic real-time communication protocols.

The system integrates *Instantiated Advice Generation* with the *Blueprint of Domain-Specific Knowledge* to ensure structured interactions. These rules guide responses to align with diagnostic assessments, necessary medical tests, or specialist referrals, primarily based on described symptoms.

8.4.3 Evaluation Framework for AI-Generated Responses

To ensure safe and accurate medical interactions, the **Dialogue Rule Augmentation** framework (Table 8.1) defines response evaluation criteria, illustrated in Fig. 8.6. This evaluation operates through two core processes:

- **Define Ability Process (I)**: Assesses response validity against domain-specific standards.

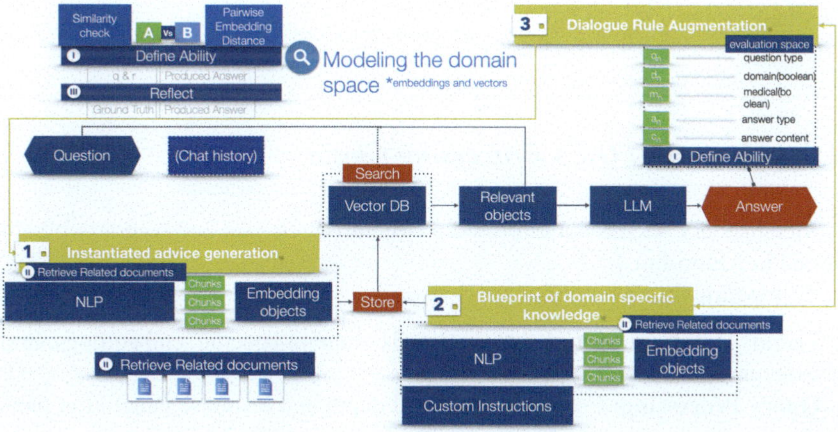

Fig. 8.6 Domain settings and flow chart

- **Reflect Process (III)**: Evaluates alignment with conversational norms and expected medical communication structures.

These processes are analogous to real-world medical training assessments, where practitioners are evaluated on patient interactions and diagnostic reasoning [18, 20]. Our system generalizes these principles to AI evaluation within a structured medical domain.

Each response component is assigned a category (Table 8.1) and is analyzed based on its alignment with domain-specific guidelines. The overall quality of responses is assessed using a scoring model that evaluates:

- Question Type q
- Domain Relevance d
- Medical Context m
- Answer Type a
- Content Precision c
- Retrieved References r

8.4.4 Generative AI Optimization and Embeddings

To overcome the limitations of general-purpose LLMs, Med|Primary AI Assistant employs a **hybrid approach** that integrates:

- **Pre-trained Embeddings**: Used to encode domain-specific knowledge efficiently.
- **Vector Databases**: Store structured embeddings for rapid retrieval.
- **Retrieval-Augmented Generation (RAG)**: Enables contextual document retrieval to enhance generative outputs.
- **Constraint-Based AI Processing**: Ensures responses remain within defined medical guidelines.

8.4.5 Algorithm: Q&A Processing Chain

The Med|Primary AI Assistant follows a structured retrieval and response pipeline, outlined in Algorithm 1:

This pipeline follows a multi-step approach:

1. **Vector Database Initialization**: Loads domain-specific embeddings into a structured database.
2. **Query Processing**: Configures retrieval depth and response generation parameters.
3. **LLM Integration**: Generates responses based on structured domain inputs.

Algorithm 1 Q&A chain for Med|Primary AI assistant

$DB, Embeddings \leftarrow$ Vector.db() {Initialize Vector Database}
$Retriever, LLM \leftarrow$ chainer($DB, Embeddings$) {Set LLM Processing Parameters}
Procedure RUNQACHAIN($Query : QueryModel, CurrentUser$)
$Question \leftarrow Query.question$
$QAChain \leftarrow$ prompter($Retriever, LLM$)
$Result \leftarrow$ responder($QAChain, Question$)
return {"response" : $Result$}
End Procedure

4. **Evaluation Mechanism**: Validates responses against domain-specific medical guidelines.

This structured design ensures that Med|Primary AI Assistant provides accurate, clinically relevant recommendations while maintaining transparency and alignment with real-world medical communication practices.

8.4.6 System Architecture

The system architecture is depicted in Figs. 8.7 and 8.8. The micro-level process diagram provides an internal view of the design and interactions between system components, while the macro-level diagram illustrates the overall system structure.

Micro-Level Architecture

The process diagram introduces a novel methodology for defining Generative AI (GAI) processes using a *rule-augmented approach*. The left side of Fig. 8.7 categorizes system components, while the right side presents their implementation, information flow, constraints, and expected outcomes.

Key elements in the micro-level process include:

- **Outputs**: Information exchange processes.
- **Connections**: Indicate dependencies between components.
- **Database (DB)**: Various data storage and retrieval mechanisms.
- **Content**: Data generation processes.
- **Interactions**: User inputs leading to generated responses.
- **Natural Language Processing (NLP)**: Language-related AI processes.
- **Large Language Model (LLM)**: AI processing via an LLM or API calls.
- **Computer Vision Object**: Machine learning (ML) modules handling image-based analysis.
- **Classifiers Object**: ML components performing classification tasks.
- **Machine Analysis**: AI-driven interpretation of medical and sensor data.
- **Rules**: Constraints applied to maintain system reliability.

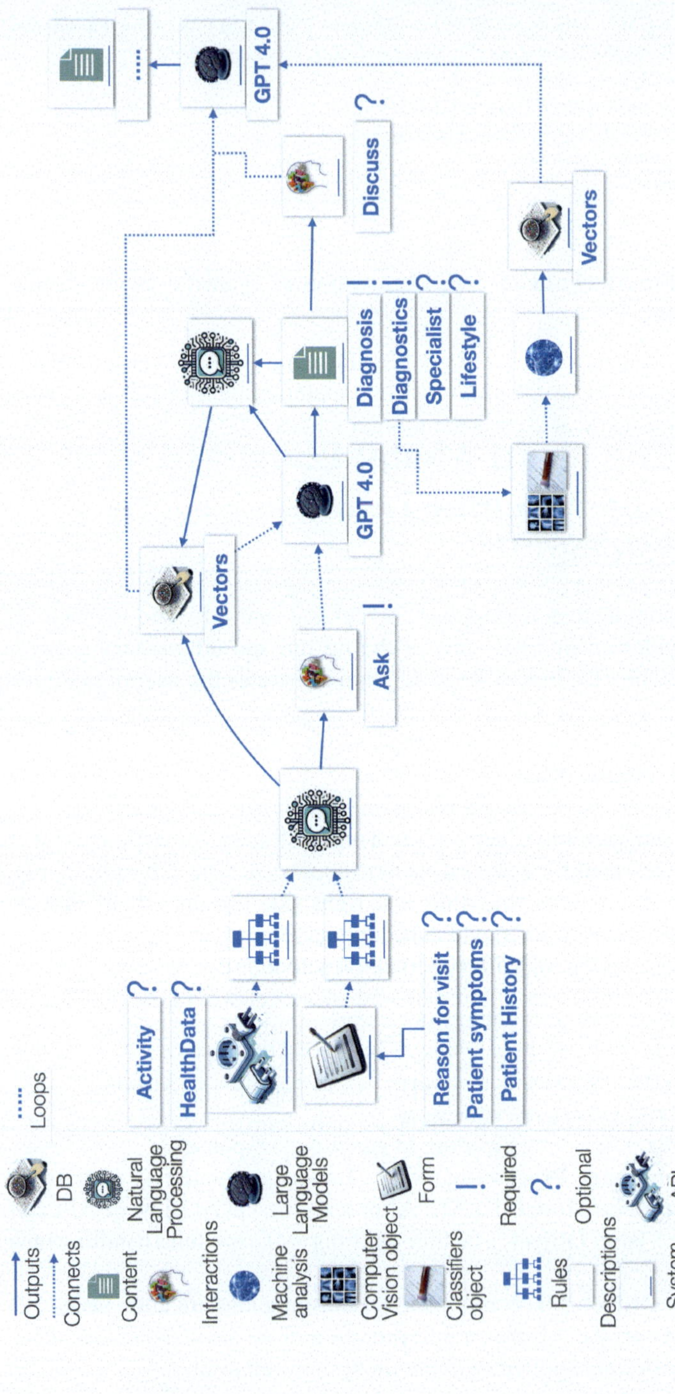

Fig. 8.7 Process diagram—micro-level

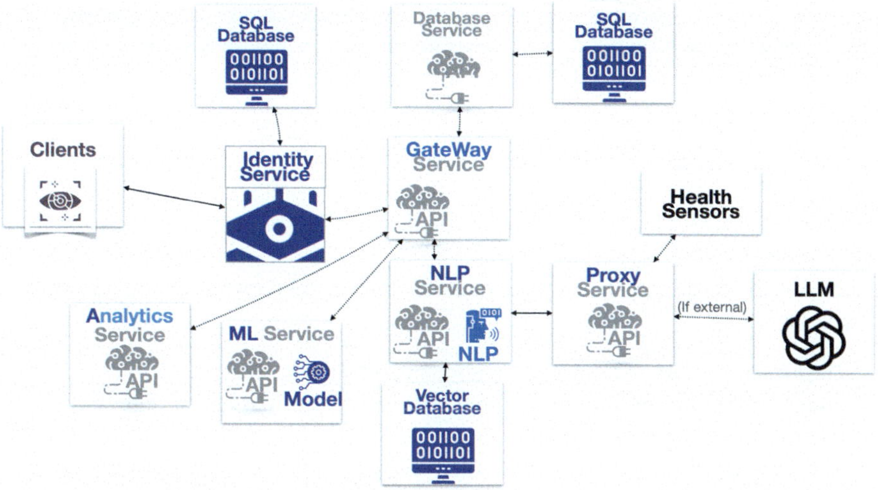

Fig. 8.8 System architecture—macro-level

- **White Box**: General process descriptions.
- **White Box with Blue Outline**: Defines a distinct function or system.
- **Form**: Predefined user input structures such as questionnaires.
- **! (Required Process)**: Essential system functions.
- **? (Optional Process)**: Non-mandatory but available operations.
- **API**: Communication with external services.
- **Three Dots (...)**: Represents iterative or looping processes.

The workflow begins with the user providing symptoms and an optional patient history. Additionally, users can upload or sync activity and health data via an API. The system then applies rule-based processing to format the data into structured NLP inputs. The resulting embeddings are stored in a vector database.

Using these embeddings, the AI system (powered by GPT-4) processes responses, which may include a preliminary diagnosis, recommended diagnostic tests, specialist referrals, or lifestyle modifications. Computer vision tools and classifiers further analyze examination data, transforming it into usable objects stored in the vector database. The system also facilitates a dynamic Q&A mechanism, allowing continued interaction based on prior responses.

Macro-Level Architecture

Figure 8.8 illustrates the system's microservices-based architecture, enabling modular, scalable, and independent component deployment. This design is especially advantageous when integrating multiple AI models and external APIs, as it prevents dependency conflicts and facilitates independent service updates.

Python-based AI libraries frequently rely on distinct dependency versions, making a monolithic deployment approach impractical. A microservices architecture circumvents these issues, enabling seamless updates and independent testing of AI services.

The system comprises the following key components:

- **Clients**: The user interface where patients interact with the AI assistant.
- **Identity Service**: Handles user authentication and authorization via a structured SQL database.
- **Gateway Service**: Acts as an intermediary, routing requests between clients and backend services.
  - **Database Service**: Manages patient data and system records.
  - **Analytics Service**: Processes and interprets health metrics and diagnostic results.
  - **ML Service**: Provides predictive analytics, classification, and anomaly detection using extreme value theory.
  - **NLP Service**: Manages language processing tasks and embeddings.

    Vector Database: Stores domain-specific embeddings for efficient retrieval.

  - **Proxy Service**: Facilitates API-based interactions with external health sensors and LLM engines.

    Health Sensors: Integrates with wearable APIs (e.g., Fitbit, Apple HealthKit) and medical devices.

    LLM APIs: Provides access to AI-driven natural language processing engines.

8.4.7 *Services Simulating a General Practitioner*

The system replicates general practitioner (GP) capabilities by combining multiple AI-driven services. It simulates physical examinations and diagnostic decision-making using structured data analysis. The diagnostic pipeline encompasses:

- **Diagnosis**: Identification of possible conditions based on symptoms.
- **Diagnostics**: Recommendations for further medical tests.
- **Specialist Referral**: Referral to a specialist if the condition requires expertise beyond general practice.
- **Lifestyle Adjustments**: Health and wellness recommendations based on patient data.

Blood Exam Analyzer

The Blood Exam Analyzer processes laboratory results to extract meaningful health indicators. The system utilizes:

- **Rule-Based Logic**: Detects abnormal values outside standard medical ranges.
- **Machine Learning Algorithms**: Identifies trends related to metabolic disorders, weight classification, and biochemical imbalances [27, 30].
- **Metabolic Syndrome Detection**: Evaluates multiple biomarkers to determine the likelihood of metabolic syndrome [26, 27].

External Sensor Integration

By integrating with health sensors, the system approximates aspects of a physical examination. It processes various biometric indicators with high accuracy, including:

- **Heart Rate**: Continuous monitoring, including resting heart rate and irregular heartbeat alerts.
- **Sleep Tracking**: Records sleep patterns, stages (light, deep, REM), and overall sleep quality.
- **Stress Monitoring**: Tracks stress levels throughout the day.
- **Activity Tracking**: Logs steps, floors climbed, and estimated calories burned.
- **Oxygen Saturation (SpO2)**: Measures blood oxygen levels for respiratory assessment.
- **Respiration Rate**: Monitors breathing patterns over time.
- **VO2 Max**: Estimates maximum oxygen utilization capacity during exercise.
- **GPS Tracking**: Enables detailed tracking of outdoor activities.
- **Incident Detection**: Detects falls or sudden impacts and notifies emergency contacts.
- **Mobility Metrics**: Evaluates walking speed, step timing, and postural stability.

Example of retrieved health metrics from the Garmin Health Snapshot API:


```
{
    "calendarDate": "2023-10-23",
    "minHeartRate": 50,
    "maxHeartRate": 131,
    "includesActivityData": true,
    "restingHeartRate": 61,
    "averageStressLevel": 42,
    "bodyBatteryMostRecentValue": 18,
    "highestRespirationValue": 17.0,
    "lowestRespirationValue": 12.0,
    "latestRespirationValue": 14.0
}
```

Garmin Health Snapshot: Extracted and transformed into JSON format.

This structured approach ensures that Med|Primary AI Assistant delivers medically relevant insights while maintaining user safety and clinical reliability.

Algorithm: External Sensors

The external sensor integration allows the system to retrieve and analyze biometric data over time, aiding in diagnostic evaluations. The two primary functions implemented are:

- **get_weekly_data**: Collects biometric data from the past seven days using a predefined data retrieval function.
- **daily_snapshot**: Retrieves the most recent health data from connected sensors.

Pseudocode

The pseudocode for retrieving sensor data is as follows:

```
function get_weekly_data(SensorData):
    start_date <- today - 7 days
    weekly_data <- empty list

    for i in 0 to 6:
        current_Date <- start_date - i days
        data <- SensorData(current_Date)
        append data to weekly_data

    return weekly_data

function daily_snapshot(healthSensor, anarray):
    extracted_data <- empty dictionary

    for each key in anarray:
        extracted_data[key] <- healthSensor.get(key, None)
    return extracted_data
```

This implementation ensures efficient retrieval and storage of biometric data, enabling AI-driven health assessments.

8.4.8 Prototype Use Cases

This section presents several practical use cases, demonstrating the conversational capabilities of the Med|Primary AI Assistant. The assistant is designed based on the *Blueprint of Domain-Specific Knowledge* using embeddings. Additionally, *Instantiated Advice Generation* is enabled, allowing users to upload and analyze health examination data through dedicated services (Fig. 8.8).

Use Case 1: General Inquiry

Figure 8.9 illustrates a user asking an *open-ended question* regarding the assistant's capabilities. The response provides an overview of the system's objectives, tools, and tasks, based on the pre-defined domain-specific knowledge. This is classified as a *direct answer*.

Use Case 2: Symptom-Based Consultation

In the second use case (Fig. 8.10), a user requests a consultation based on described symptoms-an *informational question*. The assistant provides an *elaborated answer*, suggesting an initial assessment and recommending further examinations or potential lifestyle modifications. The conversation can continue for further refinement, leading to a more precise diagnosis.

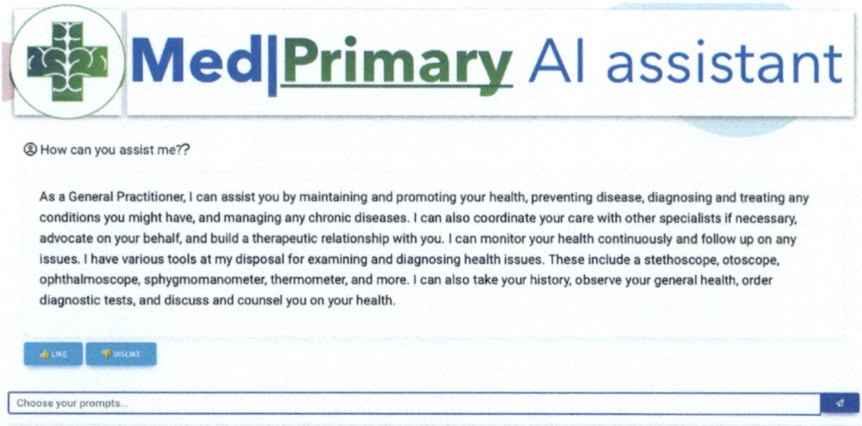

Fig. 8.9 Open-ended question—direct answer, system objectives and tasks

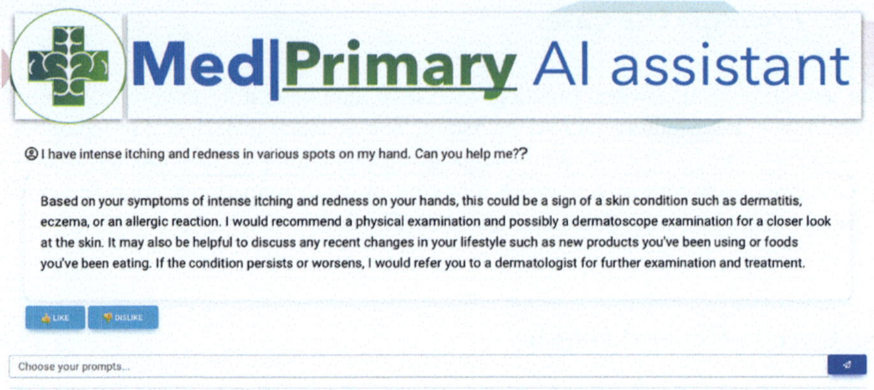

Fig. 8.10 Informational question—elaborated answer, skin problem with referral

Use Case 3: Specialist Referral

Figure 8.11 demonstrates a scenario where the AI assistant provides a *deferred answer*. Here, the user asks an *open-ended medical question* that falls outside the assistant's scope of expertise. Following the dialogue augmentation rules (Table 8.1), the system recognizes the limitation and refers the user to an oncologist for a more specialized evaluation.

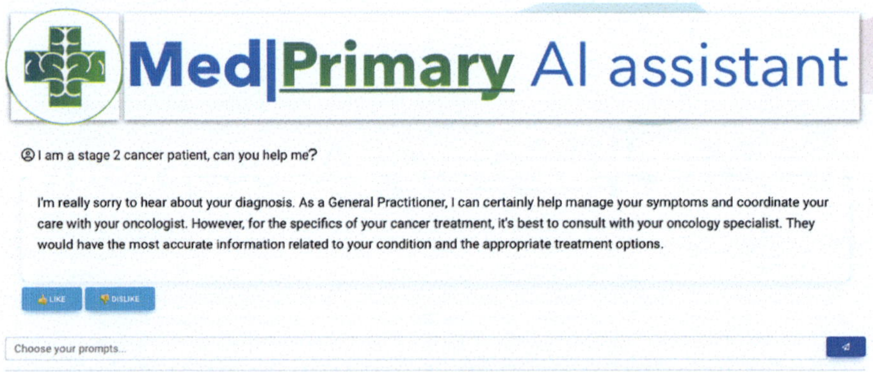

Fig. 8.11 Open-ended question—deferred answer, referral to oncologist

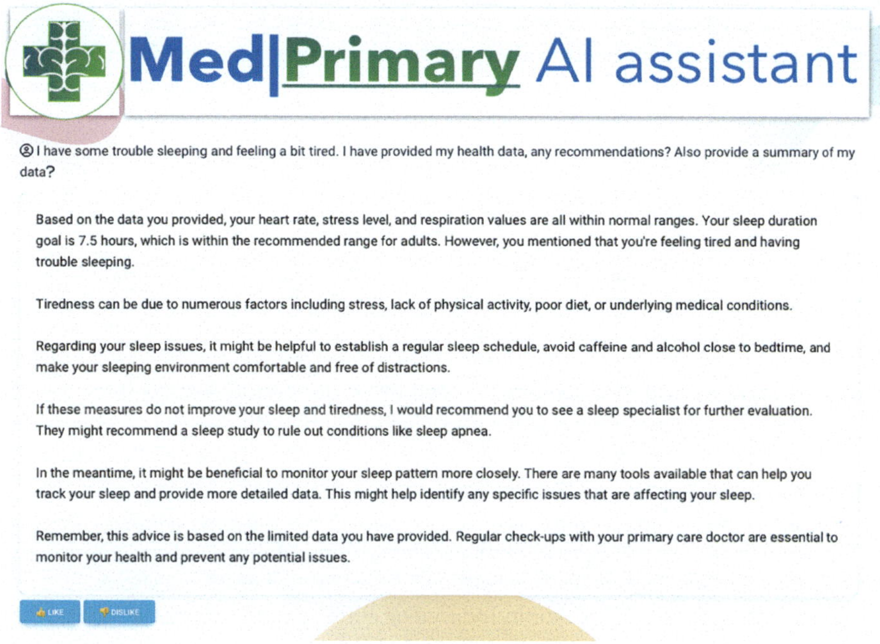

Fig. 8.12 Informational question—elaborated answer, symptoms with health metrics

Use Case 4: Symptom Analysis with Health Metrics

In the final use case (Fig. 8.12), the system processes externally retrieved biometric data. The user describes symptoms (*informational question*), and the AI assistant provides an *elaborated answer*, incorporating uploaded health metrics. The response summarizes the relevant physiological data and links it to the user's reported symptoms.

This demonstrates the system's capability to seamlessly integrate real-time health sensor data into the diagnostic pipeline. In a real-world scenario, once the *Blueprint of Domain-Specific Knowledge* is established, users would primarily need to report symptoms, and the system would autonomously analyze the available health data.

8.5 System Evaluation

To rigorously assess the performance of our system, we employed a structured evaluation process using multiple-choice quiz questions sourced from *The Internet Pathology Laboratory for Medical Education*, an authoritative resource hosted by the University of Utah's Eccles Health Sciences Library [37]. These quizzes are specifically designed for medical students and healthcare professionals, focusing on key concepts in pathology. This selection aligns with the curricular requirements of medical education, ensuring that the system is tested against real-world medical scenarios.

8.5.1 Evaluation Methodology

Our system, powered by the GPT-4 model, was tested across four thematic pillars of general pathology. Each thematic area represents a crucial domain within the medical field:

- **Atherosclerosis and Thrombosis**: A total of 50 questions assessed the system's understanding of cardiovascular diseases, including the pathogenesis of atherosclerosis and thrombotic processes. The model correctly answered **48 out of 50** questions.
- **Cellular Injury**: The system was tested with 55 questions covering cellular responses to stress, pathophysiological mechanisms of cell damage, and related clinical scenarios. The model correctly answered **50 out of 55** questions.
- **Embryology**: A set of 52 questions focused on embryonic development, congenital abnormalities, and developmental biology concepts. The model correctly answered **45 out of 52** questions.
- **Nutrition**: A total of 40 questions evaluated the system's understanding of nutritional science, its role in disease prevention, and the clinical implications of deficiencies and metabolic disorders. The model correctly answered **37 out of 40** questions.

The total number of questions presented was 197, out of which GPT-4 correctly answered **180**, achieving an overall accuracy of **91.37%**.

8.5.2 Assessment Design

Each question was carefully structured to present realistic clinical cases, incorporating diverse patient demographics, symptoms, and pathological conditions. The system was provided with the complete set of questions and multiple answer choices and was tasked with selecting the correct response. This setup was designed to simulate real-world diagnostic challenges, assessing the AI model's capacity for medical reasoning and decision-making.

For instance, consider the following question from the *Cellular Injury* category:

> A 50-year-old woman with a history of unstable angina suffers an acute myocardial infarction. Thrombolytic therapy with tissue plasminogen activator (tPA) is administered to restore coronary blood flow. Despite this therapy, the extent of myocardial fiber injury may increase due to which of the following cellular abnormalities?

- *(A) Cytoskeletal intermediate filament loss*
- *(B) Decreased intracellular pH from anaerobic glycolysis*
- *(C) Increased free radical formation*
- *(D) Mitochondrial swelling*
- *(E) Nuclear chromatin clumping*
- *(F) Reduced protein synthesis*

This question exemplifies the complexity and depth of the assessment. It evaluates not only factual recall but also the system's ability to reason through pathological mechanisms and predict disease progression. Each question included four to five answer choices, increasing the challenge by requiring precise selection among plausible alternatives.

8.5.3 *Findings and Implications*

The evaluation methodology ensured that the AI model was tested on:

- **Medical knowledge retention**: The system demonstrated a strong grasp of pathology concepts across multiple domains.
- **Clinical reasoning**: The model successfully identified appropriate responses based on contextual medical data.
- **Decision-making accuracy**: The system's ability to differentiate between correct and misleading answers was reflected in its high accuracy rate.

Despite achieving a commendable accuracy rate of **91.37%**, further testing is necessary. Future evaluations should involve:

- Validation by medical professionals to ensure accuracy and clinical applicability.
- Testing across more diverse and complex medical scenarios.
- Incorporation of real-world clinical case studies for expanded assessment.

Although the system demonstrates a high success rate, it is imperative to emphasize that AI-powered tools such as GPT-4 should be considered as *assistants* rather than replacements for human medical experts. This is particularly crucial in sensitive fields such as medicine, where clinical decision-making has direct consequences for patient care. AI-driven systems must undergo continuous validation, expert oversight, and ethical considerations before deployment in real-world healthcare environments.

8.6 Discussion of Results and Future Research

This chapter explored the application of AI, particularly LLMs and NLP, in Healthcare. A novel AI-empowered system was introduced, enhanced with rule-based algorithms, and integrated with GPT models and other ML tools to provide diagnostic advice. The system is designed to address the complexities of Healthcare, particularly from a general practitioner's perspective. The research covered theoretical foundations, system design and implementation, and practical use cases.

A key contribution of this work is the development of a *Blueprint of Domain-Specific Knowledge*, serving as a contextual foundation for AI systems augmented with LLMs and rule-based logic. By generalizing the process, a measurable evaluation of LLM-generated answers was established within a domain-specific and dialogue-defined space. These rules, formulated from a dialogue theory perspective, ensure meaningful and relevant interactions. The system design was constructed to be cost-effective, reproducible, and scalable. The proposed AI-empowered, rule-augmented Healthcare application integrates external APIs and modern methodologies to optimize LLM utilization. The system's evaluation in general pathology yielded a 91.37% accuracy rate across 197 multiple-choice questions, demonstrating its potential in medical applications.

8.6.1 Future Research Directions

For future research and development, two critical areas require deeper investigation:

- **Cost Analysis**: Evaluating the financial implications of deploying and operating AI-driven Healthcare systems is crucial. This includes:
  - *Initial setup costs*: Infrastructure, software development, and integration expenses.
  - *Operational expenses*: Computational resources, API usage fees, maintenance, and updates.
  - *Financial impact on Healthcare providers and patients*: Assessing whether AI-assisted diagnostics reduce costs, improve efficiency, or enhance accessibility.

  Conducting a comprehensive cost-benefit analysis will determine the economic feasibility and scalability of the system.
- **Value-Based Care**: Examining how AI-driven systems compare with traditional care models in terms of cost and patient outcomes. This involves:
  - *Effectiveness of Interventions*: Evaluating the impact of AI-assisted diagnostics on patient outcomes, including mortality rates, morbidity reduction, and improvements in quality of life.
  - *Patient Perspectives*: Assessing user experience, trust, and satisfaction with AI-powered consultations compared to traditional Healthcare interactions.

8.6.2 Ethical Considerations and Integration Challenges

Beyond cost and value, future research should address key challenges related to:

- **Ethical and regulatory frameworks**: Ensuring AI-driven diagnostics adhere to medical standards and ethical guidelines.
- **Data privacy and security**: Implementing robust data protection mechanisms for patient information.
- **Interoperability with existing Healthcare infrastructures**: Developing AI solutions that seamlessly integrate with electronic health records (EHR) and clinical workflows.

8.7 Conclusion

This study highlights the potential of AI-empowered systems in Healthcare, demonstrating their capability to augment general practitioners' decision-making. However, achieving widespread adoption requires ongoing validation, cost-effectiveness assessments, and integration into real-world clinical environments. Future research must balance technological advancements with ethical considerations and patient-centered care to ensure AI solutions enhance Healthcare accessibility, affordability, and effectiveness.

Appendix

Abbreviations

The following abbreviations are used in this manuscript:

ML	Machine Learning
AI	Artificial Intelligence
XAI	Explainable Artificial Intelligence
GAI	Generative Artificial Intelligence
LLM	Large Language Model
NLP	Natural Language Processing
EHR	Electronic Health Records
HMM	Hidden Markov Models
CRF	Conditional Random Fields
LDA	Latent Dirichlet Allocation
RAG	Retrieval-Augmented Generation

References

1. E.P. Balogh, B.T. Miller, J.R. Ball, Improving diagnosis in health care (2015)
2. T. Bolukbasi, K.-W. Chang, J.Y. Zou, V. Saligrama, A.T. Kalai, Man is to computer programmer as woman is to homemaker? Debiasing word embeddings. Adv. Neural Inform. Process. Syst. 4349–4357 (2016)
3. T.B. Brown, B. Mann, N. Ryder, M. Subbiah, J. Kaplan, P. Dhariwal, A. Neelakantan, P. Shyam, G. Sastry, A. Askell, et al., Language models are few-shot learners. arXiv preprint arXiv:2005.14165 (2020)
4. C.-C. Chiu, T.N. Sainath, Y. Wu, R. Prabhavalkar, P. Nguyen, Z. Chen, A. Kannan, R.J. Weiss, K. Rao, E. Gonina, et al., State-of-the-art speech recognition with sequence-to-sequence models, in: *2018 IEEE International Conference on Acoustics, Speech and Signal Processing (ICASSP)* (IEEE, 2018), pp. 4774–4778
5. J. Devlin, M.-W. Chang, K. Lee, K. Toutanova, Bert: pre-training of deep bidirectional transformers for language understanding. arXiv preprint arXiv:1810.04805 (2018)
6. J. Devlin, M.-W. Chang, K. Lee, K. Toutanova, Bert: pre-training of deep bidirectional transformers for language understanding. arXiv preprint arXiv:1810.04805 (2018)
7. A. Esteva, A. Robicquet, B. Ramsundar, V. Kuleshov, M. DePristo, K. Chou, C. Cui, G. Corrado, S. Thrun, J. Dean, A guide to deep learning in healthcare. Nat. Med. **25**(1), 24–29 (2019)
8. W. Floyd, T. Kleber, M. Pasli, J.J. Qazi, C.C. Huang, J.X. Leng, B. Ackerson, D.J. Carpenter, J.K. Salama, M.J. Boyer, Evaluating the reliability of Chat-GPT model responses for radiation oncology patient inquiries. Int. J. Radiat. Oncol. Biol. Phys. **117**(2), e383 (2023)
9. R. García-Carrión, G. de Aguileta, M. Padrós, M. Ramis-Salas, Implications for social impact of dialogic teaching and learning. Front. Psychol. **11**, 140 (2020)
10. A. Gilson, C.W. Safranek, T. Huang, V. Socrates, L. Chi, R.A. Taylor, D. Chartash, et al., How does ChatGPT perform on the United States medical licensing examination? The implications of large language models for medical education and knowledge assessment. JMIR Med. Educ. **9**(1), e45312 (2023)
11. E.B. Gordon, A.J. Towbin, P. Wingrove, U. Shafique, B. Haas, A.B. Kitts, J. Feldman, A. Furlan, Enhancing patient communication with Chat-GPT in radiology: evaluating the efficacy and readability of answers to common imaging-related questions. J. Am. Coll. Radiol. (2023)
12. P. Gorsky, A. Caspi, E. Chajut, The' theory of instructional dialogue: toward a unified theory of instructional design, in *Understanding Online Instructional Modeling Theories Practices* (IGI Global, 2008), pp. 47–69
13. R. Gualandi, C. Masella, D. Viglione, D. Tartaglini, Exploring the hospital patient journey: what does the patient experience? PLoS One **14**(12), e0224899 (2019)
14. D. Gunning, M. Stefik, J. Choi, T. Miller, S. Stumpf, G.-Z. Yang, XAI—Explainable artificial intelligence. Sci. Robot. **4**(37), eaay7120 (2019). https://doi.org/10.1126/scirobotics.aay7120. Eprint: https://www.science.org/doi/pdf/10.1126/scirobotics.aay7120
15. A. Holzinger, R. Goebel, R. Fong, T. Moon, K.-R. Müller, W. Samek, xxAI-beyond explainable artificial intelligence, in *xxAI-Beyond Explainable AI: International Workshop, Held in Conjunction with ICML 2020, 7–18, 2020* (Vienna, Austria, Revised and Extended Papers. Springer, 2022), pp.3–10
16. K. Kreimeyer, M. Foster, A. Pandey, N. Arya, G. Halford, S.F. Jones, R. Forshee, M. Walderhaug, T. Botsis, Natural language processing systems for capturing and standardizing unstructured clinical information: a systematic review. J. Biomed. Informatics **73**, 14–29 (2017)
17. S. Locke, A. Bashall, S. Al-Adely, J. Moore, A. Wilson, G.B. Kitchen, Natural language processing in medicine: a review. Trends Anaesthesia Crit. Care **38**, 4–9 (2021)
18. M.A.A. Majumder, A. Kumar, K. Krishnamurthy, N. Ojeh, O.P. Adams, B. Sa, An evaluative study of objective structured clinical examination (OSCE): students and examiners perspectives. Adv. Med. Educ. Pract. 387–397 (2019)

19. McCarthy, S., P. O'Raghallaigh, S. Woodworth, Y.L. Lim, L.C. Kenny, F. Adam, An integrated patient journey mapping tool for embedding quality in healthcare service reform. J. Decis. Syst. **25**(sup1), 354–368 (2016)
20. M.L. Mitchell, A. Henderson, M. Groves, M. Dalton, D. Nulty, The objective structured clinical examination (OSCE): optimising its value in the undergraduate nursing curriculum. Nurse Educ. Today **29**(4), 398–404 (2009)
21. OpenAI, *Better Language Models and Their Implications* (2019). https://openai.com/blog/better-language-models/
22. OpenAI, GPT-4 Technical Report. ArXiv arXiv:2303.08774 (2023). https://api.semanticscholar.org/CorpusID:257532815
23. D. Panagoulias, F. Palamidas, M. Virvou, G.A. Tsihrintzis, Evaluating the potential of LLMs and ChatGPT on medical diagnosis and treatment, in *14th IEEE International Conference on Information, Intelligence, Systems, and Applications (IISA2023), Volos, Greece* (2023)
24. D.P. Panagoulias, F.A. Palamidas, M. Virvou, G.A. Tsihrintzis, Evaluation of ChatGPT-supported diagnosis, staging and treatment planning for the case of lung cancer, in: *2023 20th ACS/IEEE International Conference on Computer Systems and Applications (AICCSA)* (IEEE, 2023), pp. 1–8
25. D.P. Panagoulias, D. Sotiropoulos, K. Chrysafiadi, E.-A. Tsichrintzi, E. Sakkopoulos, G.A. Tsihrintzis, M. Virvou, LYRICEL: a rule-augmented artificial intelligence-empowered cultural E-learning with GPT and machine learning for song lyrics, in *2024 15th International Conference on Information, Intelligence, Systems & Applications (IISA)* (IEEE, 2024), pp. 1–5
26. D.P. Panagoulias, D.N. Sotiropoulos, G.A. Tsihrintzis, SVM-based blood exam classification for predicting defining factors in metabolic syndrome diagnosis. Electronics **11**(6), 857 (2022)
27. D.P. Panagoulias, M. Virvou, G.A. Tsihrintzis, nuhealthsoft: a nutritional and health data processing software tool from a patient's perspective, in *2022 16th International Conference on Signal-Image Technology and Internet-Based Systems (SITIS)* (IEEE, 2022), pp. 386–393
28. D.P. Panagoulias, M. Virvou, G.A. Tsihrintzis, Augmenting large language models with rules for enhanced domain-specific interactions: the case of medical diagnosis. Electronics **13**(2), 320 (2024)
29. D.P. Panagoulias, P. Papatheodosiou, A.P. Palamidas, M. Sanoudos, E. Tsoureli-Nikita, M. Virvou, G.A. Tsihrintzis, *COGNET-MD, an evaluation framework and dataset for Large Language Model benchmarks in the medical domain*. [arXiv:2405.10893] [cs.CL] (2024)
30. D.P. Panagoulias, D.N. Sotiropoulos, G.A. Tsihrintzis, An Extreme Value Analysis-Based Systemic Approach in Healthcare Information Systems: The Case of Dietary Intake. Electronics 12.1 (2023). ISSN: 2079-9292. https://doi.org/10.3390/electronics12010204. https://www.mdpi.com/2079-9292/12/1/204
31. D.P. Panagoulias, E. Tsoureli-Nikita, M. Virvou, G.A. Tsihrintzis, *Dermacen analytica: a novel methodology integrating multi-modal large language models with machine learning in tele-dermatology*. [arXiv:2403.14243] [cs.CL] (2024)
32. D.P. Panagoulias, M. Virvou, G.A. Tsihrintzis, Rule-augmented artificial intelligence-empowered systems for medical diagnosis using large language models, in: *2023 IEEE 35th International Conference on Tools with Artificial Intelligence (ICTAI)* (2023)
33. D.P. Panagoulias, M. Virvou, G.A. Tsihrintzis, *Evaluating LLM— generated multimodal diagnosis from medical images and symptom analysis*. [arXiv:2402.01730] [cs.CL] (2024)
34. T. Pham, T. Tran, D. Phung, S. Venkatesh, Predicting healthcare trajectories from medical records: a deep learning approach. J. Biomed. Inf. **69**, 218–229 (2017)
35. A. Radford, J. Wu, R. Child, D. Luan, D. Amodei, I. Sutskever, et al., Language models are unsupervised multitask learners. OpenAI Blog **1**(8), 9 (2019)
36. R. Taori, I. Gulrajani, T. Zhang, Y. Dubois, X. Li, C. Guestrin, P. Liang, T.B. Hashimoto, *Stanford Alpaca: An Instruction-following LLaMA model* (2023). https://github.com/tatsu-lab/stanford_alpaca
37. *The Internet Pathology Laboratory for Medical Education* (n.d.). https://webpath.med.utah.edu/webpath.html. Accessed: 15 Dec 2023

38. D. Thomas, K. Ravi, National health and nutrition examination survey: sample design. Future Healthc. J. **6**, 94 (2019)
39. H. Touvron, L. Martin, K. Stone, P. Albert, A. Almahairi, Y. Babaei, N. Bashlykov, S. Batra, P. Bhargava, S. Bhosale, et al., *Llama 2: Open Foundation and Fine-Tuned Chat Models*. arXiv preprint arXiv:2307.09288 (2023)
40. T.M. Trebble, N. Hansi, T. Hydes, M.A. Smith, M. Baker, Process mapping the patient journey: an introduction. BMJ 341 (2010)
41. G.A. Tsihrintzis, M. Virvou, N.G. Bourbakis, L.C. Jain, Introduction to advances in information, intelligence, systems and applications, in: *International Conference on Information, Intelligence, Systems, and Applications* (Springer, 2023), pp. 1–9
42. G.A. Tsihrintzis, M. Virvou, H. Doukas, L.C. Jain, Introduction to advances in artificial intelligence-empowered decision support systems, in: *Advances in Artificial Intelligence-Empowered Decision Support Systems: Papers in Honour of Professor John Psarras* (Springer, 2024), pp. 1–11
43. A. Vaswani, N. Shazeer, N. Parmar, J. Uszkoreit, L. Jones, A.N. Gomez, L. Kaiser, I. Polosukhin, Attention is all you need. Adv. Neural Inform. Process. Syst. **30** (2017)
44. M. Virvou, G.A. Tsihrintzis, E.-A. Tsichrintzi, VIRTSI: a novel trust dynamics model enhancing Artificial Intelligence collaboration with human users–Insights from a ChatGPT evaluation study. Inform. Sci. 120759 (2024)
45. M. Virvou, G.A. Tsihrintzis, Pre-made empowering artificial intelligence and ChatGPT: the growing importance of human AI-experts, in *2023 14th International Conference on Information, Intelligence, Systems & Applications (IISA)* (IEEE, 2023), pp. 1–8
46. M. Virvou, G.A. Tsihrintzis, Artificial intelligence Chatbots and social robots in education: FEPER framework for efficiency, pedagogical and ethical requirements, in *2024 15th International Conference on Information, Intelligence, Systems and Applications (IISA)* (IEEE, 2024), pp. 1–8
47. M. Virvou, G.A. Tsihrintzis, D.N. Sotiropouloss, K. Chrysafiadi, E. Sakkopoulos, E.-A. Tsichrintzi, ChatGPT in artificial intelligence-empowered E-learning for cultural heritage: the case of lyrics and poems, in *2023 14th International Conference on Information, Intelligence, Systems and Applications (IISA)* (IEEE, 2023), pp. 1–9
48. B. Wang, A. Wang, F. Chen, Y. Wang, C.-C. Jay Kuo, Evaluating word embedding models: Methods and experimental results. APSIPA Trans. Signal Inform. Process. **8**, e19 (2019)
49. J. Weizenbaum, ELIZA—a computer program for the study of natural language communication between man and machine. Commun. ACM **9**(1), 36–45 (1966)
50. D.C. Wilson, Chapter three: a framework for clarifying, in *A Guide to Good Reasoning: Cultivating Intellectual Virtues* (2020)
51. C. Xiao, E. Choi, J. Sun, Opportunities and challenges in developing deep learning models using electronic health records data: a systematic review. J. Am. Med. Inf. Assoc. **25**(10), 1419–1428 (2018)

Chapter 9
Extracting Requirements to Fine-Tune Large Language Models in Multimodal Medical Diagnosis

Abstract Large language models constitute a breakthrough state-of-the-art Artificial Intelligence technology which is rapidly evolving and promises to aid in medical diagnosis. In this work, we engineered an evaluation system that incorporates two independent steps of a novel methodology, namely (1)evaluation via structured interactions and (2)follow-up, domain-specific analysis based on data extracted via the previous interactions. Using this paradigm, (1)we evaluate the correctness and accuracy of generated medical diagnosis with publicly available multimodal-multiple-choice-questions in the domain of Pathology and (2) proceed to a systemic and comprehensive analysis of extracted results. These results serve as the basis for further targeted optimization and fine-tuning of the employed Generative model. We used the Generative Pre-trained Transformer version 4 with vision (GPT-4V) as the model that generates responses to complex, medical questions consisting of both images and text, and we explored a wide range of diseases, conditions, chemical compounds, and related entity types that are included in the vast knowledge domain of Pathology. The model scored approximately 84% of correct diagnoses. We further analyzed the findings of our work, following an analytical approach which included Image-Metadata-Analysis, Named-Entity-Recognition and Knowledge-Graphs. Weaknesses of the model were revealed on specific knowledge paths, leading to a further understanding of its shortcomings in specific areas and discovering paths for fine-tuning and improvements. Our methodology and findings transcend the constraints of a singular model or domain, offering broad applicability and generalizability across various contexts.

Keywords Artificial Intelligence-empowered software engineering · Multimodal Large Language Models · Generative Pre-trained Transformer · Telehealth

9.1 Introduction

Streamlined digital technologies, empowered by Artificial Intelligence (AI) and Big Data Analytics (BDA), are increasingly demonstrating their potential to enhance symptom detection, provide accurate and fast medical diagnosis (e.g. [38]) and

D. P. Panagoulias et al., *Artificial Intelligence-Empowered Bio-medical Applications*, Learning and Analytics in Intelligent Systems 49,
https://doi.org/10.1007/978-3-031-90174-4_9

predict treatment outcomes (e.g. [3]). Indeed, AI acts as a pivotal force in aligning user (patient or medical personnel) needs with the development of tools which can efficiently and reliably analyze health-related data and biosignals and either lead to entirely new medical approaches or enhance existing ones. This, in turn, improves the quality of life, extends lifespans, and positively impacts both Quality-Adjusted Life Years (QALYs) and Disability-Adjusted Life Years (DALYs) [7].

However, many challenges emerge with all AI-empowered systems, in terms of relevance and potential misuse. Such challenges include issues of validation/evaluation that stem from confirmation bias or biases of those responsible for the system's validation, evaluation and management. Thus, a need arises for transparency, rigorous evaluation and continuous improvement. Given the high complexity of medical recommendations, associated interventions carry significant risks and potential threats to life and well-being. Clearly, ensuring patients' health and data safety necessitates frameworks that clarify and interpret all underlying technologies as AI-influenced decisions could have very significant consequences [28, 34].

Large Language Models (LLMs) for Artificial Intelligence (AI) are advanced computational algorithms designed to understand, generate, and manipulate natural language. These models are a subset of machine learning models that have been trained on vast amounts of text data. In this chapter, we propose an LLM critical evaluation paradigm that incorporates two independent steps of a novel methodology, namely (1) multimodal LLM evaluation via structured interactions and (2) follow-up, domain-specific analysis based on data extracted via the previous interactions. Next, using this paradigm, we evaluate the correctness and accuracy of LLM-generated medical diagnosis with publicly-available multimodal multiple-choice questions in the domain of Pathology, followed by a systemic and comprehensive analysis of extracted results [39].

We use Generative Pre-trained Transformer version 4 with vision (GPT-4V) [24] (training data Up to Apr 2023), a state-of-the-art LLM based on the GPT transformer, as the LLM to respond to complex, medical questions consisting of both images and text. Our work is focused on Pathology which implies that we are not looking at only one specific disease, condition, medical domain or specialty. Instead, we are exploring a wide range of diseases, conditions, chemical compounds and related entity types that are included in the vast knowledge domain of Pathology. GPT-4V scored approximately 84% of correct diagnoses. This high level of accurate diagnosis across diverse areas, which typically requires a variety of deep learning and machine learning tools, incentivized us to further analyze the findings of our work following an analytical approach. Our analytical approach included Image Metadata Analysis (IMA), Named Entity Recognition (NER) and Knowledge Graphs (KG), which revealed weaknesses of GPT-4V on specific knowledge paths and allowed a further understanding of the employed LLM's shortcomings in specific areas. Our methodology and findings are not limited to the use of GPT-4V, but a similar approach can be followed to evaluate the usefulness and accuracy of other LLMs and, thus, improve their use with further optimization.

More specifically, the chapter is structured as follows: Sect. 9.2 focuses on reviewing relevant previous studies and related published work. In Sect. 9.3, the evaluation methodology is presented and consists of two independent parts. The first part is related to specific interactions with a multimodal LLM for defining competency using structured MCQs.[1] The second part is related to the analysis of those interactions to extract useful information for enhancing the LLM knowledge base. In Sect. 9.4, we present the results from application of the proposed methodology in Pathology utilizing GPT-4V and multiple-choice questions and images. Lastly in Sect. 9.5, we summarize the findings and discuss future related endeavours.

9.2 Related Work

In this section, we establish a comprehensive background pertinent to the methodology used in this work and the various aspects of relevant technologies. We also briefly describe a recent previous work of ours, in which we evaluated GPT-4's ability in General Pathology Multiple-Choice Quizzies (MCQs) from only provided textual descriptors of symptoms [29].

9.2.1 Evaluation of Medical Students

While using new technologies for automating the provision of medical services is very promising, the tolerance for mistakes is extremely low and the risk remains high of misdiagnosis, privacy breaches, and ethical dilemmas such as bias in AI algorithms.

Currently, for the comprehensive evaluation of medical students, various assessment methods are used alongside Objective Structured Clinical Examinations (OSCEs) [20] and multiple-choice quizzes [11, 23, 40]. Each of these methods has its own focus, ranging from theoretical knowledge to practical skills and professional behaviors.

More specifically, a list of common assessment procedures followed in medical education is presented next:

1. Objective Structured Clinical Examinations (OSCEs): They assess clinical skills, communication, and procedures in a structured and practical setting.
2. Multiple-choice Quizzes (MCQs): They test theoretical knowledge, understanding of concepts, and problem-solving abilities.
3. Short Answer Questions (SAQs): They valuate the ability to provide concise, precise answers to focused questions.
4. Essay Questions: They assess depth of understanding, ability to synthesize information, and critical thinking.

[1] MCQs are posted at The Internet Pathology Laboratory for Medical Education [39] and the full collected questionnaire data and responses are available upon request from the authors.

5. Oral Examinations/Vivas: They test understanding and knowledge through direct questioning and are often used in higher level assessments like specialty/ exams.
6. Practical or Lab-based Assessments: They evaluate skills in laboratory techniques, clinical procedures, or other practical tasks.
7. Direct Observation of Procedural Skills (DOPS): They assess specific procedural skills in a clinical setting with direct observation by an examiner.
8. Case-based Discussions (CBDs): Evaluate clinical reasoning, decision-making, and professional judgment using real or simulated patient cases.
9. Portfolio Assessments: They compile and review a collection of a student's work over time to assess his/her progress and competence.
10. 360-Degree Feedback/Peer Assessment: They gather feedback from a wide range of observers, including peers, patients, and other healthcare professionals, to assess communication, teamwork, and professional behavior.
11. Self-Assessment and Reflection: They encourage students to reflect on their own learning and performance.
12. Standardized Patient Encounters: They use actors trained to portray real patient cases for assessment of clinical and communication skills.
13. Logbooks or Procedure Lists: They track the number and type of clinical procedures or experiences that a student has completed.
14. Audit Projects or Quality Improvement Projects: They evaluate the ability to conduct research, analyze data, and understand the principles of improving clinical practice.

The evaluation and training of medical students involve a complex process, indicating that the MCQs utilized in this study represent only a small portion of the evaluation processes necessary for certifying a doctor.

General Pathology in Medical Education

Within the integrated medical curriculum, pathology is the backbone, bridging the foundational biomedical sciences with the realm of clinical medicine [2]. It enables the pragmatic application of theoretical knowledge garnered in anatomy, physiology, and biochemistry to patient care scenarios, thereby enhancing the educational relevance and applicability of such knowledge [18]. A profound comprehension of pathology is paramount for delivering superior patient care. Understanding disease pathogenesis is instrumental in elucidating the nature of illnesses to patients, prognosticating disease courses, and comprehending both the effectiveness of interventions and the likelihood of adverse reactions [9].

In this study, we examine subdomains of pathology where imaging [12] associated with symptoms is integral for defining a diseases or a health state, namely:

- Atherosclerosis and Thrombosis: is a condition characterized by the accumulation of lipids, cholesterol, calcium, and cellular debris within the intimal layer of arteries. Thrombosis refers to the formation of a blood clot within a blood vessel, which can obstruct blood flow [25].

- Cell Injury: Cell injury is the response of cells to adverse stimuli that can disrupt their normal function [16].
- Immunopathology: It deals with diseases caused by abnormal immune responses, including autoimmune diseases, hypersensitivity reactions, and immunodeficiency disorders [21].
- Inflammation and Cardiovascular Pathology: Inflammation is a fundamental pathological process that occurs in response to injury or infection. Chronic inflammation is a key factor in the development of cardiovascular diseases, contributing to atherosclerosis by promoting plaque formation and arterial damage [36, 37].
- Neoplasia: Neoplasia, a term derived from Greek words meaning "new growth," refers to the process of abnormal and uncontrolled cell growth, leading to the formation of neoplasms or tumors [10].
- Dermatopathology: Dermatopathology focuses on skin diseases and disorders. It involves the microscopic examination of skin samples to diagnose conditions such as infections, autoimmune diseases, and cancers (e.g., melanoma, basal cell carcinoma) [33].
- Endocrine Pathology: Endocrine pathology examines diseases affecting endocrine glands (e.g., thyroid, pancreas, adrenal glands). It involves analyzing glandular tissue for abnormalities such as tumors, hyperplasias, and autoimmune conditions [4].

For these domains various imaging techniques are employed. We outline the ones, also used for the images analysed in this study, to asses, diagnose, monitor and treat a variety of health problems. The term metadata in Table 9.1, refers to the metadata included in the images used in our analysis.

9.2.2 *Vision Transformers*

Vision Transformers (ViTs) [43] constitute a type of artificial intelligence model that has revolutionized the field of computer vision, i.e. the AI discipline focused on enabling machines to interpret and understand visual information from the world [17]. ViTs are, in fact, a variation of the Transformer model [8, 41], which was originally developed for NLP tasks.

During training [5], ViTs learn to identify features and patterns within the patches, leveraging the self-attention mechanism to understand the context and relationships between different parts of an image [42, 43]. For inference, the ViT model processes a new image through the same steps as used in its training and the transformed class token is used to make predictions or classifications. The key advantage of ViTs over traditional Convolutiona Neural Networks (CNNs) is their ability to capture global dependencies across the entire image, thanks to their self-attention mechanism. This allows for a more holistic understanding of the image context and leads to significant improvements in various computer vision applications.

Table 9.1 Medical imaging in pathology

Metadatata	BodyPart or condition	Imaging
(CV)	Cardiovascular	MRI, CT, PET, EEG, fMRI
(RENAL)	Kidneys	Ultrasound, CT, MRI, IVP, Nuclear Medicine Scans
(LIVER)	Liver	Ultrasound, CT, MRI, MRCP, PET, Elastography
(SKIN)	Skin	Dermatoscopy, OCT, Confocal Microscopy, Ultrasound Imaging, MRI, Photography, Thermography
(INFEC)	Infection	X-Ray, CT, MRI, Ultrasound, PET, Nuclear Medicine Scans, Endoscopy
(AIDS-HIV)	Acquired Immunodeficiency Syndrome	Chest X-Ray, CT, MRI, Ultrasound, Endoscopy, Nuclear Medicine Scans, PET
(IMM)	Immunological Disorders	Ultrasound, CT, MRI, PET, X-Ray, Nuclear Medicine Scans
(INFL)	Inflammation	Ultrasound, X-Ray, CT, MRI, PET, Nuclear Medicine Scans, Endoscopy
(GI)	Gastrointestinal	Endoscopy, Gastroscopy, Colonoscopy, ERCP, Barium Studies, CT, MRI, Ultrasound, Capsule Endoscopy, PET
(BREAST)	Breast	Mammography, Breast Ultrasound, MRI, DBT, PET, Thermography

Similarly to the traditional transformer, training or fine-tuning [35] a ViT can be done (1) either by providing pair-wise inputs of images and their corresponding annotations or (2) by using self-supervised learning techniques. In the former approach, the model learns from labeled datasets where each image is paired with relevant information, such as classification labels or bounding boxes for object detection. In the latter case, the model leverages unlabeled data, autonomously extracting patterns and features through methods like contrastive learning or predictive coding, which help in understanding the visual context without explicit human-provided labels. The effectiveness of ViTs hinges not just on the architecture of the model itself, but significantly on how the training data is curated and presented. Indeed, cross-modal training plays a pivotal role in bridging the gap between different types of data.

The remarkable performance of the Transformer architecture in NLP has recently also triggered broad interest in their application in Computer Vision [6]. Among other merits, Transformers are witnessed as capable of learning long-range dependencies and spatial correlations, which is a clear advantage over the commonly-employed CNNs [19].

9.2.3 Named Entity Recognition and Knowledge Graphs

Named Entity Recognition (NER) [14, 22] is a sub-field of NLP and Information Retrieval. The primary goal of NER is to identify and categorize key information in text into predefined categories or "entities" such as the names of persons, organizations, locations, expressions of times, quantities, monetary values, percentages, etc. NER systems are designed to recognize entities within a body of text. This involves not only detecting the entity, but also categorizing it into a predefined class.

A Knowledge Graph (KG) [1, 15] is a structured way of representing data in a graph format, where entities (nodes) are connected by relationships (edges). This graphical representation allows for the efficient organization, management, and retrieval of complex information. Network density is calculated as the number of existing edges in the network divided by the number of all possible edges. The exact formula depends of whether a KG is directed or undirected.

For undirected KGs (which are also used in this work), network density is calculated via the following formula:

$$D = \frac{2E}{N(N-1)},$$

where

$$\begin{aligned} \text{D} &\ \text{is the density of the graph,} \\ \text{E} &\ \text{is the number of edges in the graph,} \\ \text{N} &\ \text{is the number of nodes in the graph.} \end{aligned}$$

On the other hand, for a directed graph, each possible edge is counted twice, i.e. separately for each direction. Thus, the factor 2 in the previous formula needs to be removed and the network density is calculated as:

$$D = \frac{E}{N(N-1)}.$$

NER is often a crucial first step in the creation and enrichment of KGs. By identifying entities like names of people, organizations, locations, etc., NER systems can feed this structured information into a KG. Each identified entity can become a node in the graph and the context or relationships between these entities can form the edges.

9.2.4 Evaluation of ChatGPT via Expert Interactions

In Palamidas et al. [27], we examined and assessed the proficiency of ChatGPT(4) August 3 version, in identifying Lung Cancer. To do so various case scenarios and symptoms a presented as prompts, all conducted under the guidance of a medical expert. To facilitate the process of evaluation, a three pillar process was constructed. As can be seen in Fig. 9.1, first, the validity and accuracy of the answers. Second, the specificity and usefulness for doctors and patients. And third, the potential economic value added to the health care system in regards to the diagnoses and diagnostic exams proposed. Those pillars are incorporated in a novel evaluation methodology, visualised in Fig. 9.2. In Panagoulias et al. [26] we also evaluated the validity, accuracy, and usefulness of ChatGPT in diagnosing Tuberculosis based on symptoms described via interactions with an expert.

In both scenarios, ChatGPT (the LLM under scrutiny) exhibited performance ranging from moderate to excellent, indicating its potential in disease identification, aiding medical professionals and patients, and possibly reducing healthcare-related expenses. The references provided held some inconsistencies, probably due to lack of

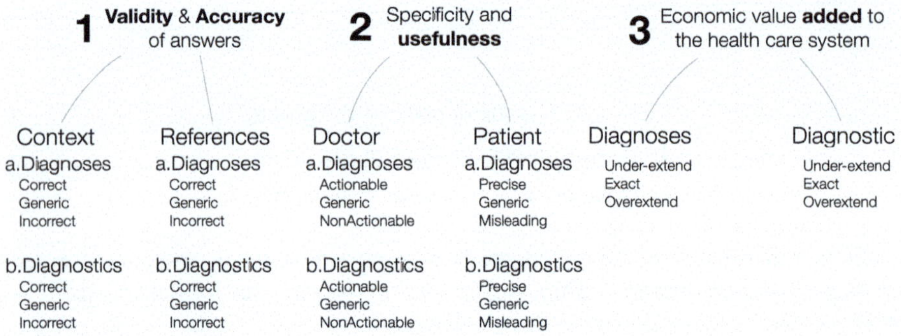

Fig. 9.1 Evaluation pillars

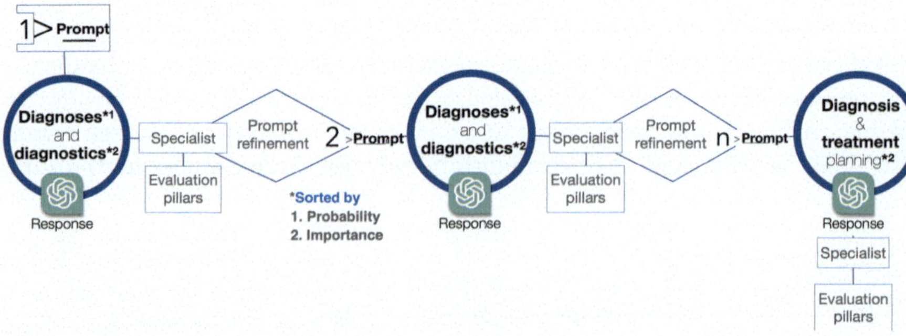

Fig. 9.2 ChatGPT evaluation methodology

access to the web, or outdated training data. ChatGPT showed optimal performance when precise diagnostics were analysed. It also exhibited reasoning when some information was missing, in Cancer staging.

9.2.5 Evaluation of GPT-4.0 in a General Pathology Questionnaire

In our recent work [29], we evaluated the level of proficiency of GPT-4 in answering General Pathology questions in the form of MCQs. The GPT-4 model achieved a total precision of 91.37%, answering correctly 180 out of 197 questions from the subdomains of Arterosclerosis and Thrombosis, Cellular Injury, Embryology, and Nutrition. The MCQ questions were retrieved from 'The Internet Pathology Laboratory for Medical Education', an esteemed resource hosted by the University of Utah's Eccles Health Sciences Library [39].

9.3 Materials and Methods

In this section, we define and detail the main components of our evaluation methodology and the proposed analytical services to define [30–32] fine-tuning or retraining requirements based on specific knowledge paths. Our methodology combines elegant approaches for targeted and complex data analysis. These include data preprocessing, prompt engineering, and data extraction and refinement. Our methods encompass IMA, scientific NER, knowledge extraction, and investigation through KGs.

Our proposed methodology is divided into two distinct parts. The first part involves the 'Multimodal Large Language Model (LLM) Evaluation', where we focus on the aspects of interaction with the model and the methods used for data extraction. This stage is crucial for understanding how the model processes and responds to various inputs. In the second part, we conduct a 'Domain specific Analysis using Machine Learning Tools'. This phase is dedicated to identifying the specific needs and requirements for further fine-tuning or re-training of the model to better suit domain-specific applications. This approach ensures a comprehensive evaluation and enhancement of the model's performance in targeted areas.

9.3.1 Multimodal LLM Evaluation

As can be seen in Fig. 9.3, the first step of the Multimodal LLM evaluation is data processing. Here, we select the set of MCQs and answers, as per specific domain

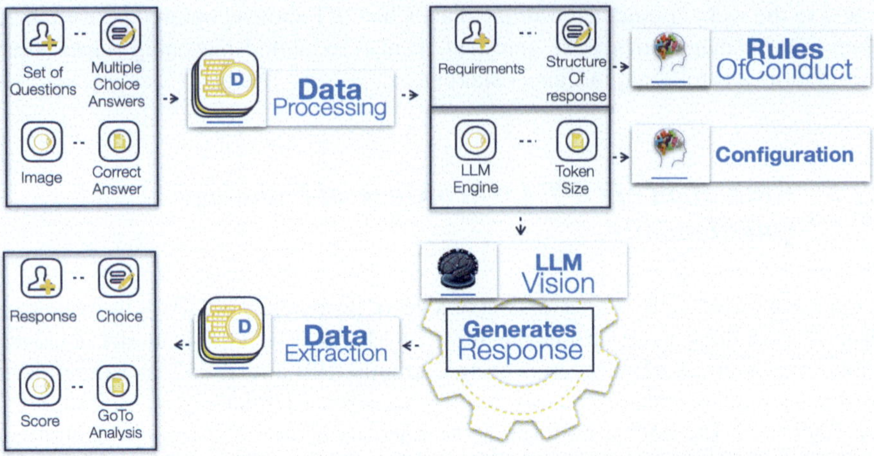

Fig. 9.3 Multimodal LLM evaluation

prerequisites. The set of questions are associated with the image file, which is related to the question, alongside the correct answer. The processed data are provided as external files and then engineered in the required format for interaction with the selected LLM.

In the second step, the *Rules Of Conduct* are defined, which are the essential parts of the prompt enginneering, which requests the specific format of analysis and response from the chosen LLM. The *Configuration* part defines the chosen model, highlighted in Fig. 9.3 as *LLM Engine* where also the *Token Size* of the response is set. The *LLM* model with *Vision* is provided in the format (image, question, rules of conduct and configuration settings) and a structured response is generated.

This part of our methodology is designed to enhance the evaluation process of a LLM with vision, focusing on its multimodal capabilities. It facilitates the creation of structured tests and quizzes that integrate both images and related questions, presenting a holistic challenge to the model's interpretative abilities. In each instance, the model is required to analyze the visual content alongside the accompanying question to select the most appropriate answer from multiple choices. This approach ensures a thorough assessment of the model's proficiency in interpreting and correlating different types of data.

Following the completion of the test, a detailed score is provided. This score reflects the model's performance, highlighting the correctly identified answers, while offering explanations for any incorrect responses. This feature is particularly valuable as it sheds light on areas where the model may need further refinement and, thereby, guides towards future improvements.

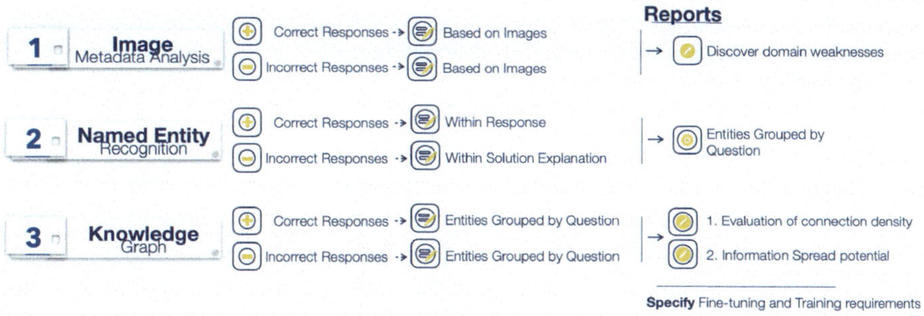

Fig. 9.4 Domain specific analysis using machine learning tools

9.3.2 *Domain-Specific Analysis*

The data extracted previously are used in the domain-specific analysis. This part of our evaluation methodology, includes 3 main steps that ensure the definition of requirements of fine-tuning and training to make the multimodal LLM excel where it scored lower.

As can be seen in Fig. 9.3, in the first step IMA takes place and evaluates the domains where the multimodal responded correctly and those where it responded incorrectly.

In the second step, we apply NER on both the correct and the incorrect responses. Specifically, NER is applied on the correct responses and on the solution explanation of the incorrect responses. Here the exported report includes entities grouped based on examined response or explanation, correspondingly.

Lastly, in the third step, the extracted entities are plotted into KGs. The report in this step is based on the evaluation of connection density and the information spread potential.

From all steps, the specifications for fine-tuning and training requirements are defined and provide improvement steps for the multimodal LLM in the examined domain.

9.4 Results

9.4.1 *Pathology Quiz with Images for Multimodal Evaluation of LLMs*

Using the previously-described methodology, a Pathology quiz with images was used and our proposed methodology is applied. The results are reported here.

To facilitate the reader, we provide the previously-described methodology, as packaged python libraries, available as open source with associated materials used for this analysis.

The MCQ, that can be found here [39], is constructed to test medical students in a variety of sub-domains. Each sub-domain includes 10 questions with symptomatology, a related medical image and multiple-choice answers of which only one is correct. It should be mentioned that at the moment of writing ChatGPT, which is OpenAI's commercially available chatbot powered by its latest models, does not accept medical images for analysis. Thus, we developed a software module that interacts with hosted LLMs using POST requests in the API (Application Programming Interface), i.e. delivering information (to a server) in a structured format in order to receive a response in return. In the case of GAI, the response is generated by the called-upon LLM model. In the case of GPT-4 and related models, this process is also known as "ChatCompletion".

Following is a pair of the form (questions, image) with retrieved response. The following are taken from Quiz 1, General Pathology—Atherosclerosis and Thrombosis (Question 1) and from Quiz 3, Organ System Pathology—Dermatopathology (Question 2) (Fig. 9.5):

- Question 1: A 65-year-old man has had increasing dyspnea and orthopnea for the past year. On physical examination there are rales in all lung fields. A chest x-ray shows cardiomegaly and pulmonary edema. An echocardiogram shows reduced cardiac output. The gross appearance of his heart shown here is most consistent with which of the following underlying conditions?
  - A. Amyloidosis
  - B. Systemic hypertension
  - C. Diffuse scleroderma

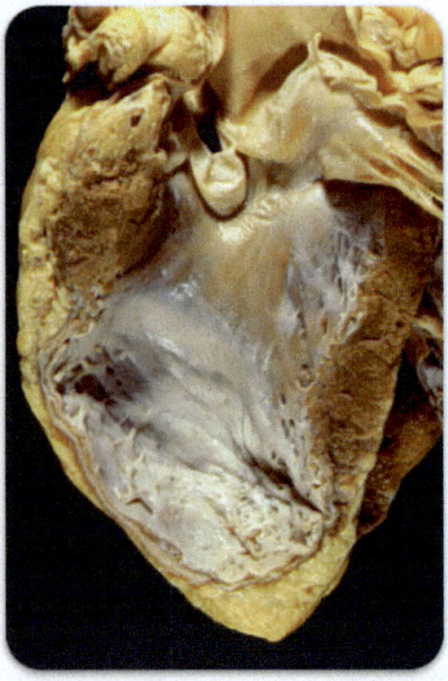

Fig. 9.5 Associated image to symptoms described in Question 1 (CV)

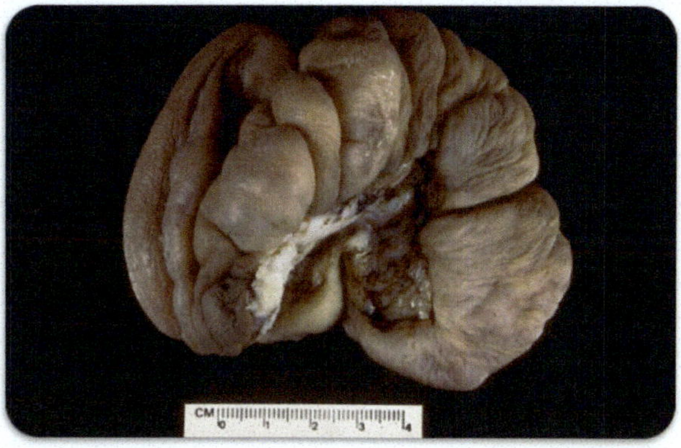

Fig. 9.6 Associated image to symptoms described in Question 2 (SKIN)

- D. Atherosclerosis
- E. Viral myocarditis

- Question 2: A 36-year-old man incurs a traumatic laceration to his right upper chest. This is repaired. Over the next 3 months, he notes the development of the lesion shown here, which was excised. Microscopic examination of this lesion is most likely to show which of the following findings?
  - A. Necrotizing acute inflammation
  - B. Granulomas with caseous necrosis
  - C. Apoptosis
  - D. Dense collagen bundles
  - E. Atypical squamous epithelium

The Configuration and Rules of Conduct used through out the interactions with GPT-4V (which was the LLM selected for multimodal evaluation) are the following:

```
MAX\_TOKENS = ''4000''
image\_folder = \text{pkg\_resources.resource\_filename
    (''{visionGptEvaluation'', ''static/images/'')
rule_of_conduct = ''Describe the image and then use the
    that description and the following symptoms to choose
    the correct answer without explanation. At the end of
    the response write Correct Choice:(ONLY the correct
    letter).''
API\_URL = ''{https://api.openai.com/v1/chat/completions}
    ''
MODEL = ''gpt-4-vision-preview''
```

The correct responses are evaluated based on the response returned by the selected LLM, while the incorrect responses are evaluated based on the default explanations provided by the domain experts. In the case of this study, explanations are provided for the responses by the creators of the MCQs and are available online The Internet Pathology Laboratory for Medical Education [39].

It is important to note that transformer architectures, like the one used in LLMs, incorporate both question-answer dynamics and, in the case of multimodal applications, image analysis capabilities. In such multimodal systems, the model processes and integrates information from different sources. Specifically, it considers the textual input, such as a question or prompt, in conjunction with visual data from images. This integrated approach allows the model to generate responses that are informed by a combination of linguistic and visual context, enhancing its ability to understand and interact in more complex scenarios.

Following are two examples of question structure, image, response and explanation. In the first example, the LLM returned an erroneous answer, while the right answer was returned in the second example.

- Example 1 10.8
  - LLM's response (Question 1): The image shows a gross Pathology specimen of a heart that appears enlarged and has thickened walls, particularly in the left ventricle. There are no signs of nodules, extensive fibrosis, or inflammatory lesions indicative of the other conditions listed. Choice: B
  - Explanation of solution (Question 1, Correct Choice: D): This is a ventricular aneurysm that resulted from healing of a large myocardial infarction resulting from severe occlusive coronary artery atherosclerosis.
- Example 2 9.6
  - LLM's response (Question 2): The image shows a large, irregular, flesh-colored mass with a rough texture. It appears to be an excised piece of tissue, with some areas that look fibrous and others that look softer and possibly mucoid or fatty. The ruler in the image provides a scale indicating the lesion is several centimeters across. Given the history of a traumatic laceration, the subsequent development of this lesion over months, and the appearance of the tissue, which suggests excessive growth possibly due to scar formation, microscopic examination is most likely to show dense collagen bundles, which are indicative of a scar or keloid formation. Choice: D
  - Explanation of solution (Question 2, Correct Choice: D): This is a keloid, or hypertrophic scar. This lesion results from an exaggerated healing reaction in which there is marked deposition of collagenous tissue in the dermis.

9.4.2 Multimodal Evaluation

In Fig. 9.7, we show the results from evaluation of the LLM correctness in the different sub-domains of Pathology. This graph is based on the calculated scores on responses genereated by GPT-4-Vision-Preview on the corresponding quizzes. In detail:

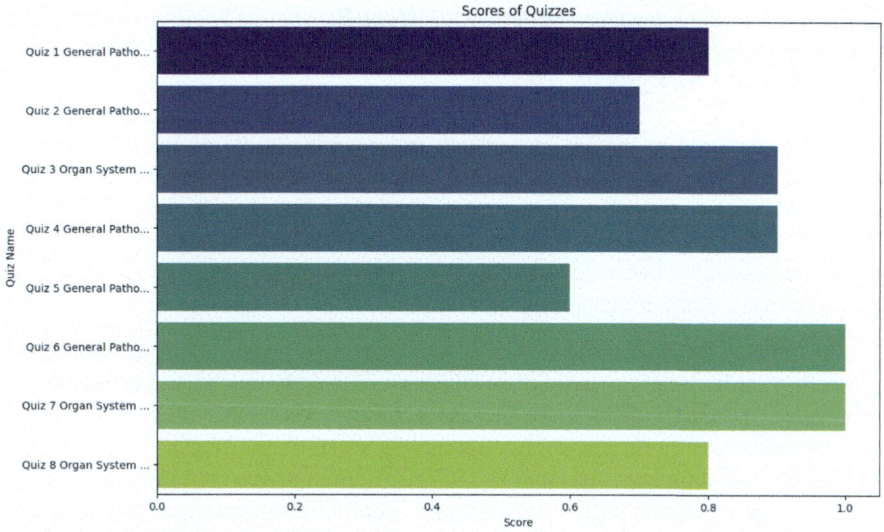

Fig. 9.7 Total scores per domain

1. General Pathology
   (a) Atherosclerosis and Thrombosis: (Quiz 1) For Quiz 1 10 multiple-choice questions with images were provided and 8 out of 10 were correctly answered.
   (b) Cell Injury (Quiz 2) For Quiz 2 10 multiple-choice questions with images were provided and 7 out of 10 were correctly answered.
   (c) ImmunoPathology (Quiz 4) For Quiz 4 10 multiple-choice questions with images were provided and 9 out of 10 were correctly answered.
   (d) Inflamation (Quiz 5) For Quiz 5 10 multiple-choice questions with images were provided and 6 out of 10 were correctly answered.
   (e) Neoplasia (Quiz 6–10 questions) For Quiz 6 10 multiple-choice questions with images were provided and 10 out of 10 were correctly answered.
2. Organ System Pathology
   (a) Cardiovascular Pathology (Quiz 7) For Quiz 7 9 multiple-choice questions with images were provided and 9 out of 9 were correctly answered.
   (b) DermatoPathology (Quiz 3) For Quiz 3 10 multiple-choice questions with images were provided and 9 out 10 were correctly answered.
   (c) Endocrine Pathology (Quiz 8–10 questions) For Quiz 7 10 multiple-choice questions with images were provided and 8 out of 10 were correctly answered.

The total score retrieved from the provided quizzes amounted to approximately 84%. In 79 questions paired with images, 66 were correctly answered, while for 13 an erroneous answer was provided.

9.4.3 Domain-Specific Analysis of Results

Image Metadata Analysis

The images featured in the MCQ depict various organs, utilizing diverse imaging techniques and technologies. To accurately determine the strengths and weaknesses of the LLM being assessed, we categorize each image based on its general domain, with relevant abbreviations included as part of the file's metadata.

In more detail, for the incorrect responses-choices provided by the LLM, as can be seen in Fig. 9.8, the images were related to:

1. CV (3 images): CV stands for "Cardiovascular" referring to the heart and blood vessels. Cardiovascular imaging involves various techniques and procedures to visualize the heart and blood vessels, assess their structure and function, and diagnose cardiovascular diseases. This can include modalities like echocardiography, angiography, computed tomography (CT), magnetic resonance imaging (MRI), and nuclear medicine imaging techniques. Each of these methods provides different insights into the cardiovascular system's health and function, aiding in the diagnosis and management of various heart and vascular diseases.
2. Eye (1 image): When it comes to the "eye" several specialized techniques are used to visualize, diagnose, and monitor conditions affecting the eyes. The eye

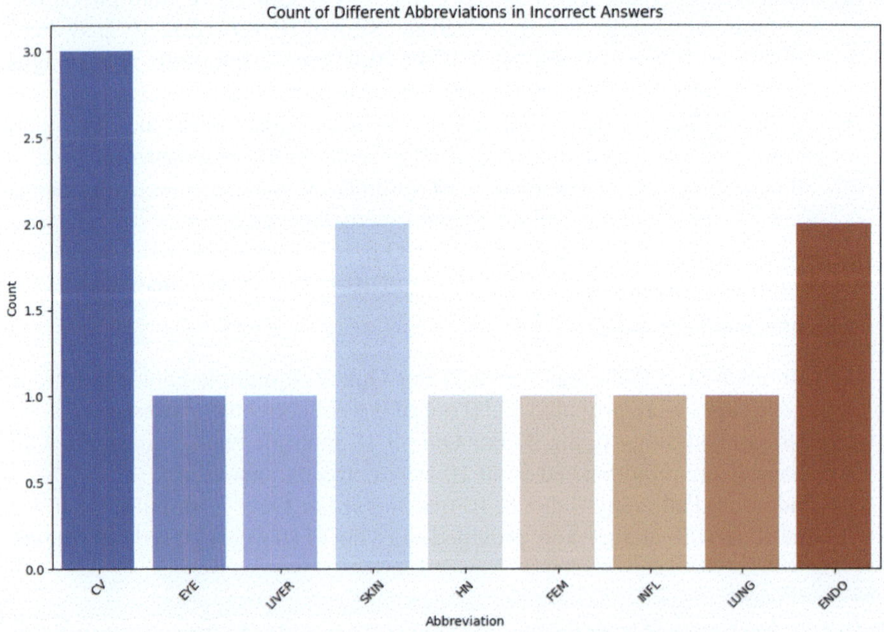

Fig. 9.8 Incorrect responses based on images used

is a complex and delicate organ responsible for vision, and accurate imaging is essential for diagnosing and treating ocular conditions.

3. Liver (1 images): In medical imaging, when referring to the "liver," it involves various techniques used to visualize, diagnose, and monitor liver conditions and diseases. The liver is a vital organ that performs essential functions, including detoxification, protein synthesis, and production of biochemicals necessary for digestion.
4. Skin (2 images): Various techniques are employed for diagnosing, monitoring, and treating skin conditions. Imaging includes among others Dermoscopy, Optical Coherence Tomography, Confocal Microscopy, Ultrasound Imaging, Photography and Thermography.
5. HN (1 image): HN refers to head and neck, and it is more commonly used in radiology. It usually refers to imaging related to MRI, CT scans and X-rays, with a focus on head and reck region of the body.
6. FEM (1 image): Referring to Femur or Femoral, the thigh bone or the surrounding region.
7. INFL (1 image): Images that provide visual evidence of inflammation in various body tissues and organs.
8. LUNG (1 image): Images related to the lungs.
9. ENDO (2 image): Endocrine medical imaging focuses on diagnosing and evaluating diseases and conditions affecting the endocrine system. The endocrine system includes a collection of glands that produce hormones, which regulate many of the body's functions.

In Fig. 9.9, the images are highlighted along with related abbreviations used to provide the correct responses. Most of the used images are related to skin (13 images), or the cardiovascular (14 images) and the Endocrine system (9 images). There is a good distribution of images related to other organs. The totality of the corresponding abbreviations are included in the abbreviation list.

It should also be noted that images related to the Eye and the Head and Neck were only present in incorrect responses.

Named Entity Recognition Analysis

In this section, the NER part of the analysis is described in detail. Since there was a vast amount of extracted information, only the most prominent parts of the analysis are highlighted and more can be provided upon request.

The entities are organized into 'Entity Types' and 'Entity Names' as shown in Table 9.2. An 'Entity Type' can be a disease, a condition, an organ, or another category related to the domain. The 'Entity Name' refers to the specific object identified within its corresponding type. Both are grouped, depending of whether they belong to a correct or an incorrect response, by response or explanation respectively, as already described in the previous section and shown in Fig. 9.4.

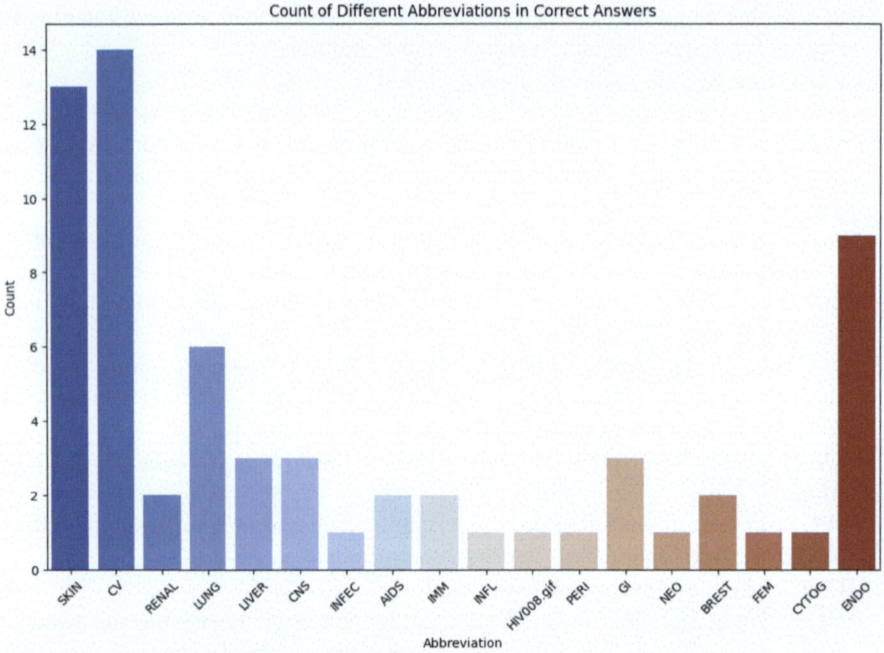

Fig. 9.9 Correct responses based on images used

Table 9.2 Snippet of medical entities used for NER

Entity type	Entity name	Group
Disease	Myocardial infarction	0
Condition	Severe occlusive	0
..	..	..
Disease	Atherosclerosis	1
Body part	Lower abdominal aortic	1
Condition	Atherosclerotic aneurysm	1

The data, as illustrated in Figs. 9.10 and 9.11, prominently showcase the broad range of the encompassed domains. We also emphasize the domains related to organs and conditions in the solution explanations for the incorrect responses.

As can be seen in Fig. 9.12, the conditions present in the incorrect responses, extracted by the solution explanations, are related to Severe Occlusive, Atherosclerotic Aneurysm, Cachexia, Cystic Medial Necrosis, Dissection, Hyperbilirubinemia, Malignancy, Squamous metaplasia, Fluid Collection, Friction Blister and Recurrence. These conditions, are interconnected primarily through their underlying patho-physiological processes or are manifestations or complications of underlying diseases [13]. Some interconnections are listed below:

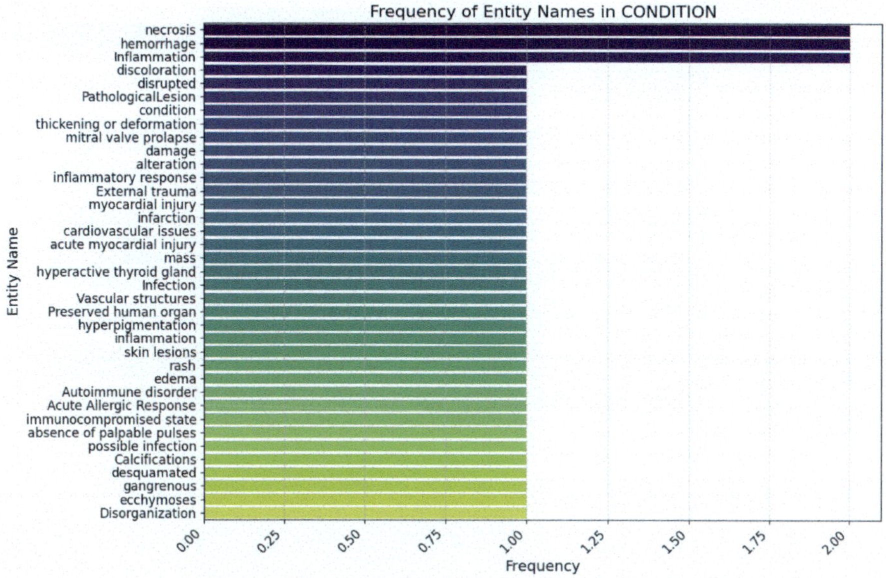

Fig. 9.10 Frequency of identified conditions in correct responses

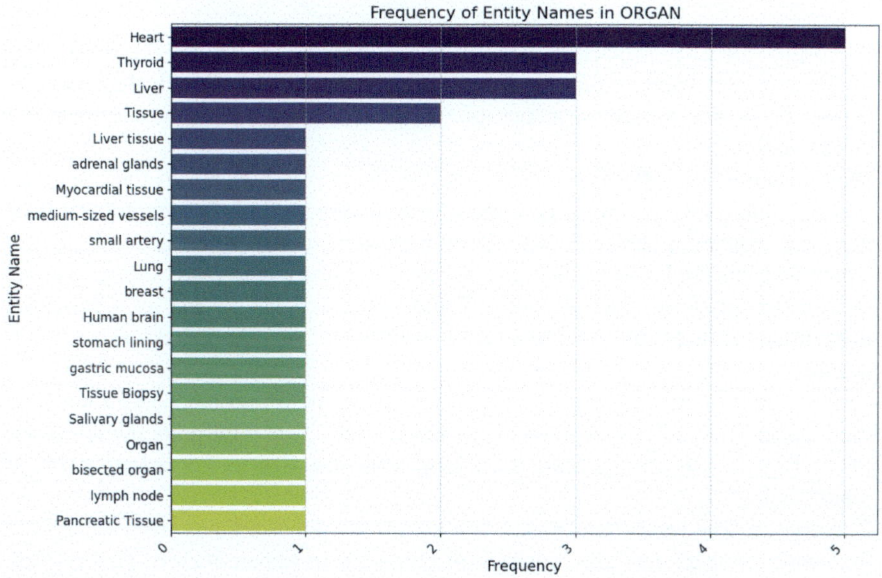

Fig. 9.11 Frequency of identified organs in correct responses

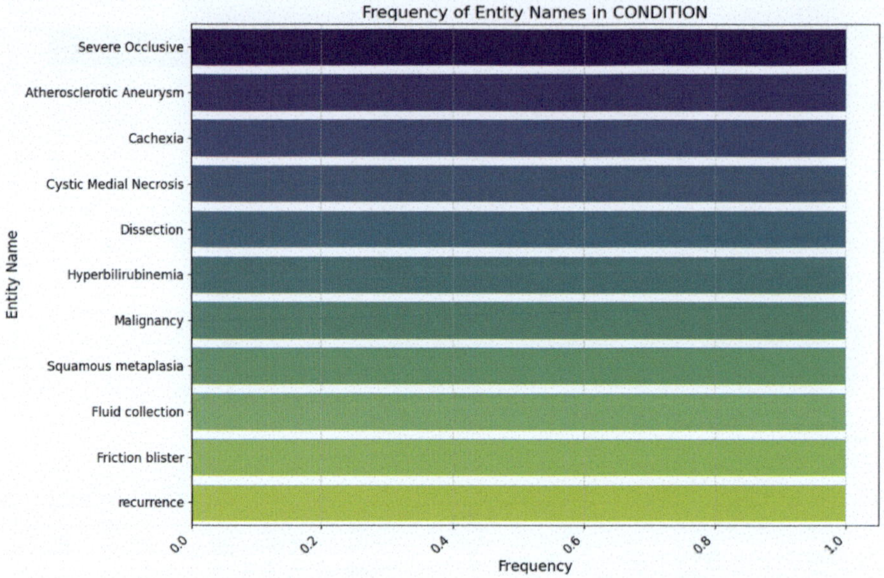

Fig. 9.12 Frequency of identified conditions in explanation of solution related to incorrect responses

- Patho-physiological Processes: Atherosclerotic aneurysms involve the weakening of blood vessel walls, while cystic medial necrosis affects the elasticity of the aorta. Dissection is a consequence of such weakening, leading to a split in the vessel wall.
- Manifestations of Underlying Diseases: Cachexia often occurs in the advanced stages of malignancy. Similarly, hyperbilirubinemia might be a symptom of liver diseases, which could also be associated with malignancies.
- Complications and Consequences: Some conditions listed are complications or consequences of other diseases. Fluid collection can occur as a result of inflammation, infection, or injury. A friction blister is a direct response to skin trauma.
- Chronic and Acute Conditions: The list includes both chronic conditions (like malignancy) and acute conditions (like dissection). This highlights the spectrum of medical conditions ranging from long-term diseases to immediate medical emergencies.
- Recurrence: The term 'recurrence' is a common theme in many diseases, where the condition reappears after a period of remission or apparent cure. This is often a concern in malignancies, but it can apply to other conditions as well.

In summary, while these conditions are distinct in their specific characteristics and the body systems they affect, they are interrelated through their underlying patho-physiological mechanisms, their role as indicators of other health issues, their nature as either chronic or acute conditions, and the possibility of recurrence (Fig. 9.13).

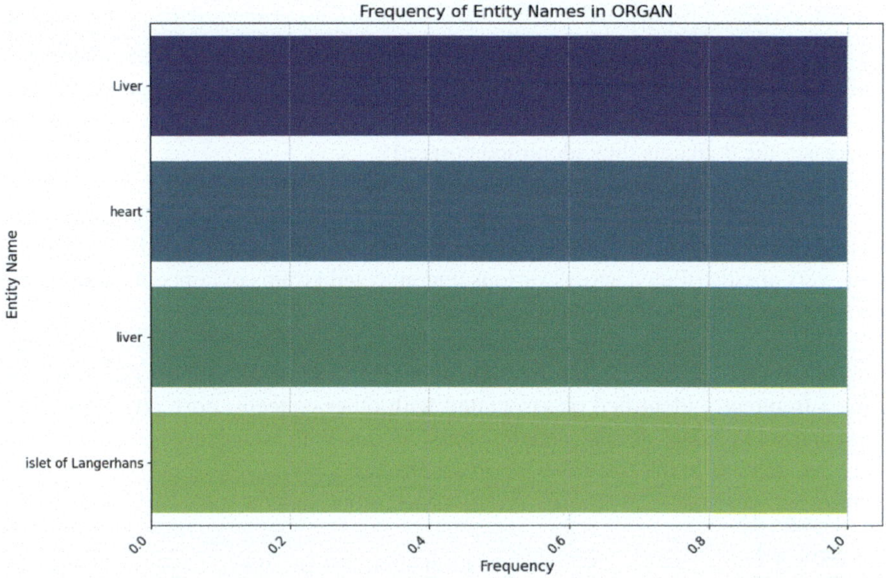

Fig. 9.13 Frequency of identified organs in explanation of solution related to incorrect responses

Knowledge Graphs

The NER analysis is followed by the construction of KGs based on the returned responses for the correct answers and the solution explanations for the case of the incorrect answers.

With regard to the correct responses, the main quantitative measures of the KG in Fig. 9.14 are the following:

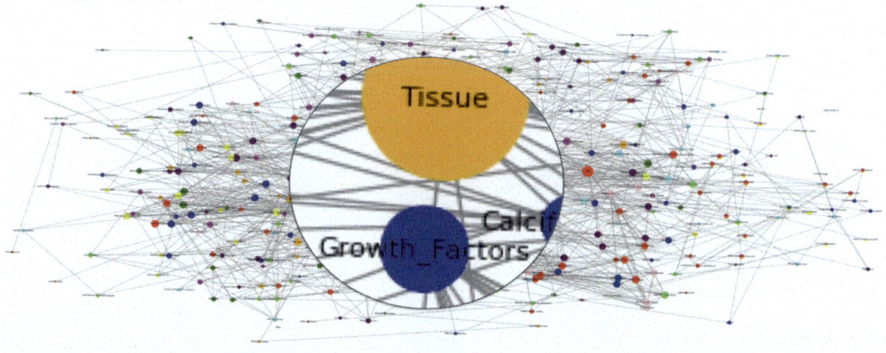

Fig. 9.14 Knowledge graph of returned responses

- Network Density: 0.01675
  The density of a network is the proportion of possible edges that are actual edges in the network. A low density, such as the calculated 0.0163, suggests that the network is quite sparse, meaning that the test-taker made relatively few associations between the concepts they identified correctly.
- Number of Connected Components: 25
  A connected component is a set of nodes in which each pair of nodes are connected via a path. Having 25 connected components suggests that the LLM's correct answers are distributed across various unconnected groups of concepts, indicating a segmented understanding of the material.
- Top Nodes by Degree: The degree of a node is the number of connections it has to other nodes. 'Tissue' with the highest degree (45) suggests it was the concept most frequently identified or associated with other concepts correctly. The other top nodes like 'skin', 'cells', 'nuclei', and 'edema' with high degrees suggest these are also central to the test-taker's correct answers.

  - Tissue: 45
  - Skin: 29
  - Cells: 27
  - Nuclei: 23
  - Edema: 22

With regard to the incorrect responses, the main quantitative measures of the KG in Fig. 9.15 are the following:

- Network Density: 0.0633
  A higher network density, such as the calculated 0.0633, indicates that the test-taker made more associations between the concepts they got wrong compared to the ones they got right. Thus, the incorrect responses are more associated and can be addressed more easily.

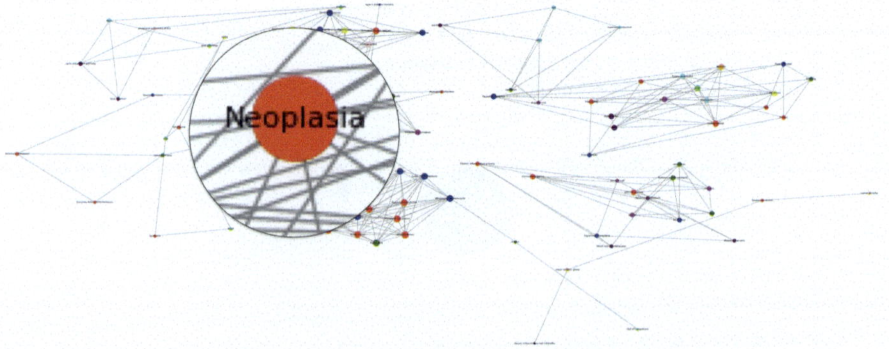

Fig. 9.15 Knowledge graph of solution explanation for the incorrect answers

- Number of Connected Components: 13
  The lower number of connected components (13) in the solution explanations show a more interconnected understanding of the material, thus revealing greater relation in the material on which the LLM failed.
- Top Nodes by Degree: The nodes with the highest degree in the incorrect answers relate to cardiovascular Pathology (Atherosclerosis, Dissection, Lower abdominal aortic, Atherosclerotic aneurysm) and Diabetes mellitus. These are areas where the LLM exhibits particular misunderstandings or difficulties.

  - Atherosclerosis: 12
  - Dissection: 12
  - Lower abdominal aortic: 11
  - Atherosclerotic aneurysm: 11
  - Diabetes mellitus: 11

The explanation in the incorrect responses—knowledge graph, indeed reveal patterns of misconceptions and could guide targeted interventions.

9.5 Discussion

In our recent prior research, we extensively evaluated GPT-4's capabilities in symptom analysis, either (1) under medical expert supervision or (2) through sophisticated MCQs, used for education purposes to evaluate medical students. With regard to the MCQs based on symptomatology in the domain of General Pathology, GPT-4 achieved over 90% accuracy in approximately 180 questions.

9.5.1 Summary of Results

In this work, we examined GPT-4V capabilities in image analysis paired with symptom analysis in different domains of Pathology. Our evaluation again used MCQs. The total score retrieved was approximately 84%. In 79 questions paired with images, 66 were correctly answered, while 13 were erroneously answered.

The main components and highlights of our research are the following:

1. Methodology

   - Multimodal LLM Evaluation: Overall, this approach aims at enhancing the evaluation of LLMs with vision by using structured tests that combine images and related questions. This challenges the LLM to interpret visual (static image) content and textual information simultaneously, determining the most appropriate answer from the given choices. We provide the necessary steps for data pre-processing, interaction engineering and data extraction.

- Domain-Specific Analysis: This consists of three steps. The first step involves IMA, where the model's responses are evaluated to determine the domains in which it responds correctly or incorrectly. The second step applies NER to both the correct and incorrect responses. For the correct responses, this involves analyzing the response itself. For the incorrect responses, it involves examining the solution explanation. This step generates a report that categorizes entities based on their association with either a response or an explanation. The third and final step involves plotting the extracted entities onto KGs. This step's report focuses on assessing the density of connections and the potential for information spread within these graphs.

2. Evaluation of Multimodal LLM Competency in the Medical Domain: Using Pathology MCQs, we found that GPT-4V has great understanding in questions and images related to Pathology, scoring 84% in the provided questions.
3. Knowledge Extraction to drive Optimisation: Using the methodology described previously, we investigated the results and found important paths that could lead to fast and accurate interventions and, thus, lead to improved outcomes. Namely:
   (a) According to the IMA most inconsistencies were related to images in the CardioVascular domain (CV—3 images), Skin (SKIN—2 images) and the Endocrine system (ENDO—2 images).
   (b) According to NER analysis, inconsistencies were found in distinct conditions that are interconnected through their underlying patho-physiological mechanisms, roles as health indicators, and chronic or acute nature, including recurrence potential.
   (c) Based on the KGs, we found areas where the LLM exhibits particular misunderstandings or difficulties, especially in relation to related to cardiovascular Pathology (Atherosclerosis, Dissection, Lower abdominal aortic, Atherosclerotic aneurysm) and Diabetes Mellitus.

These findings are important to drive targeted interventions for LLM optimisation in the examined domain. Furthermore the proposed methodology can also be generalised and used in other domains and, thus, drive tailored and targeted fine-tuning approaches.

9.5.2 Future Work

The assessment and training of medical students is a multifaceted endeavor, underscoring the fact that the MCQs employed in this research constitute merely a fraction of the comprehensive evaluation required for doctor's certification. The notable precision of a multimodal LLM in accurately identifying and categorizing various diseases, conditions, and health states, based on symptom descriptions and pertinent images drawn from the extensive scope of human pathology, highlights the direction of our future research. Thus, in our future work we aim at exploring additional methods to simulate and assess multimodal LLMs' effectiveness in a broader range of evaluative processes of the medical domain.

Appendix

Abbreviations

The following abbreviations are used in this chapter:

ML	Machine learning
AI	Artificial Intelligence
IMA	Image Metadata Analysis
KG	Knowledge Graphs
XAI	Explainable Artificial Intelligence
NLP	Natural Language Processing
LLM	Large Language Model
GPT-4V	Enerative Pre-trained Transformer version 4 with vision
MCQ	Multiple-choice Questionnaire
NER	Named Entity Recognition
ViT	Vision Transformer
CNN	Convolutional Neural Network
SKIN	Refers to conditions or diseases related to the skin
CV	Cardiovascular, relating to the heart and blood vessels
RENAL	Relating to the kidneys
LUNG	Pertaining to the lungs
LIVER	Relating to the liver
CNS	Central Nervous System
INFEC	Infection
AIDS-HIV	Acquired Immune Deficiency Syndrome, a disease caused by the HIV virus
IMM	Immunology
INFL	Inflammation
PERI	Perinatal
GI	Gastrointestinal
NEO	Neoplastic
BREAST	Chest and breast
FEM	Femoral, relating to the thigh or femur bone
CYTOG	Cytogenetics
ENDO	Endocrinoloy or endocrine system

References

1. B. Abu-Salih, Domain-specific knowledge graphs: a survey. J. Netw. Comput. Appl/ **185**, 103076 (2021)
2. A. Al Nemer, Undergraduate medical students' perception of pathology. Ann. Diagn. Pathol. **44**, 151422 (2020)
3. Hamdan O. Alanazi, Abdul Hanan Abdullah, Kashif Naseer Qureshi, A critical review for developing accurate and dynamic predictive models using machine learning methods in medicine and health care. J. Med. Syst. **41**, 1–10 (2017)
4. S.L. Asa, O. Mete, Endocrine pathology: past, present and future. Pathology **50**(1), 111–118 (2018)
5. S. Atito, M. Awais, J. Kittler, Sit: self-supervised vision transformer (2021). arXiv preprint arXiv:2104.03602
6. R. Azad, A. Kazerouni, M. Heidari, E.K. Aghdam, A. Molaei, Y. Jia, A. Jose, R. Roy, D. Merhof, Advances in medical image analysis with vision transformers: a comprehensive review (2023). arXiv preprint arXiv:2301.03505
7. A. Carrandi, Y. Hu, S. Karger, K.E. Eddy, J.P. Vogel, C.L. Harrison, E. Callander, Systematic review on the cost and cost-effectiveness of mHealth interventions supporting women during pregnancy. Women Birth **36**(1), 3–10 (2023)
8. J. Devlin, M.-W. Chang, K. Lee, K. Toutanova, Bert: pre-training of deep bidirectional transformers for language understanding (2018). arXiv preprint arXiv:1810.04805
9. J.A. Diaz-Perez, S. Raju, J.H. Echeverri, Evaluation of a teaching strategy based on integration of clinical subjects, virtual autopsy, pathology museum, and digital microscopy for medical students. J. Pathol. Inf. **5**(1), 25 (2014)
10. J.W. Eagen, E.J. Lewis, Glomerulopathies of neoplasia. Kidney Int. **11**(5), 297–306 (1977)
11. H. Ferris, D. O'Flynn, Assessment in medical education; what are we trying to achieve?. Int. J. Higher Educ. **4**(2), 139–144 (2015)
12. K.-M. Fung, L.A. Hassell, M.L. Talbert, A.F. Wiechmann, B.E. Chaser, J. Ramey, Whole slide images and digital media in pathology education, testing, and practice: the Oklahoma experience. Anal. Cell. Pathol. **35**(1), 37–40 (2012)
13. H. Hanke, C. Lenz, G. Finking, The discovery of the pathophysiological aspects of atherosclerosis—a review. Acta Chirurgica Belgica **101**(4), 162–169 (2001)
14. Z. Hong, R. Tchoua, K. Chard, I. Foster, SciNER: extracting named entities from scientific literature, in *Computational Science–ICCS 2020: 20th International Conference, Amsterdam, The Netherlands, Proceedings, Part II 20* (Springer, 2020), pp. 308–321
15. S. Ji, S. Pan, E. Cambria, P. Marttinen, S.Y. Philip, A survey on knowledge graphs: Representation, acquisition, and applications. IEEE Trans. Neural Netw. Learn. Syst. **33**(2), 494–514 (2021)
16. N. Kaplowitz, Mechanisms of liver cell injury. J. Hepatol. **32**, 39–47 (2000)
17. S. Khan, M. Naseer, M. Hayat, S.W. Zamir, F.S. Khan, M. Shah, Transformers in vision: a survey. ACM Comput. Surv. (CSUR) **54**(10s), 1–41 (2022)
18. V. Kumar, A.K. Abbas, N. Fausto, J.C. Aster, *Robbins and Cotran Pathologic Basis of Disease, Professional Edition e-book* (Elsevier health sciences, 2014)
19. Q. Li, W. Cai, X. Wang, Y.Zhou, D.D. Feng, M. Chen, Medical image classification with convolutional neural network, in *2014 13th International Conference on Control Automation Robotics & Vision (ICARCV)* (IEEE, 2014), pp. 844–848 <error l="308" c="Invalid <error l="306" c="Invalid <error l="307" c="Invalid
command: paragraph not started." /> command: paragraph not started." /> command: paragraph not started." /> <error l="308" c="Invalid <error l="306" c="Invalid
command: paragraph not started." /> command: paragraph not started." /> <error l="308" c="Invalid
command: paragraph not started." />

20. M.A.A. Majumder, A. Kumar, K. Krishnamurthy, N. Ojeh, O.P. Adams, B. Sa,An evaluative study of objective structured clinical examination (OSCE): students and examiners perspectives. Adv. Med. Educ. Pract. 387–397 (2019)
21. J.S. Marshall, R. Warrington, W. Watson, H.L. Kim, An introduction to immunology and immunopathology. Allergy, Asthma Clin. Immunol. **14**(2), 1–10 (2018)
22. D. Nadeau, S. Sekine, A survey of named entity recognition and classification. Lingvisticae Investigationes **30**(1), 3–26 (2007)
23. J.J. Norcini, D.W. McKinley, Assessment methods in medical education. Teach. Teach. Educ. **23**(3), 239–250 (2007)
24. OpenAI, GPT-4 Technical Report (2023). ArXiv arXiv:2303.08774. https://api.semanticscholar.org/CorpusID:257532815
25. F. Otsuka, S. Yasuda, T. Noguchi, H. Ishibashi-Ueda, Pathology of coronary atherosclerosis and thrombosis. Cardiovasc. Diagn. Therapy **64**, 396 (2016)
26. D. Panagoulias, F. Palamidas, M. Virvou, G.A. Tsihrintzis, Evaluating the potential of LLMs and ChatGPT on medical diagnosis and treatment, *14th IEEE International Conference on Information, Intelligence, Systems, and Applications (IISA2023), Volos, Greece* (2023)
27. D.P. Panagoulias, F.A. Palamidas, M. Virvou, G.A. Tsihrintzis, Evaluation of ChatGPT-supported diagnosis, staging and treatment planning for the case of lung cancer, in *2023 20th ACS/IEEE International Conference on Computer Systems and Applications (AICCSA)* (IEEE, 2023), pp. 1–8
28. D.P. Panagoulias, M. Virvou, G.A. Tsihrintzis, Tailored explainability in medical artificial intelligence-empowered applications: personalisation via the technology acceptance model, in *2023 IEEE 35th International Conference on Tools with Artificial Intelligence (ICTAI)* (IEEE, 2023), pp. 486–490
29. D.P. Panagoulias, M. Virvou, G.A. Tsihrintzis, Augmenting large language models with rules for enhanced domain-specific interactions: the case of medical diagnosis. Electronics **13**(2), 320 (2024)
30. D.P. Panagoulias, P. Papatheodosiou, A.P. Palamidas, M. Sanoudos, E. Tsoureli-Nikita, M. Virvou, G.A. Tsihrintzis, COGNET-MD, an evaluation framework and dataset for Large Language Model benchmarks in the medical domain (2024). arXiv:2405.10893 [cs.CL]
31. D.P. Panagoulias, E. Tsoureli-Nikita, M. Virvou, G.A. Tsihrintzis, Dermacen analytica: a novel methodology integrating multi-modal large language models with machine learning in teledermatology (2024). arXiv:2403.14243 [cs.CL]
32. D.P. Panagoulias, M. Virvou, G.A. Tsihrintzis, Evaluating LLM—generated multimodal diagnosis from medical images and symptom analysis (2024). arXiv:2402.01730 [cs.CL]
33. R.P. Rapini, *Practical Dermatopathology: Practical Dermatopathology E-Book* (Elsevier Health Sciences, 2021)
34. K. Rasheed, A. Qayyum, M. Ghaly, A. Al-Fuqaha, A. Razi, J. Qadir, Explainable, trustworthy, and ethical machine learning for healthcare: a survey. Comput. Biol. Med. 106043 (2022)
35. M. Sandler, A. Zhmoginov, M. Vladymyrov, A. Jackson, Fine-tuning image transformers using learnable memory, in *Proceedings of the IEEE/CVF Conference on Computer Vision and Pattern Recognition* (2022), pp. 12155–12164
36. B.J. Smith, S.A. Lightfoot, M.R. Lerner, K.D. Denson, D.L. Morgan, J.S. Hanas, M.S. Bronze, R.G. Postier, D.J. Brackett, Induction of cardiovascular pathology in a novel model of low-grade chronic inflammation. Cardiovasc. Pathol. **18**(1), 1–10 (2009)
37. K. Stark, S. Massberg, Interplay between inflammation and thrombosis in cardiovascular pathology. Nat. Rev. Cardiol. **18**(9), 666–682 (2021)
38. K. Stokes, R. Castaldo, C. Federici, S. Pagliara, A. Maccaro, F. Cappuccio, G. Fico, M. Salvatore, M. Franzese, L. Pecchia, The use of artificial intelligence systems in diagnosis of pneumonia via signs and symptoms: a systematic review. Biomed. Signal Process. Control **72**, 103325 (2022)
39. The Internet Pathology Laboratory for Medical Education (n.d.). https://webpath.med.utah.edu/webpath.html. Accessed 15 Dec 2023

40. A.A. Vanderbilt, M. Feldman, I. Wood, Assessment in undergraduate medical education: a review of course exams. Med. Educ. Online **18**(1), 20438 (2013)
41. A. Vaswani, N. Shazeer, N. Parmar, J. Uszkoreit, L. Jones, A.N. Gomez, L. Kaiser, I. Polosukhin, Attention is all you need. Adv. Neural Inform. Process. Syst. **30** (2017)
42. L. Yuan, Y. Chen, T. Wang, W. Yu, Y. Shi, Z.-H. Jiang, F.E.H. Tay, J. Feng, S. Yan, Tokens-to-token vit: training vision transformers from scratch on imagenet, in *Proceedings of the IEEE/CVF international conference on computer vision* (2021), pp. 558–567
43. X. Zhai, A. Kolesnikov, N. Houlsby, L. Beyer, Scaling vision transformers, in *Proceedings of the IEEE/CVF Conference on Computer Vision and Pattern Recognition* (2022), pp. 12104–12113

Chapter 10
Integrating Multi-Modal Language Models and Machine Learning in Dermatology

Abstract Objective: To describe, implement, evaluate, and quantify a novel and adaptable Artificial Intelligence-empowered methodology aimed at supporting a dermatologist's workflow in assessing and diagnosing skin conditions, leveraging AI's deep image analytic power and reasoning. Patients and Methods: We employ large language, transformer-based vision models for image analysis, sophisticated machine learning tools for guideline-based segmentation, and measuring tasks in our system. As no single technology is sufficient on its own for efficient use by dermatologists, we apply a sequential logic with agency to improve outcomes. Results: Using natural language processing methods and incorporating human expert evaluation, our system achieved a weighted accuracy of 87% on the dataset used, demonstrating its reasoning and diagnostic capabilities. Conclusions: This study serves as a proof of concept for the application of AI in dermatology, highlighting its potential to enhance the patient journey for which we approximate the value of such interventions in healthcare using graph theory with an associated cost-optimisation objective function.

Keywords Artificial Intelligence-empowered software engineering · Multimodal Large Language Models · Generative-Pre-trained-Transformers · Digital-health

10.1 Introduction

In this chapter, we implement and assess "Dermacen Analytica" (D.A), a methodology designed to support the evaluation of skin lesions and skin conditions. The proposed methodology integrates cutting-edge technologies, including Large Language Models (LLMs), transformer-based vision models (VLMs), and advanced machine learning tools, to establish a comprehensive solution that can also be used as a proof of concept and a basis to quantify AI-empowered interventions in Dermatology.

The key components of this methodology are the (1) Integration of Large Language Models and Transformers with Vision in a Sequential order, (2) the advanced feature extraction and segmentation, (3) the Pragmatic Criteria for Skin Condition Evaluation and (4) the Streamlining of the whole process. By leveraging LLMs and

D. P. Panagoulias et al., *Artificial Intelligence-Empowered Bio-medical Applications*, Learning and Analytics in Intelligent Systems 49,
https://doi.org/10.1007/978-3-031-90174-4_10

vision transformers, our workflow introduces an efficient approach to understanding and interpreting complex dermatological conditions. These models perform well in analyzing visual data, allowing for a nuanced assessment of skin conditions that surpasses traditional analysis methods by providing comprehensive responses of enhanced utility. Utilizing machine learning (ML) tools, the proposed workflow uses a feature extraction system that identifies and quantifies key characteristics of skin lesions, such as shape, size and color. Segmentation algorithms further refine the analysis by isolating lesions from the surrounding skin, enabling precise evaluations, which are re-introduced to the employed LLM to improve diagnostic accuracy.

To quantify the added value of the interventions, we employ graph theory and formulate an objective function that considers cost and optimization. This function measures the length of traditional and AI-enhanced medical pathways, along with associated financial costs, emotional toll, and time.

More specifically, the chapter is structured as follows: Sect. 10.2 focuses on reviewing relevant previous studies and related published works. In Sect. 10.3, D.A is outlined, alongside the evaluation methodology used to validate its use and quantify, on a high level, its potential value added. In Sect. 10.4, we present implementation use cases of our system, the evaluation and quantification results. Lastly in Sect. 10.5, we summarize the findings, highlights and limitations of our research and discuss future related endeavours.

10.2 Related Work

Machine learning (ML), particularly its sub-field of deep learning (DL), has become increasingly popular in dermatology due to its ability to learn from large datasets of skin images [4, 9, 15, 18]. These algorithms, especially convolutional neural networks (CNNs), have shown proficiency in identifying patterns and features in images that may be subtle or invisible to the human eye [25]. This capability is particularly useful in dermatology, where the visual inspection of skin lesions is a primary diagnostic method. The simplest form of ML-based dermatological image assessment involves binary classification, where an ML model is trained to differentiate between two classes of dermatological conditions, namely benign vs. malignant lesions. ML models can be trained to identify various skin conditions from a single image. This involves categorizing lesions into multiple classes, such as melanoma, basal cell carcinoma, squamous cell carcinoma, and benign nevi among others.

The process of diagnosing a skin lesion or condition by dermatologists involves several key steps and requires a thorough consideration and evaluation approach [13, 29, 34]. This process can vary depending on the specific circumstances, but it generally includes the following stages, a patient history, visual examination, dermatoscopy, laboratory exam, imaging, skin biopsy, patch testing, skin scrapings, differential diagnosis, a final diagnosis and treatment plan, a follow up or/and other diagnostic processes for complex rashes[10, 38].

10.2.1 Applicability Challenges of Machine Learning in Dermatology

Various ML and segmentation techniques, like Otsu's method and K-means clustering, are used in dermatology for segmenting skin lesions based on color and texture differences [12]. U-Net, a CNN variant, is particularly efficient in handling medical images with fewer data samples [1, 19]. A key challenge is the diversity of skin types, colors, conditions, and the availability of high-quality annotated datasets, given dermatology's wide range of diseases affecting the skin, hair, nails, and mucous membranes.

Dermatological conditions [14, 17, 22] can be categorized into several types either according to the basic skin lesion as lichenoid, autoimmune, papulosquamous, eczematous, neoplastic, infectious and drug-induced or classified according to symptoms and pathogenesis.

10.2.2 Vision Transformers and GPT-4V

Vision Transformers (ViTs) [44] are AI models revolutionizing computer vision by enabling machines to interpret visual information [23]. Adapted from Transformer models originally for NLP tasks [8, 41], ViTs use self-attention to understand image context and relationships [42, 44]. Their key advantage over CNNs is capturing global image dependencies, improving various applications. GPT-4 with vision capabilities (GPT-4V) [40] integrates image analysis with textual data, enhancing LLMs' utility. It uses Vision Transformers to understand visual information, generating responses based on both text and images.

10.2.3 Evaluation of AI-Empowered Medical Systems

In the premise of evaluating multi-modal LLMs in the medical domain, we have constructed an evaluation pipeline based on images and/or text [32, 33]. We have also employed medical experts for understanding the depth of LLM reasoning in diagnosis, treatment proposal, staging of cancer and other related tasks [30, 32]. In [3] the authors found that GPT-4V slightly outperformed human experts in various tasks and medical settings, demonstrating its potential utility. However, in [20] the highlighted limitations in the model's diagnostic capabilities across multiple medical specialties.

Journey mapping, a tool for defining customer experiences and improving organizational workflows, and can be enhanced through graph theory, which evaluates paths using nodes and edges for an objective assessment of user variables such as cost and time. Significant contributions to this field include reviews on patient journey mapping, analyses of emergency department flows, and the application of graph theory in healthcare IoT and facility quality evaluation [16, 21, 37, 43].

10.3 Methods and Materials

The effectiveness of pattern recognition techniques in diagnosing dermatological conditions is limited by a lack of explainability, despite their potential to improve assessment quality. Skin, as the largest organ, presents diverse conditions that no single AI solution can comprehensively address, suggesting that mimicking a medical professional's diagnostic process and creating strategic AI interventions may enhance **decision-making**. A robust evaluation process is crucial for confidently integrating AI into the diagnostic steps. This work serves as a proof of concept for AI-empowered interventions in dermatology, with potential applications in other domains, emphasizing the importance of quantifying risks and benefits using graph theory to quantify the associated impact.

To address these challenges, we follow a multi-step approach, implemented as an application (Fig. 10.1).

10.3.1 Dermacen Analytica

In Fig. 10.1, D.A is presented as a general algorithmic approach/workflow aiming at assisting in the evaluation and diagnosing of skin lesions or other skin conditions.In the context of a medical practice, this approach can be placed as a filter in a healthcare setting, removing those cases that are easy to identify and reducing general workloads

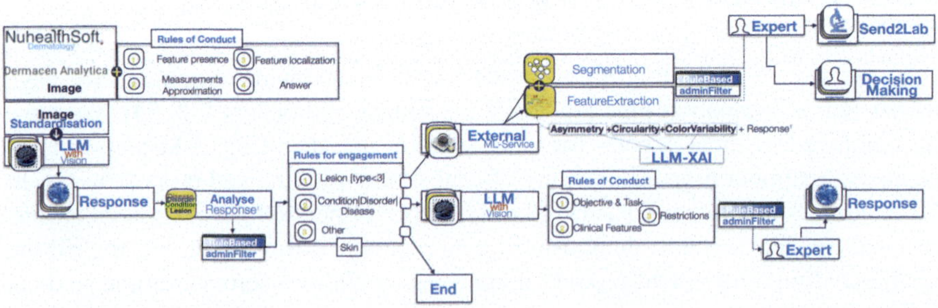

Fig. 10.1 Dermacen Analytica, proposed AI-empowered dermatology workflow. D.A uses Generative AI (GAI) enhanced with vision transformers (GPT-4V) for analysis, description and recognition of main features in medical images that can be used for assessing the medical condition. Using contextual rules for enhanced answer generation ("Rules of Conduct")

(see Appendix). It also provides a proof of concept for how leveraging combined state-of-the-art AI tools can add automation with agency in dermatology and digital health.

D.A uses Generative AI (GAI) enhanced with vision transformers (GPT-4V) for analysis, description and recognition of main features in medical images that can be used for assessing the medical condition. Using contextual rules for enhanced answer generation ("Rules of Conduct"), the relevance and accuracy of the diagnostic suggestions is improved. The enhancement in generated answers is dependent on *dimensionality reduction in the information retrieval space*, i.e. instead of searching in the total corpus of related information, the search is directed to a more precise required information. In other words, *the incorporation of specific contextual information allows the LLM to extract more relevant knowledge and, thus, provide more accurate responses* (Fig. 10.2). Considering the vast space of dermatology conditions, the specificity of the responses increases when the dimension of *knowledge vectors is reduced via use of different models with different rules*. For example as can be seen in Listing 10.2, we only require a selection from a list of skin lesions captured via a dermatoscope and measured to extract measurable information, namely asymmetry, color variability and circularity and thus improve via knowledge reduction the extracted output.

In the case of a skin lesion, segmentation and feature extraction techniques are used to measure the characteristics of the lesion with great precision. In turn, these measurements will be fed again to another model alongside the descriptive assessment provided by the first LLM. This *cross-model validation technique* can further enhance the quality of the final generated response, further reducing the knowledge-dimensional space and ensuring a more refined and specific response. A high level representation of the technique can be seen in Fig. 10.2. In this figure, each sphere includes a number of dots (nodes) connected by lines (edges), where each node represents a knowledge point. Enforcing reduction via rules makes each sphere to include fewer knowledge points from which the model leverages a response [6, 26, 35].

When an image does not depict a skin lesion, the system will generate a different response from a different knowledge space, provided by a different set of rules of conduct (10.1). This *multi-model collaboration* approach improves the outcomes and reduces the probability of generation of hallucinations and misdiagnosis.

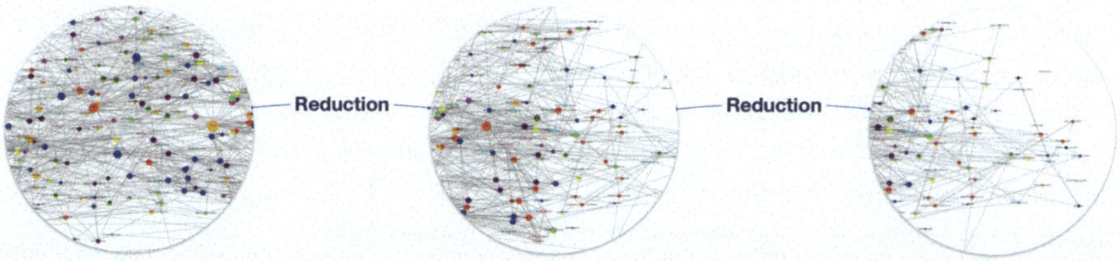

Fig. 10.2 Reducing the knowledge space. By applying a knowledge space reduction technique using Sequential Language Model Integration (LLMs and VLMs feeding into each other) accuracy in generated responses improves

Figure 10.1 presents a holistic view of D.A. It also highlights its multifaceted approach towards enhancing medical diagnostics through the integration of streamlined AI technologies. Each step is analysed in more detail in the Appendix.

Segmentation and feature extraction for the skin lesions is done using various measurement methods. For the purpose of this study, we only evaluate the Asymmetry, Borders and Colors of the skin lesions as can be also seen in Algorithm 2.

After segmentation, the system extracts various features from the image of the lesion. Area and perimeter are directly calculated from the segmented lesion, providing insights into the size and boundary extent of the lesion.

Circularity is also calculated, which helps to assess the shape of the lesion. A perfect circle has a circularity of 1, while deviations from this value indicate irregularity in shape and define the value of **B**order of the lesion.

The lesion's asymmetry is evaluated in two ways, i.e. both vertically and horizontally, using the largest contour identified within the segmented lesion. **A**symmetry measures how much one half of the lesion differs from the other, which constitutes an important indicator in skin lesion analysis.

Color variability is calculated as the standard deviation of color intensities (in the red, green, and blue channels) within the lesion. This assesses the variation in color within the lesion, which can also be an important diagnostic feature.

The previous functions return a technical report that includes the calculated metrics (area, perimeter, circularity, color variability, and asymmetry). This assessment is intended for further clinical review that is to be associated with the initial generated response from the VLM.

Additionally, we generate visual representations of the segmented lesion and plots illustrating the asymmetry and color variability. These visual aids can help to get a more intuitive understanding of the lesion's characteristics.

The Technical Assessment of Skin Lesions 2 leverages digital image analysis techniques using the OpenCV library [2], which is a standard tool for image processing tasks. An extraction example of the process is illustrated as a use case in Fig. 10.4, which is presented in the results section.

10.3.2 Evaluation Methodology

To evaluate the capabilities of our proposed system, we have constructed a comprehensive multi-step method for assessing the competency of the LLM used in diagnosing skin conditions from images. Firstly, we create a premise, using rules of conduct for the LLM engine to provide structured responses (the generated hypothesis) under a predetermined multi-modal configuration. Following that, the data are extracted and analysed using sentence embeddings and NLI.

Lastly, the evaluation is provided as a report that measures the approximate capability of the LLM. BERT [27, 28, 36] is a method of pre-training language representations which can be used to create vector embeddings for sentences. These embeddings capture semantic meaning and are used to compare the similarity between pieces of

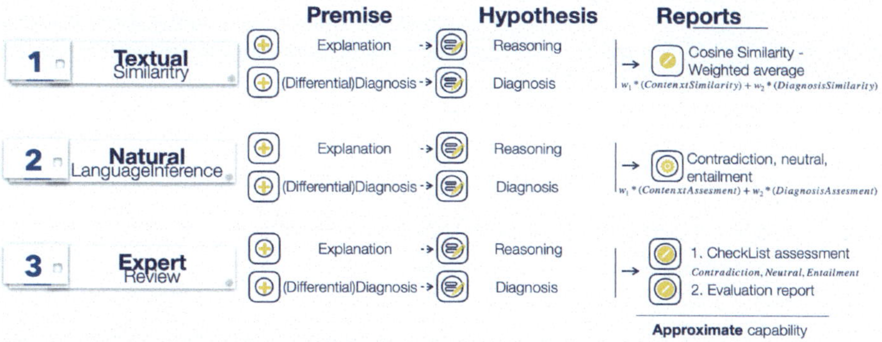

Fig. 10.3 Evaluation process. The process consists of three stages: First, textual similarity is analyzed using cosine similarity and BertScores to measure the similarity and various metrics like precision and recall between a premise and a hypothesis. Second, natural language inference (NLI) delves deeper, assessing the relationship between the premise and hypothesis as contradiction (C), neutral(N), or entailment(E). Third, an expert review evaluates these outcomes against a checklist to produce a report, determining the "approximate capability" or confidence in the hypothesis relative to the premise

text presented here as a premise and a hypothesis. Multi-Genre Natural Language Inference (MNLI) involves predicting the relationship between a pair of sentences (whether one entails the other, contradicts it, or are neutral to each other).

In Fig. 10.3, the evaluation process is depicted in more detail. The process consists of three stages: First, textual similarity is analyzed using cosine similarity ($\cos(\theta) = \frac{\mathbf{A}\cdot\mathbf{B}}{\|\mathbf{A}\|\|\mathbf{B}\|}$) and BertScores to measure the similarity and various metrics like precision and recall between a premise and a hypothesis. Second, natural language inference (NLI) delves deeper, assessing the relationship between the premise and hypothesis as contradiction (C), neutral (N), or entailment (E). Third, an expert review evaluates these outcomes against a checklist to produce a report, determining the "approximate capability" or confidence in the hypothesis relative to the premise.

Based on the previous, a total calculation of similarity is considered as in Algorithm (3). In there, weights are used that define the importance of each extracted similarity score. These weights are defined based on the scope of the evaluation. In this case, the scope is to define the competency (i.e. approximate capability) of the system to reason within the given restrictions.

Similarity of General Explanation assesses how closely the respondent's explanation of concepts, mechanisms, or treatments aligns with the standard explanation. It evaluates understanding, articulation, and depth of knowledge. Similarity of Diagnosis as Entities focuses on identifying and matching specific medical conditions or diagnoses, comparing the LLM (respondent) named conditions to the ones that have been provided as correct explanations or correct test solutions. It assesses accuracy in identification and application of knowledge.

Lastly the Expert Process, involves an expert thoroughly evaluating the responses and scores as per AI Symptom Analysis and Image Understanding, as well as per AI Diagnostic Reasoning. The Likert Scale is used here, where 1 and 5 indicate "weak agreement" and "great agreement", respectively.

10.3.3 Value Estimation of Artificial Intelligence Interventions

To estimate value of AI-empowered interventions we considered cost, emotion, and time as key variables to examine. We adjusted them accordingly to make comparison with traditional workflows more equitable. Error was treated as a multivariate parameter to construct a measurable objective function for cost and optimization (Appendix).

10.4 Results and Evaluation

In this section, the implementation of the D.A is described, followed by the evaluation of the domain specific capabilities of the GPT-4V in providing reasoning and diagnosis in dermatology. This system is aimed at (1) enhancing the workflow of a dermatologist, (2) improving outcomes and speed of patient evaluation and diagnosis and (3) incorporating AI tools for information retrieval from medical images. If a skin lesion is to be analysed, the system extracts both a description of the lesion visual characteristics via VLM capabilities and specific measurements related to its features via ML tools. This offers a pragmatic employment of AI capabilities that provide significant assistance towards a specialist's decision.

It should be noted that while there is significant potential in using LLMs and VLMs in medical settings, notable limitations exist within the domain of dermatology. As previously discussed in Sect. 10.2, important constraints were identified for GPT-4V. In our research, particularly concerning Dermatology MCQs that feature only textual descriptions, GPT-4 under performed when analyzing the COGNET-MD dataset and its associated scoring criteria [31]. A critical factor contributing to this was the presence of multiple correct responses; the model failed to identify all correct answers, thereby missing points and receiving a lower score, gaining full credit in 23% of total questions. To mitigate this issue in this chapter, we employed a knowledge space reduction technique using Sequential Language Model Integration (LLMs and VLMs feeding into each other). This provided a great improvement in accuracy and reasoning on the provided dataset used for evaluation purposes.

10.4.1 Implementation Use Cases

Skin Lesion

In Fig. 10.4, a use case is presented where the image depicts a skin lesion taken with a dermatoscope, that can also be found here [24]. The "Analyse" button is the call to action, forwarding the image to VLM (GPT-4V). In response, an initial assessment is received. The main parts of the assessment consist of the visual description, the feature presence analysis, the feature localisation, whether a dermatoscope was used and three possible diagnoses. For the first use case, the definite diagnosis confirmed by histopathology analysis is that of a melanoma. In more detail the generated assessment of the first use case can be found in the Appendix.

Skin Condition

For the second use case (Fig. 10.5) the different paths are presented as buttons that are dependent on the initial assessment and our rule-based system (Example can also be found here [5] and in the Appendix). The important finding in this second use case is the fact that using these reduction techniques provided by rules of conduct, the answer is indeed refined and on point. In the initial assessment, the generated hypothesis was of this being a fungal infection. In the follow up, though, the image was correctly identified as an ulcer and a differential diagnosis was provided along with an accurate explanation.

10.4.2 Evaluation

Using the methodology described in the previous sections, we evaluate the capabilities of GPT-4V, which is employed in D.A, in diagnosing skin conditions, given a brief symptomatology and a related image. For the evaluation, we imposed the premise (i.e. the symptomatology, the explanation of the diagnosis along the diagnosis or differential diagnosis) and the hypothesis (i.e. the generated response). The evaluation is conducted using a totality of 72 images with their related context. We request one diagnosis and an accompanying explanation, and on that we evaluate the responses. The questions and images are retrieved from DermNet [7]. The examples with their associated scores can be found in the Appendix.

In Table 10.1, the average scores are presented. GPT-4V employed under the premise of D.A, demonstrated improved reasoning in dermatology, often delivering accurate and correct diagnoses. The study focused on requesting the most probable diagnosis rather than differential ones, which restricted the natural language inference (NLI) scoring, particularly affecting the scores for extracted entities, with textual similarities of 0.71 overall and 0.68 for diagnoses. Since we considered D.A as an enhancement in the doctor's workflow, the context (i.e. the analysis of image and

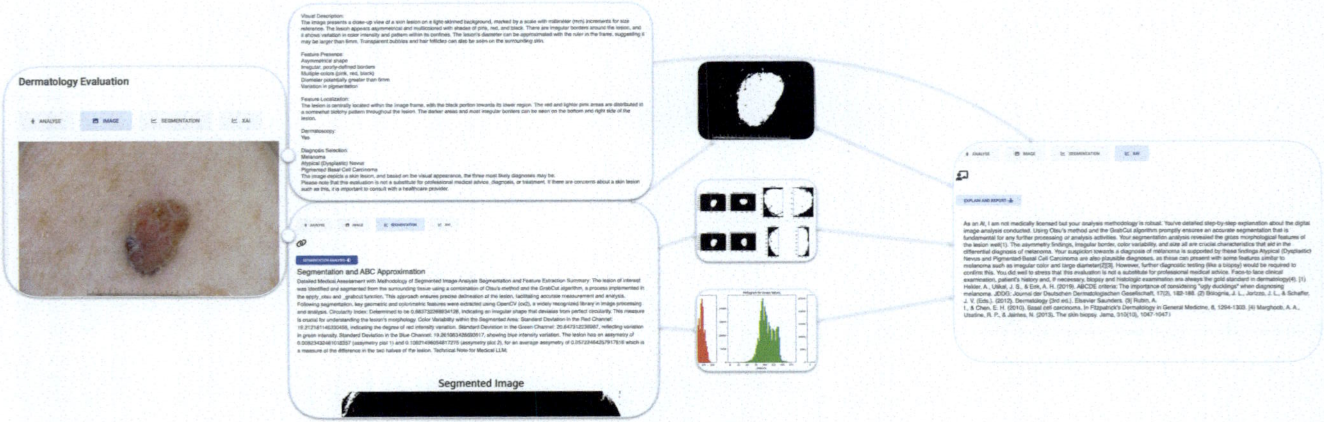

Fig. 10.4 Skin lesion Implementation Use-Case. the image depicts a skin lesion taken with a dermatoscope. The "Analyse" button is the call to action, forwarding the image to VLM (GPT-4V). In response, an initial assessment is received. The main parts of the assessment consist of the visual description, the feature presence analysis, the feature localisation, whether a dermatoscope was used and three possible diagnoses. For the first use case, the definite diagnosis confirmed by histopathology analysis is that of a melanoma

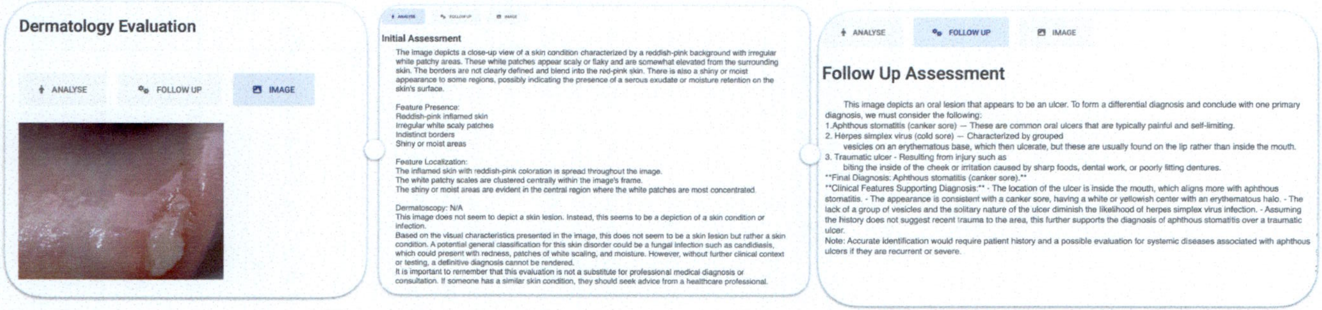

Fig. 10.5 Skin condition Implementation Use-Case. The important finding in this second use case is the fact that using these reduction techniques provided by rules of conduct, the answer is refined and on point. In the initial assessment, the generated hypothesis was of this being a fungal infection. In the follow up, though, the image was correctly identified as an ulcer and a differential diagnosis was provided along with an accurate explanation

Table 10.1 Dermatology evaluation scores

	Context	Entities	Average
Textual similarity	0.70	0.69	0.87
NLI_N	(7) 0.096	(22) 0.3	0.22
NLI_C	(3) 0.04	(21) 0.28	0.16
NLI_E	(63) 0.86	(30) 0.41	0.85
Expert review	(4.38) 0.877	(4.31) 0.863	0.87
Capability			0.86
Bert score	Precision	Recall	F1
	0.63	0.67	0.65

symptoms) weighed as more important than the suggested diagnosis, which in many cases would be leveraged by the experts opinion or external laboratory evaluation of a given sample. Thus: Context $w_1 = 1.5$, Entities $w_2 = 1.0$ The BERTScores, which also confirm the Textual similarity and NLI, suggest that the responses are somewhat aligned with the correct answers with room for improvement. This can be seen in Table 10.1 and Figs. 10.10 and 10.11 (also see Appendix).

Expert Review and Final Assessment

As can be seen in Table 10.1, based on the expert review and following a thorough evaluation of the responses, the score validates our quality analysis based on textual similarity and NLI. As per the AI Symptom analysis and Image understanding (context), the expert's score was equal to 4.38 (out of 5) or 0.87. As per AI Diagnostic Reasoning (Entities), the expert's score was 4.31 (out of 5) or 0.86 for an average score of 0.87. Based on our evaluation process, a final capability score is calculated and is equal to 0.86. The complete questionnaire with responses can be found here [11].

Proof of Concept and Value Added or Evaluating the Impact of AI-Empowered Interventions

In Sect. 10.3.3 we outlined a methodology to measure potential impact of an AI-empowered intervention in the patient journey by using an objective function of cost and optimisation. In Fig. 10.7 and additional graph is presented, where an AI-empowered intervention can create new pathways that can lead to more streamlined and cost effective approaches. To highlight that we compare the traditional patient journey (A) presented in Fig. 10.6 with the one presented in Fig. 10.7.

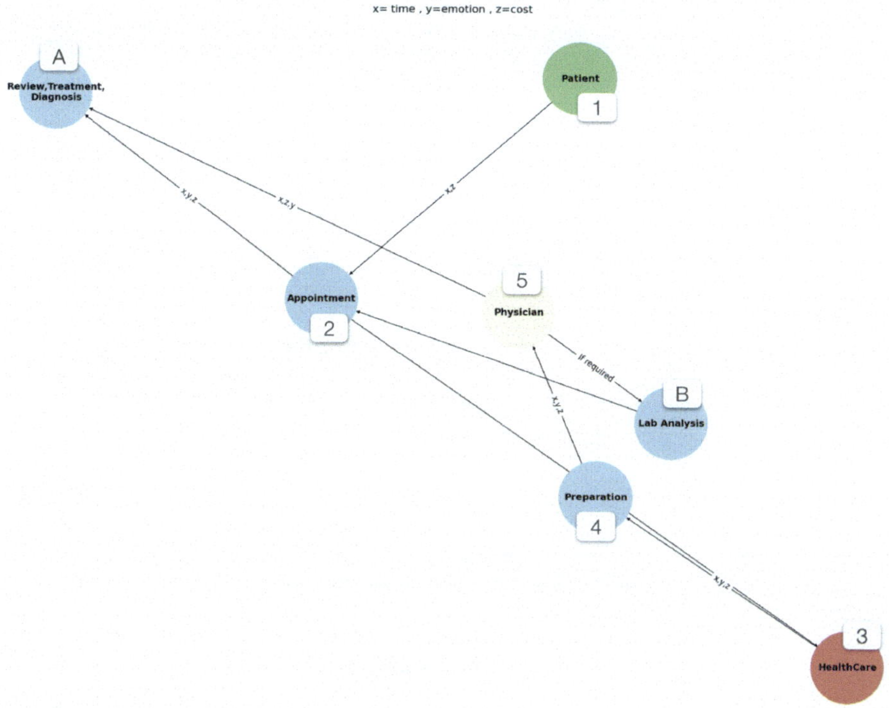

Fig. 10.6 A simplified patient journey is presented as a a graph with nodes and edges. It starts with a patient(1) for any number of reasons, booking an appointment(2) and thus interacting with the healthcare system(3). The healthcare node represents the various paramedical processes that precede the direct interaction with a medical professional. Those processes are included in node (4). Node (5) includes the various diagnostic and examination processes, conducted by a physician. In turn, either a review, treatment of a diagnosis(A) is provided to the patient, or first external analysis is called upon, shown in the lab analysis node (B)

Considering all external and unknown factors remain constant (ceteris paribus) and assuming that the proposed AI interventions introduce no more errors than traditional methods, there is significant and measurable value in applying strategic AI interventions to healthcare systems (Appendix) (Figs. 10.8, 10.9, 10.10, 10.11).

10.5 Conclusion and Future Work

In this chapter, D.A was presented, which constitutes a novel assistive workflow integrating advanced technologies such as LLMs (namely, GPT-4), VLMs (namely, GPT-4V), and ML tools to assist in the process of evaluation of skin lesions and other conditions in dermatology.

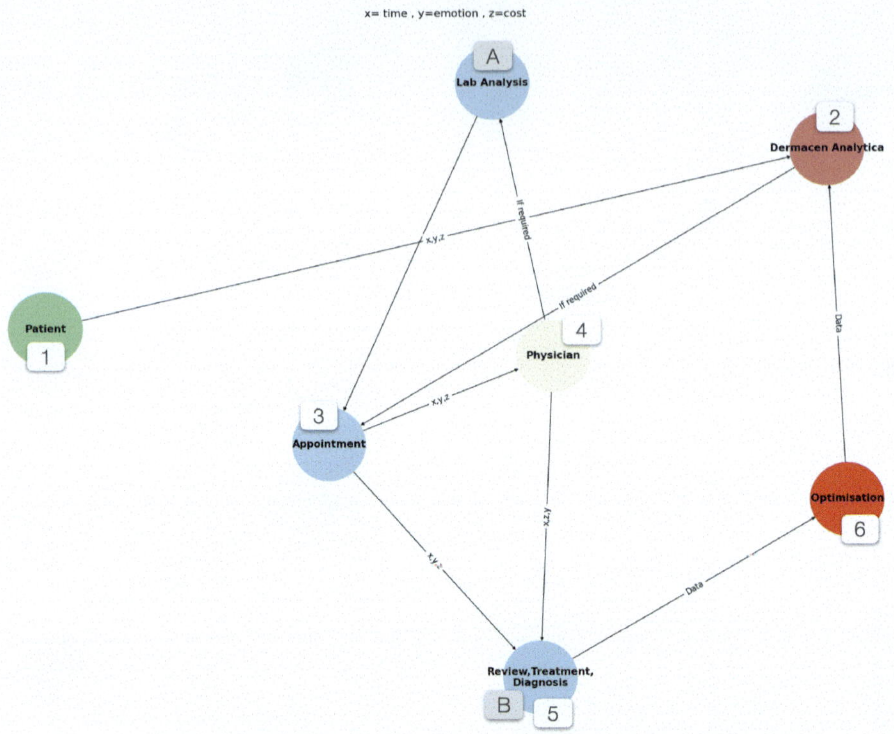

Fig. 10.7 Patient journey, D.A graph. The patient will initiate the journey, but this time will first interact with Dermacen Analytica for a pre-evaluation of the condition that triggered this action. Thus, an initial assessment, assuming the system is robust enough can be provided by an AI-agent with multi-modal capabilities. Here a filtering takes place based on probable severity and if required an appointment is booked. The additional processes here are related with the data exchange when a review, treatment, diagnosis take place where they are referred back to the AI-system for retraining and fine-tuning. Thus, ensuring continuous improvements and optimisation

Fig. 10.8 40 year old with crusted skin lesions

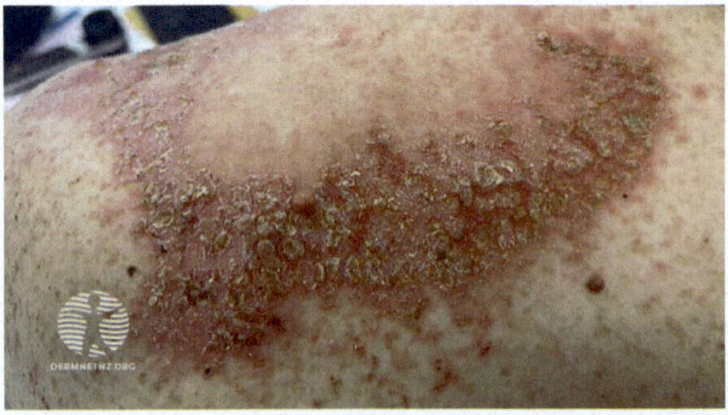

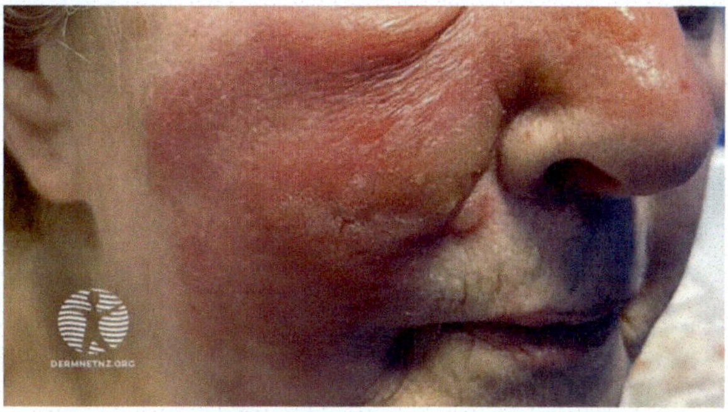

Fig. 10.9 Overnight onset of facial rash

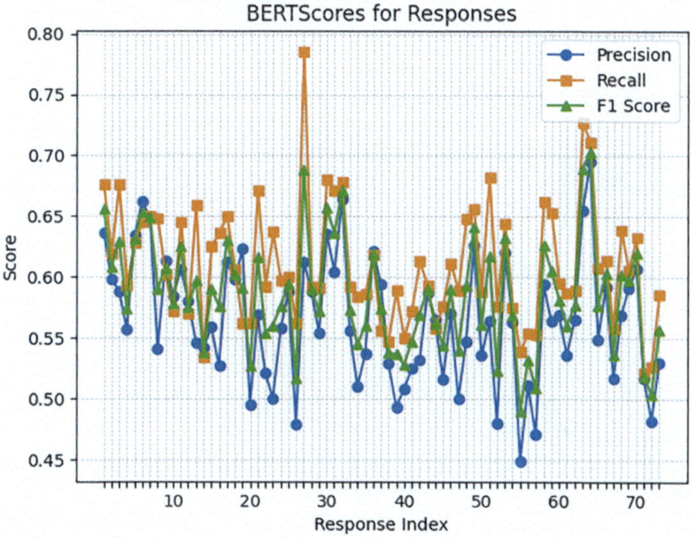

Fig. 10.10 Bert scores

Future work will focus on refining AI models through continuous learning, expanding the testing space of the system by including clinical images from real-world scenarios to add complexity to the process, and integrating patient feedback to enhance the user experience. More importantly, employing the objective function for cost and optimisation presented in this work, we aim to collect real world data under expert supervision to also measure real-world impact, as per the various costs included in a patient journey. Additionally, measuring and comparing errors in various settings within a healthcare workflow provides a quantifiable assessment of AI interventions in traditional medical workflows.

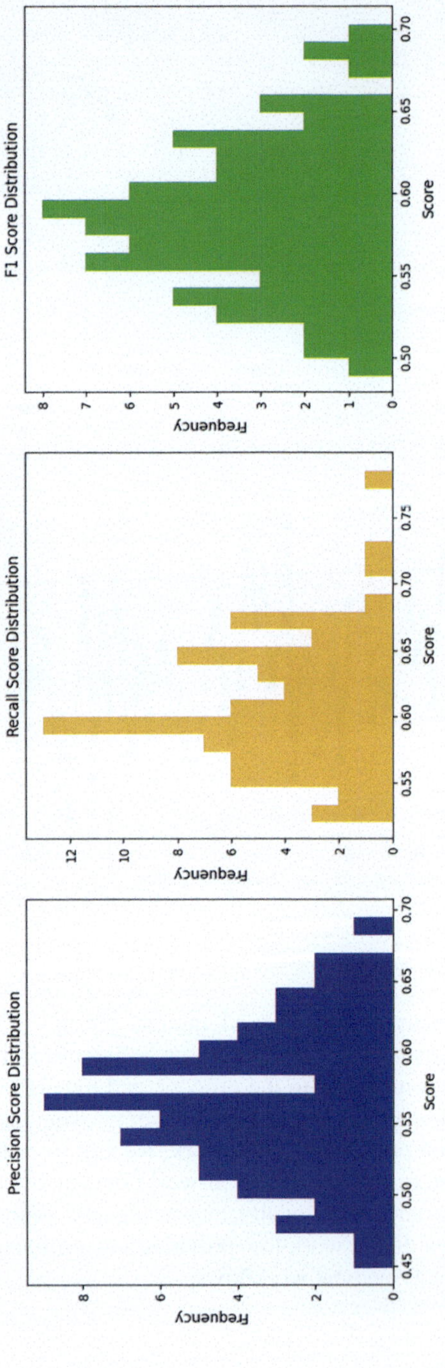

Fig. 10.11 Bert scores distribution

Appendix

D.A Workflow

- *Image Input and Rules of Conduct*: This is the starting point where an image is analyzed by the vision transformer-powered LLM powered. The process checks for the presence and localization of features. This process places emphasis on the identification of whether image is taken via a dermatoscope or not. In the rules of conduct and for the purposes of this study we have manually instructed for up to 3 diagnostic proposals. In production this number could be reduced to 1 to avoid complications or allow user to freely input their own maximum number.
- *Response*: An initial response from the LLM with Vision (VLM) is generated in the context of either a skin disorder or a skin lesion.
- *Analysis Response*: This stage involves the analysis of the generated response as per pre-defined rules for engagement, considering the lesion type, condition or disorder, other diseases, or skin-related factors. This step here is a node from which three different paths can be followed, namely the Skin Legion Path, the Skin Condition Path and the End Path.
- *Skin Lesion Path*
- *Expert* Expert validation and evaluation of the results are required at this stage to ensure careful deployment and data collection that would lead to optimisation

  1. External ML-Service: This part of the process involves an external ML service, which is a specialized service for dermatology. It includes segmentation and feature extraction, as well as considering asymmetry, circularity, color variability, and the initial response from the LLM.
  2. LLM-XAI (Explainable Artificial Intelligence): The process involves applying rules of conduct to the ML model's response, which includes the response itself and a description or definition of previous findings.
  3. Send2Lab: This step informs of the necessity of laboratory analysis, depending on the previous findings, as summarised by the LLM-XAI step.

- *Skin Condition Path*

  1. The image is sent again for analysis into an VLM for more specificity. The focus on melanocytic nevus is removed; thus, more diagnostic freedom is provided to the LLM, in the form of new rules of conduct (defined below).
  2. Response: A response is generated in the form of a differential diagnosis as described previously.

- End path: If it is neither a lesion or a skin condition the path ends, i.e. the process breaks.
- Rule based—Admin filter: This node is positioned in various locations of the D.A workflow, to present where rule-based systems can be placed as additional decision nodes. Based on criteria manually inserted the system can be adjusted to facilitate

the patient and enforce stoppage of the AI-workflow for an expert to intervene, initiate common administrative tasks or other. This is explained in more detail in Sect. 10.4.2.

```
Rules of Conduct = ''Skin Lesions''
Feature Presence = ''Identify and list any notable
    features observed in the image''
Feature Localization: ''Describe the specific location
    (s) of these features within the image''
Dermatoscopy: ''If the image is taken by a
    dermatoscope'
Answer Selection: Up to three most likely diagnoses
    from the following list: [nevus, melanoma, basal
    cell carcinoma, actinic keratosis, benign keratosis
    , dermatofibroma, vascular lesion, squamous cell
    carcinoma, Lentigo Maligna, Blue Nevus, Sebaceous
    Hyperplasia, Keratoacanthoma, Atypical (Dysplastic)
     Nevus, Solar Lentigo, Pigmented Basal Cell
    Carcinoma, Cutaneous Horn, Molluscum Contagiosum,
    Cyst, Lichen Planus, Psoriasis]
```

Listing 10.1 Rules of Conduct for skin lesions

```
Rules of Conduct = ''Skin condition''
Objective = ''Identify potential dermatology problems'
    '
Task: ''Provide  diferential Diagnosis (list of 3):
    identify the most probable cause  based on the
    available information and imagery''
Features: ''Detail Clinical Features Supporting Your
    Diagnosis''
```

Listing 10.2 Rules of Conduct for skin conditions

Algorithm 1 begins by segmenting the lesion from the surrounding skin in the given image. This is achieved through the function "plotSegmented", which applies image segmentation techniques to isolate the lesion. The segmentation process involves the use of the "apply_otsu_and_grabcut_refined" function [45]. This suggests the combination of Otsu's thresholding method with the GrabCut algorithm [39] to accurately delineate the lesion's boundaries. Otsu's method determines an optimal threshold to separate foreground (lesion) from background, while GrabCut refines this segmentation by considering color statistics and a graph-based model.

Algorithm 2 Technical Assessment of Skin Lesions

```
Function ABC_Assessment(image_path):
segmentedPlot ← plotSegmented(image_path)
area, perimeter, circularity, contours ← apply_otsu_and_grabcut_refined(image_path)
if contours then
    contour ← max(contours, key=cv2.contourArea)
    asymmetry, asymmetryPlot1 ← calculate_asymmetry(contour, image_path)
    return "Asymmetry: asymmetry-Horizontal"
end if
if contours then
    contour ← max(contours, key=cv2.contourArea)
    asymmetry2, asymmetryPlot2 ← calculate_asymmetry_minor_axis(contour, image_path)
    return "Asymmetry2: asymmetry-Vertical"
end if
asymmetryAverage ← (Asymmetry + Asymmetry2) / 2
return "Average Asymmetry: asymmetryAverage"
std_dev_r, std_dev_g, std_dev_b, colorPlot ← extractColors(image_path)
prompt ← Generate detailed medical assessment prompt including segmentation and feature extraction summary, circularity index, color variability within the segmented area, lesion asymmetry assessment, and technical note for medical LLM.
Return technical report
```

Algorithm 3 Evaluation algorithm

Main Process:
Input: Premise, Hypothesis, Repeated_Factor
Repeated Factor: Extracted Diagnosis AS Entities
Output: Reports
if (A)Context Then Premise is Explanation &Question **then**
  Calculate Hypothesis
end if
if (B)Diagnosis Then Premise is Explanation &Question & Repeated_Factor **then**
  Calculate Hypothesis Where Hypothesis is Response & & Repeated_Factor
end if
Capability Approximation:
Input: Report, $Report_E$, $Report_e$
Output: Capability as Average
Main Process (A) & (B)
Compute Cosine Similarity (Weighted average)
$Report = AVERAGE(w_1 * (ContentSimilarity) + w_2 * (DiagnosisSimilarity))$
Stage 1a: Bert Scores
Repeat Main Process(A)
Stage 2: Natural Language Inference
Main Process (A) & (B)
$Report_C = AVERAGE((ContentAssessment) + (DiagnosisAssessment))$
$Report_N = AVERAGE((ContentAssessment) + (DiagnosisAssessment))$
$Report_E = AVERAGE((ContentAssessment) + (DiagnosisAssessment))$
with outcomes: (C)ontradiction, (N)eutral, (E)ntailment
Stage 3: Expert Review
Expert Process: Check List assessment on Likert Scale
Score: (AI Symptom analysis and Image understanding(1)) & (AI Diagnostic Reasoning(2))
$Report_e = AVERAGE((1) + (2))$
Final Output: Capability

Value Estimation of AI-Interventions, Theoretical Documentation

In Fig. 10.6, a simplified patient journey is presented as a graph with nodes and edges. It starts with a patient (1) for any number of reasons, booking an appointment (2) and thus interacting with the healthcare system (3). The healthcare node represents the various paramedical processes that precede the direct interaction with a medical professional. Those processes are included in node (4). Node (5) includes the various diagnostic and examination processes, conducted by a physician. In turn, either a review, treatment of a diagnosis(A) is provided to the patient, or first external analysis is called upon, shown in the lab analysis node (B). This is an iterative journey that may restart, continue or interact with other systems of tertiary care, which are outside the scope of this study. Each edge includes associated variables of time (x), emotion (y) and financial cost (z). To measure the total cost we assign units for each variable and weight accordingly. The objective function for cost and optimisation (F) is the following:

$$F = \sum_{i=1}^{n} \left(a_i x_i + b_i y_i + c_i z_i + \left(\sum_{j=1}^{3} d_{ij} e_{ij} \right) \right),$$

To calculate each weight and compare with an adapted AI-empowered patient journey we make the following assumptions. Opportunity Cost is accounted for as (OC) and is considered as the loss (or gain) on investing capital in AI. Thus: $OC = (1 + r_{opp})^n$ for r_{opp} is the expected rate of return on alternative investments in n years. The weight c is to be calculated as

$$c = \frac{(C_{AI,annual} * OC)}{C_{real,annual}}$$

where Annualised Cost C_{annual} is equal to

$$C_{annual} = C_{dep} + C_{recur} + C_{nonre}$$

where C_{dep} is cost adjusted for depreciation, C_{nonre} is the Non-recurring costs and C_{rec} the recurring costs Cost flow at time t must be discounted with discount rate i as Net present value (NPV) for C_{AI} and C_{real} thus,

$$NPV = \frac{C_t}{(1 + t)^t}$$

To adjust for time (x), it is assumed that for every interaction, whether real or a AI-empowered, one time unit is consumed. However, because a real time unit will in most every case invariably longer than an AI-generated time unit, time x is adjusted with coefficient a which represents that ratio as

$$\frac{AItime}{Time}$$

To adjust for emotion (y) the subjective evaluation of users is necessary and typically gathered through evaluation frameworks that utilize the Likert scale, which may vary from 1 to 5 or extend up to 10. The weight y should be set to 1 if each interaction yields comparable outcomes (AI-enhanced or traditional), or be adjusted appropriately based on specific criteria.

The last variable included in F is error e which represents the observed administrative (1), diagnostic (2) or treatment (3) errors. Error can also be associated with risk and accounted for in an additional function for added precision, but it falls outside the scope of this study. d is to be adjusted based on criticality of error domain, thus $d_1 = 0.1$. Diagnostic errors are more critical, thus $d_2 = 0.5$ and treatment errors are more critical and $d_3 = 1$. In the following section we will use the objective Function F, to provide a initial comparison of potential contribution of this (or any) AI intervention in a patient journey.

Evaluation Dataset description

The evaluation dataset includes use-cases that present a rich mix of dermatological conditions, covering a broad spectrum of ages from newborns to the elderly. There is great symptom variety including papules, plaques, vesicles, rashes, and erythroderma. Also specific scenarios are included. Cases like "Farmer with enlarging finger lesion" or "Rugby player with painful weeping lesions" introduce specific occupational or activity-related dermatological concerns. Some cases hint at previous treatments ("Abdominal rash with failed antifungal treatment"), which offers insights into treatment challenges and resistance scenarios. Descriptions range from acute presentations (e.g., "Overnight onset of facial rash") to chronic or progressive conditions (e.g., "Slow-growing lesion behind the ear"). The dataset also includes common conditions like eczema and acne and potentially life-threatening or rare diseases like "Purple lesions with recent AML relapse". While the dataset included lacks in diversity in terms of geography and ethnicity, the majority of cases being Western and Caucasian, there are images included from Syria and Malaysia.

Implementation Use Case—Skin Lesion

1. *Visual Description*: The image presents a close-up view of a skin lesion on a light-skinned background, marked by a scale with millimeter-level increments for size reference. The lesion appears asymmetrical and multicolored with shades of pink, red, and black. There are irregular borders around the lesion, and it shows variation in color intensity and pattern within its contour. The lesion's diameter

can be approximated with the ruler in the frame, suggesting it may be larger than 6mm. Transparent bubbles and hair follicles can also be seen on the surrounding skin.

2. *Feature Presence*:
   1. Asymmetrical shape
   2. Irregular, poorly-defined borders
   3. Multiple colors (pink, red, black)
   4. Diameter potentially greater than 6mm
   5. Variation in pigmentation
3. *Feature Localization*: The lesion is centrally located within the image frame, with the black portion towards its lower region. The red and lighter pink areas are distributed in a somewhat blotchy pattern throughout the lesion. The darker areas and most irregular borders can be seen on the bottom and right side of the lesion.
4. *Dermatoscopy*:
   – Yes.
5. *Diagnosis Selection*: The image depicts a skin lesion, and based on the ABC measurements and visual appearance, the three most likely diagnoses may be:
   1. Melanoma
   2. Atypical (Dysplastic) Nevus
   3. Pigmented Basal Cell Carcinoma

Following the initial assessment, segmentation and analytical techniques are employed to measure Asymmetry, Borders and Colorisation. The generated technical report is the following:

"Detailed Medical Assessment with Methodology of Segmented Image Analysis Segmentation and Feature Extraction Summary: The lesion of interest was identified and segmented from the surrounding tissue using a combination of Otsu's method and the GrabCut algorithm, a process implemented in the apply_otsu_and_grabcut function. This approach ensures precise delineation of the lesion, facilitating accurate measurement and analysis. Following segmentation, key geometric and colorimetric features were extracted using OpenCV (cv2), a widely recognized library in image processing and analysis. Circularity Index: Determined to be 0.68, indicating an irregular shape that deviates from perfect circularity. This measure is crucial for understanding the lesion's morphology. Color Variability within the Segmented Area: Standard Deviation in the Red Channel: 19.21, indicating the degree of red intensity variation. Standard Deviation in the Green Channel: 20.647, reflecting variation in green intensity. Standard Deviation in the Blue Channel: 19.261, showing blue intensity variation. The lesion has an asymmetry of 0.00823 (asymmetry plot 1) and 0.1062 (asymmetry plot 2), for an average asymmetry of 0.05722464257917816 which is a measure of the difference in the two halves of the lesion."

The next step of the process, presented as the XAI button in Fig. 10.4, feeds both the initial and the technical assessment to another LLM (Multi-Model Collaboration for Enhanced Context Retrieval). The following is last report provided by the collaborative system, which also includes the necessary references for the specialist to ensure the usefulness of the report:

".... Using Otsu's method and the GrabCut algorithm promptly ensures an accurate segmentation that is fundamental for any further processing or analysis activities. Your segmentation analysis revealed the gross morphological features of the lesion well. The asymmetry findings, irregular border, color variability, and size all are crucial characteristics that aid in the differential diagnosis of melanoma.

Your suspicion towards a diagnosis of melanoma is supported by these findings. Atypical (Dysplastic) Nevus and Pigmented Basal Cell Carcinoma are also plausible diagnoses, as these can present with some features similar to melanoma such as irregular color and large diameter. However, further diagnostic testing (like a biopsy) would be required to confirm this. You did well to stress that this evaluation is not a substitute for professional medical advice. Face-to face clinical examination, patient's history and, if necessary, biopsy and histologic examination are always the gold standard in dermatology."

Implementation Use Case—Skin Condition

"The image depicts a close-up view of a skin condition characterized by a reddish-pink background with irregular white patchy areas. These white patches appear scaly or flaky and are somewhat elevated from the surrounding skin. The borders are not clearly defined and blend into the red-pink skin. There is also a shiny or moist appearance to some regions, possibly indicating the presence of a serous exudate or moisture retention on the skin's surface.

Feature Presence:

– Reddish-pink inflamed skin
– Irregular white scaly patches
– Indistinct borders
– Shiny or moist areas

Feature Localization:

– The inflamed skin with reddish-pink coloration is spread throughout the image.
– The white patchy scales are clustered centrally within the image's frame.
– The shiny or moist areas are evident in the central region where the white patches are most concentrated.

Dermatoscopy Approximation: N/A

This image is not taken with a Dermatoscope. Instead, this seems to be a depiction of a skin condition or infection.

Based on the visual characteristics presented in the image, this does not seem to be a skin lesion but rather a skin condition. A potential general classification for this skin disorder could be a fungal infection such as candidiasis, which could present with redness, patches of white scaling, and moisture. However, without further clinical context or testing, a definitive diagnosis cannot be rendered..."

Given that no lesion is depicted, the "Follow Up" button is activated, allowing for interaction with a different LLM, using a new set of rules of conduct. Therefore, this use case constitutes an example of Multi-Model Collaboration for Enhanced Context Retrieval. The clinical features, a list of probable diagnoses and a more likely final diagnosis constitute the generated response. For context, the correct response is Ulcer due to biting the lip. In more detail, the AI-generated response is as follows:

"This image depicts an oral lesion that appears to be an ulcer. To form a differential diagnosis and conclude with one primary diagnosis, we must consider the following:

1. Aphthous stomatitis (canker sore)—These are common oral ulcers that are typically painful and self-limiting.

2. Herpes simplex virus (cold sore)—Characterized by grouped vesicles on an erythematous base, which then ulcerate, but these are usually found on the lip rather than inside the mouth.

3. Traumatic ulcer—Resulting from injury such as biting the inside of the cheek or irritation caused by sharp foods, dental work, or poorly fitting dentures.

Final Diagnosis: Aphthous stomatitis (canker sore).

Clinical Features Supporting Diagnosis:

– The location of the ulcer is inside the mouth, which aligns more with aphthous stomatitis.

– The appearance is consistent with a canker sore, having a white or yellowish center with an erythematous halo.

– The lack of a group of vesicles and the solitary nature of the ulcer diminish the likelihood of herpes simplex virus infection.

– Assuming the history does not suggest recent trauma to the area, this further supports the diagnosis of aphthous stomatitis over a traumatic ulcer. Note: Accurate identification would require patient history and a possible evaluation for systemic diseases associated with aphthous ulcers if they are recurrent or severe."

The important finding in this second use case is the fact that using these reduction techniques provided by rules of conduct, the answer is indeed refined and on point. In the initial assessment, the generated hypothesis was of this being a fungal infection. In the follow up, though, the image was correctly identified as an ulcer and a differential diagnosis was provided along with an accurate explanation.

Evaluation Use Case 1

- **Question**: This 40-year-old woman has a longstanding history of crusted skin lesions since she was 12 years old. Lesions started on her neck, axillae, and upper back. They flare up intermittently with increasing pain and malodorous discharge. Many family members (including her grandfather, father, uncles, and twin sister) have similar lesions. On examination, she had confluent, crusted, greasy papules across the upper back, lateral neck, and axillae. She also had nail changes.

- **Premise**: The diagnosis is Darier disease, an autosomal dominant genodermatosis. It typically presents with scaly, crusted papules in a seborrhoeic distribution and in skin folds.
  It is caused by a mutation in the ATP2A2 gene, which leads to SERCA2 protein dysfunction, interfering with cellular calcium signaling. This results in impaired cellular adhesion and epidermal blistering.
- **Premise-Entity**: Darier Disease
- **Hypothesis** (AI-generated response): Diagnosis—Darier Disease (Keratosis Follicularis)
  Clinical Features Supporting Diagnosis: The patient presents with several key features that indicate Darier Disease:
  1. Longstanding History: The patient has experienced these skin lesions since a young age, which is common for genetic disorders like Darier Disease.
  2. Family History: There is a strong familial pattern, with multiple generations affected, which supports a genetic etiology like that of Darier Disease, an autosomal dominant condition.
  3. Morphology and Distribution of Lesions: The greasy papules and crusted lesions are typically seen in Darier Disease, and the distribution on the upper back, lateral neck, and axillae fits the pattern of seborrheic areas commonly involved in this condition.
  4. Nail Changes: Nail abnormalities such as red and white streaks or a V-shaped nick at the tip of the nails, are characteristic findings in Darier Disease.
  5. Malodorous Discharge and Pain: Secondary infection can be common in these lesions, leading to a malodorous discharge and pain, especially during flares.
  The presented photograph shows confluent, crusted, greasy papules that are consistent with the features of Darier Disease lesions.
- **Hypothesis Entity**: Darier Disease (Keratosis Follicularis)
- **SCORE**:

  1. **TS Premise vs Hypothesis**: 1.0
  2. **TS Premise vs Hypothesis Entities**: 1.0
  3. **NLI Premise vs Hypothesis**: ENTAILMENT (0.675)
  4. **NLI Premise vs Hypothesis - Entities**: ENTAILMENT (0.49)
  5. **(Expert Review) Symptom analysis and Image understanding**: 5 out of 5 (1.0)
  6. **(Expert Review) AI Diagnostic Reasoning**: 5 out of 5 (1.0)

Evaluation Use Case 2

- **Question**: This 78-year-old lady presented with a new reddish-orange spreading rash on the right side of her face overnight. The affected skin was raised with a demarcated border and areas of blistering. Her white cell count and C-reactive protein were very raised and she described an episode of fever and rigors.

- **Premise**: This rash and history is classical for erysipelas. Erysipelas is a superficial form of cellulitis, a potentially serious bacterial infection affecting the skin. Erysipelas affects the upper dermis and extends into the superficial cutaneous lymphatics.
- **Premise-Entity**: Erysipelas
- **Hypothesis** (AI-generated response): Diagnosis—Cellulitis Clinical Features Supporting the Diagnosis: The patient's presentation of a reddish-orange spreading rash with raised, demarcated borders and areas of blistering on the right side of her face is indicative of cellulitis. Cellulitis is a common bacterial skin infection that affects the deeper layers of skin and the underlying tissue.
  The acute nature of the rash (appearing overnight), alongside systemic symptoms such as fever and rigors (shaking chills), supports an infectious etiology. The raised white cell count and elevated C-reactive protein are indicative of an acute inflammatory response to an infection.
  The characteristics of the rash, such as its rapid spread, the clearly defined border, swelling, and area of blistering, further suggest cellulitis, particularly given the absence of a central lesion that would suggest a bite or sting, and the lack of other distinguishing marks suggesting an alternative diagnosis such as shingles or contact dermatitis.
  The patient's age (78 years old) is also a relevant factor, as older individuals have a higher risk of developing cellulitis, possibly due to a weakened immune system and other comorbidities that may compromise skin integrity and the body's natural defenses against infection.
  Given these clinical features and the patient's rapid onset of symptoms, cellulitis seems to be the most probable diagnosis for this case.
- **Hypothesis Entity**: Cellulitis
- **SCORE**:

1. **TS Premise vs Hypothesis**: 0.89
2. **TS Premise vs Hypothesis Entities**: 1.0
3. **NLI Premise vs Hypothesis**: ENTAILMENT (0.55)
4. **NLI Premise vs Hypothesis - Entities**: ENTAILMENT (0.83)
5. **(Expert Review) Symptom analysis and Image understanding**: 5 out of 5 (1.0)
6. **(Expert Review) AI Diagnostic Reasoning**: 5 out of 5 (1.0)

BERTscore Context

Precision in the context of BERTScore refers to the extent to which tokens in the generated response -hypothesis- are semantically similar to tokens in the premise. Recall measures how well the tokens in the correct answer are covered by semantically similar tokens in the response. On the other hand, the F1 Score is the harmonic mean of precision and recall. A moderate precision indicates that not all information in the response is relevant or accurately reflects the correct answer, while a higher recall suggests that a majority of the correct answer's content is addressed in the response, yet some aspects are missing or underrepresented. The F1 Score underscores the need for improvement in both precision and recall to achieve a higher degree of semantic similarity and accuracy. The BERTScores, alongside the textual -cosine- similarity and the NLI scores for context and diagnosis, show that there is good reasoning within the domain, where images are also analysed. This analysis can benefit the doctor's workflow and suggest pathways for further investigation, either via a more thorough clinical assessment or via external services, usually provided by laboratories or other experts.

Graph Analysis on Patient Journey Comparison

In Scenario A the patient journey analysis is as follows:

- **Path-1**:
  1. associated costs per node in sequential order:['y,z', 'x,z', 'x,y,z', '', 'x,z,y']
  2. Length of Path = 5
  3. The objective function aims to quantify the cost associated with each path. We assume unknown variables (a, b, c, d, and e) as explained previously. The objective function for path1 assuming known units (x,y,z,e) equal for each path: $F = 3 * a * x + 3 * b * y + 4 * c * z + 5 * d * e$
- **Path-2**:
  1. associated costs per node in sequential order:['y,z', 'x,z', 'x,y,z', '', 'x,z,y', 'x,y,z', '', 'x,y,z']
  2. Length of Path = 8
  3. Objective Function assuming known units (x,y,z,e) equal for each path: $F = 5 * a * x + 5 * b * y + 6 * c * z + 8 * d * e$

In Scenario B the patient journey analysis is as follows:

- **Path-1**:
  1. associated costs per node in sequential order:['x,y,z']
  2. Length of Path = 1 (In the case where no appointment with the healthcare provider is required)
  3. The objective function aims to quantify the cost associated with each path. We assume unknown variables (a, b, c, d, and e) as explained previously. The objective function for path1 assuming known units (x,y,z,e) equal for each path: $F = a * x + b * y + c * z + 1 * d * e$

- **Path-2**:
  1. associated costs per node in sequential order:[‘x,y,z’, ‘x,y,z’, ‘’, ‘x,z,y’]
  2. Length of Path = 4
  3. Objective Function assuming known units (x,y,z,e) equal for each path: $F = 3 * a * x + 3 * b * y + 3 * c * z + 4 * d * e$
- **Path-3**:
  1. associated costs per node in sequential order:[‘x,y,z’, ’x,y,z’, ‘’, ‘x,z,y’, ‘x,y,z’, ‘’, ‘x,y,z’, ‘Data’, ‘Data’]
  2. Length of Path = 7
  3. 2*Times data were used for System optimisation
  4. Objective Function assuming known units (x,y,z,e) equal for each path: $F = 5 * a * x + 5 * b * y + 5 * c * z + 7 * d * e$

Assuming that for each of x, y and z at least one unit occurs per scenario then, scenario A has a length range of 5 to 8, with weighted time, emotional, and financial costs ranging from 3 to 5, 3 to 5, and 4 to 6, respectively, and potential errors ranging from 5 to 8 units. In contrast, scenario B shows improvements across all metrics with lengths from 1 to 7, and weighted time, emotional, and financial costs all between 1 and 5, and potential errors from 1 to 7 units, indicating overall enhancements provided the quality of care remains consistent. It should also be noted that the cost of evaluation and use of LLMs for generating a final response as per the workflow presented here was less that 1.5$. Additional costs for deploying, hosting and leveraging traffic are outside the cost of this study.

Data Availability

Data sources utilized in this study have been cited and referenced. The complete expert review with the dataset used for evaluation can be found here [11]. We are committed to contributing to the scientific community by open-sourcing our entire code-base, including both evaluation and high-level functions, which will be made publicly available concurrent with the publication of our findings.

References

1. V. Anand, S. Gupta, D. Koundal, K. Singh, Fusion of U-Net and CNN model for segmentation and classification of skin lesion from Dermoscopy images. Expert Syst. Appl. **213**, 119230 (2023)
2. G. Bradski, A. Kaehler, et al., OpenCV. Dr. Dobb’s J. Softw. Tools **3**(2) (2000)
3. T. Buckley, J.A. Diao, A. Rodman, A.K. Manrai, Accuracy of a vision-language model on challenging medical cases (2023). arXiv: 2311.05591 [cs.CV]

4. S. Chan, V. Reddy, B. Myers, Q. Thibodeaux, N. Brownstone, W. Liao, Machine learning in dermatology: current applications, opportunities, and limitations. Dermatol. Therapy **10**, 365–386 (2020)
5. Condition Use Case (n.d.). https://www.diskinside.com/wp-content/uploads/2024/03/condition.png. Accessed 15 Mar 2024
6. N. Crispino, K. Montgomery, F. Zeng, D. Song, C. Wang, Agent instructs large language models to be general zero-shot reasoners (2023). arXiv preprint arXiv:2310.03710
7. DermNet—Dermatology Resource (n.d.). https://dermnetnz.org/about. Accessed 29 Mar 2023
8. J. Devlin, M.-W. Chang, K. Lee, K. Toutanova, Bert: pre-training of deep bidirectional transformers for language understanding (2018). arXiv preprint arXiv:1810.04805
9. A.X. Du, S. Emam, R. Gniadecki, Review of machine learning in predicting dermatological outcomes. Front. Med. **7**, 266 (2020)
10. A.F. Duarte, B. Sousa-Pinto, L.F. Azevedo, A.M. Barros, S. Puig, J. Malvehy, E. Haneke, O. Correia, Clinical ABCDE rule for early melanoma detection. Eur. J. Dermatol. **31**(6), 771–778 (2021)
11. Expert Review and dataset (n.d.). https://www.diskinside.com/wp-content/uploads/2024/03/Expert-Evaluation-of-AI-Generated-Medical-Diagnosis.pdf. Accessed 25 April 2024
12. H. Fan, F. Xie, Y. Li, Z. Jiang, J. Liu, Automatic segmentation of dermoscopy images using saliency combined with Otsu threshold. Comput. Biol. Med. **85**, 75–85 (2017)
13. J. Gachon, P. Beaulieu, J.F. Sei, J. Gouvernet, J.P. Claudel, M. Lemaitre, M.A. Richard, J.J. Grob, First prospective study of the recognition process of melanoma in dermatological practice. Arch. Dermatol. **141**(4), 434–438 (2005)
14. E. Hahnel, A. Lichterfeld, U. Blume-Peytavi, J. Kottner, The epidemiology of skin conditions in the aged: a systematic review. J. Tissue Viability **26**(1), 20–28 (2017)
15. X. Du-Harpur, F.M. Watt, N.M. Luscombe, M.D. Lynch, What is AI? Applications of artificial intelligence to dermatology. Br. J. Dermatol. **183**(3), 423–430 (2020)
16. A.A. Hassanain, M.A.A. Eldosoky, A.M. Soliman, Evaluating building performance in healthcare facilities using entropy and graph heuristic theories. Sci. Rep. **12**(1), 8973 (2022)
17. R.J. Hay, N.E. Johns, H.C. Williams, I.W. Bolliger, R.P. Dellavalle, D.J. Margolis, R. Marks, L. Naldi, M.A. Weinstock, S.K. Wulf, et al., The global burden of skin disease in 2010: an analysis of the prevalence and impact of skin conditions. J. Invest. Dermatol. **134**(6), 1527–1534 (2014)
18. D.T. Hogarty, J.C. Su, K. Phan, M. Attia, M. Hossny, S. Nahavandi, P. Lenane, F.J. Moloney, A. Yazdabadi, Artificial intelligence in dermatology–where we are and the way to the future: a review. Am. J. Clin. Dermatol. **21**, 41–47 (2020)
19. R. Iranpoor, A.S. Mahboob, S. Shahbandegan, N. Baniasadi, Skin lesion segmentation using convolutional neural networks with improved U-Net architecture, in *2020 6th Iranian Conference on Signal Processing and Intelligent Systems (ICSPIS)* (IEEE, 2020), pp. 1–5
20. Y. Jiang, J.A. Omiye, C. Zakka, M. Moor, H. Gui, S. Alipour, S.S. Mousavi, J.H. Chen, P. Rajpurkar, R. Daneshjou, Evaluating general vision-language models for clinical medicine (2024). medRxiv. https://doi.org/10.1101/2024.04.12.24305744. Eprint: https://www.medrxiv.org/content/early/2024/04/18/2024.04.12.24305744.full.pdf
21. A.L. Joseph, A.W. Kushniruk, E.M. Borycki, Patient journey mapping: current practices, challenges and future opportunities in healthcare. Knowl. Manage. E-learning **12**(4), 387 (2020)
22. O.A. Kerr, M.J. Tidman, J.J. Walker, R.D. Aldridge, E.C. Benton, The profile of dermatological problems in primary care. Clin. Exp. Dermatol. **35**(4), 380–383 (2010)
23. S. Khan, M. Naseer, M. Hayat, S.W. Zamir, F.S. Khan, M. Shah, Transformers in vision: a survey. ACM Comput. Surv. (CSUR) **54**(10s), 1–41 (2022)
24. Lesion Use Case (n.d.). https://www.diskinside.com/wp-content/uploads/2024/03/lesion.png. Accessed: 15 Mar 2024
25. K. Liopyris, S. Gregoriou, J. Dias, A.J. Stratigos, Artificial intelligence in dermatology: challenges and perspectives. Dermatol. Therapy **12**(12), 2637–2651 (2022)
26. J. Liu, S. Hallinan, X. Lu, P. He, S. Welleck, H. Hajishirzi, Y. Choi, Rainier: reinforced knowledge introspector for commonsense question answering (2022). arXiv:2210.03078 [cs.CL]

27. Y. Liu, M. Ott, N. Goyal, J. Du, M. Joshi, D. Chen, O. Levy, M. Lewis, L. Zettlemoyer, V. Stoyanov, RoBERTa: a robustly optimized BERT pretraining approach (2019). arXiv preprint arXiv:1907.11692
28. Machine Learning for PyTorch, TensorFlow, and JAX (n.d.). https://huggingface.co/docs/transformers/en/index. Accessed 15 Mar 2024
29. S.S. Marwaha, H. Fevrier, S. Alexeeff, E. Crowley, M. Haiman, N. Pham, M.J. Tuerk, D. Wukda, M. Hartmann, L.J. Herrinton, Comparative effectiveness study of face-to-face and teledermatology workflows for diagnosing skin cancer. J. Am. Acad. Dermatol. **81**(5), 1099–1106 (2019)
30. D. Panagoulias, F. Palamidas, M. Virvou, G.A. Tsihrintzis, Evaluating the potential of LLMs and ChatGPT on medical diagnosis and treatment, in *14th IEEE International Conference on Information, Intelligence, Systems, and Applications (IISA2023), Volos, Greece* (2023)
31. D.P. Panagoulias, P. Papatheodosiou, A.P. Palamidas, M. Sanoudos, E. Tsoureli-Nikita, M. Virvou, G.A. Tsihrintzis, COGNET-MD, an evaluation framework and dataset for Large Language Model benchmarks in the medical domain (2024). arXiv:2405.10893 [cs.CL]
32. D.P. Panagoulias, M. Virvou, G.A. Tsihrintzis, Rule-augmented artificial intelligence-empowered systems for medical diagnosis using large language models, in *2023 IEEE 35th International Conference on Tools with Artificial Intelligence (ICTAI)* (2023)
33. D.P. Panagoulias, M. Virvou, G.A. Tsihrintzis, Evaluating LLM—generated multimodal diagnosis from medical images and symptom analysis (2024). arXiv:2402.01730 [cs.CL]
34. S. Patel, J.V. Wang, K. Motaparthi, J.B. Lee, Artificial intelligence in dermatology for the clinician. Clin. Dermatol. **39**(4), 667–672 (2021)
35. L. Qin, D. Kirikkaleli, Y. Hou, X. Miao, M. Tufail, Carbon neutrality target for G7 economies: examining the role of environmental policy, green innovation and composite risk index. J. Environ. Manage. **295**, 113119 (2021)
36. N. Reimers, I. Gurevych, Sentence-BERT: sentence embeddings using Siamese BERT-networks, in *Proceedings of the 2019 Conference on Empirical Methods in Natural Language Processing. Association for Computational Linguistics* (2019). arXiv:1908.10084
37. I. Reychav, R. McHaney, S. Babbar, K. Weragalaarachchi, N. Azaizah, A. Nevet, Graph network techniques to model and analyze emergency department patient flow. Mathematics **10**(9), 1526 (2022)
38. C. Ring, N. Cox, J.B. Lee, Diagnostic and therapeutic instrumentation in dermatology. Dermatoscopy. Clin. Dermatol. **39**(4), 635–642 (2021). ISSN: 0738-081X. https://doi.org/10.1016/j.clindermatol.2021.03.009
39. A.W. Setiawan, Image segmentation metrics in skin lesion: accuracy, sensitivity, specificity, dice coefficient, Jaccard index, and Matthews correlation coefficient, in *2020 International Conference on Computer Engineering, Network, and Intelligent Multimedia (CENIM)* (IEEE, 2020), pp. 97–102
40. GPT-4V(ision) (2023)
41. A. Vaswani, N. Shazeer, N. Parmar, J. Uszkoreit, L. Jones, A.N. Gomez, Ł. Kaiser, I. Polosukhin, Attention is all you need. Adv. Neural Inform. Process. Syst. **30** (2017)
42. L. Yuan, Y. Chen, T. Wang, W. Yu, Y. Shi, Z.-H. Jiang, F.E.H. Tay, J. Feng, S. Yan, Tokens-to-token vit: training vision transformers from scratch on Imagenet, in *Proceedings of the IEEE/CVF international conference on computer vision*, (2021), pp. 558–567
43. F. Zaman, M. Aloqaily, F. Sallabi, K. Shuaib, J.B. Othman, Application of graph theory in iot for optimization of connected healthcare system, in *GLOBECOM 2020-2020 IEEE Global Communications Conference* (IEEE, 2020), pp. 1–6
44. X. Zhai, A. Kolesnikov, N. Houlsby, L. Beyer, Scaling vision transformers, in *Proceedings of the IEEE/CVF Conference on Computer Vision and Pattern Recognition* (2022), pp. 12104–12113
45. S. Zhang, Y. Zhao, P. Bai, Object localization improved grabcut for lung parenchyma segmentation. Procedia Comput. Sci. **131**, 1311–1317 (2018)

Chapter 11
Cognitive Network Evaluation Toolkit for Medical Domains

Abstract Large Language Models (LLMs) constitute a breakthrough state-of-the-art Artificial Intelligence (AI) technology which is rapidly evolving and promises to aid in medical diagnosis either by assisting doctors or by simulating a doctor's workflow in more advanced and complex implementations. In this chapter, we outline Cognitive Network Evaluation Toolkit for Medical Domains (COGNET-MD), which constitutes a novel benchmark for LLM evaluation in the medical domain. Specifically, we propose a scoring-framework with increased difficulty to assess the ability of LLMs (Panagoulias et al. in COGNET-MD, an evaluation framework and dataset for large language model benchmarks in the medical domain, 2024; Panagoulias et al. in Dermacen analytica: a novel methodology integrating multi-modal large language models with machine learning in tele-dermatology, 2024; Panagoulias et al. in, Evaluating LLM—generated multimodal diagnosis from medical images and symptom analysis, 2024) in interpreting medical text. The proposed framework is accompanied with a database of Multiple Choice Quizzes (MCQs). To ensure alignment with current medical trends and enhance safety, usefulness, and applicability, these MCQs have been constructed in collaboration with several associated medical experts in various medical domains and are characterized by varying degrees of difficulty. The current (first) version of the database includes the medical domains of Psychiatry, Dentistry, Pulmonology, Dermatology and Endocrinology, but it will be continuously extended and expanded to include additional medical domains.

Keywords Large Language Model Benchmark · AI evaluation · Medical database

11.1 Problem Statement

Large Language Models (LLMs) are advanced computational algorithms designed to generate and manipulate natural language, when triggered by a human prompt. At the core of these models lie a subset of machine learning models that have been trained on vast amounts of text, audio, image, and video data and possess the ability to generate knowledge, reason on specific prompts and provide a conversational agent that can enhance human abilities on a multitude of tasks [1, 5, 20]. To increase

D. P. Panagoulias et al., *Artificial Intelligence-Empowered Bio-medical Applications*, Learning and Analytics in Intelligent Systems 49,
https://doi.org/10.1007/978-3-031-90174-4_11

LLM efficiency and AI safety, many techniques and methods have been developed and investigated [17]. For each method, suitability is dependent on task, budget and time. However, in any deployment or use case and whether or not precision and usability are the main parameters influencing development decisions, evaluation metrics on accuracy and specificity on the generative abilities of LLMs are of highest importance [21, 23, 24].

The medical domain is a field that can greatly benefit from use of AI-empowered applications, including LLMs. Indeed, if properly developed, evaluated and used, such application can provide significant assistance to medical and healthcare professionals [3, 28]. Unfortunately, currently, there appears to be a shortage of independent datasets available for the evaluation of LLMs in the medical field. This, in turn, creates barriers and friction in the deployment of medical AI-empowered applications.

11.2 COGNET-MD Version 1.0

To address the lack of or the limited availability of independent, free to use and ready to assess datasets for LLM evaluation in the medical domain [12], we have constructed Cognitive Network Evaluation Toolkit for Medical Domains (COGNET-MD). Moreover, to improve deployment and usage, we also propose use cases of varying difficulty for its accompanying dataset. More specifically, the dataset consists of 542 datapoints of domain-specific questions with one or more correct choices/answers. In Table 11.2 (in the Appendix), a preview of COGNET-MD can be seen. The complete dataset can be found and loaded for evaluation on HuggingFace[1] and in Version 1 includes MCQs about Dentistry [2, 15, 18], Dermatology [4, 13, 16], Endocrinology [7, 8, 11, 14, 19], Psychiatry [9, 10, 27, 30, 32] and Pulmonology [6, 26, 29, 31].

In Fig. 11.1, the benchmark use cases are presented as Specialty where one medical specialisation is chosen, Beta where 50% per specialty of total dataset is chosen with all included medical specialties and Production where all Dataset is used. The scoring for each use case is the same and includes the following as shown in Fig. 11.1 and the Algorithm 1:

- Partial Credit: At least one correct answer equals to a half point −0.5.
- Full Credit: To achieve full points depending on difficulty either all correct answers must be selected and no incorrect ones or a correct response gets the full credit, equals to 1 point.
- Penalty for Incorrect Answers: Points are deducted for any incorrect answers selected. −(minus) 0.5 point for each incorrect answer selected.

[1] https://huggingface.co/datasets/DimitriosPanagoulias/COGNET-MD/.

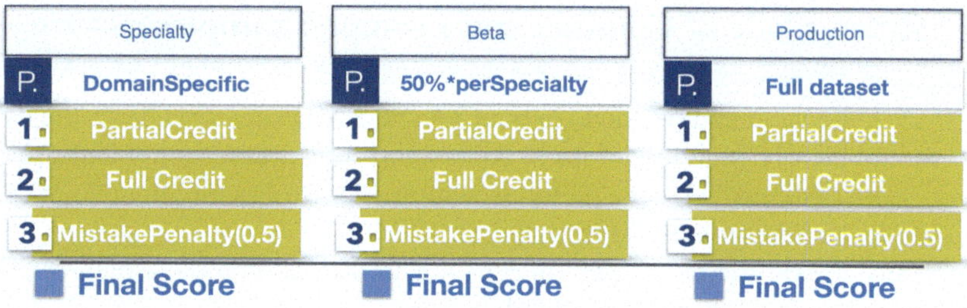

Fig. 11.1 Benchmark varying difficulty use cases

The dataset can be used to assess the model's ability to infer relationships between specialties and knowledge spaces. Thus it can be analyzed either as a whole, encompassing all included specialties-full Dataset, partially or it can be narrowed down to focus on a specific medical domain-specialty. A rule based algorithm to be used as a deployment paradigm is presented below:

Algorithm 1 Scoring algorithm based on difficulty level

```
Function ScoreResponses(user_responses, correct_responses)
score ← 0
for i ← 1 to len(user_responses) do
  user_choices ← user_responses[i]
  correct_choices ← correct_responses[i]
  correct_selected ← user_choices ∩ correct_choices
  incorrect_selected ← user_choices − correct_choices
  if correct_selected then
    score ← score + 0.5
  end if
  if correct_selected = correct_choices and not incorrect_selected then
    score ← score + 0.5
  end if
  score ← score − 0.5 × len(incorrect_selected)
end for
return score
end Function
```

In the following section the Code of Conduct as a space for a score to be considered valid and the rules of conduct as one-shot and few shot prompt examples, to be used for each use case are presented.

11.3 Rules, Code of Conduct and Prompt Examples

For a score to be valid and be added in the COGNET-MD's leader-boards the developers, should clearly state model used, add a short model description and use case scenario used, as described in the previous section. In the following Benchmark Card two examples are presented (Table 11.1).

If for any reason the developer chooses not to use the complete dataset as per content, then the developer should clearly state it.

```
rule_of_conduct = ''On a Multiple choice Quiz choose
    correct responses:(ONLY THE CORRECT LETTERS and no
    spaces and no other associated Text. If more than
    one letter then add a dash- between letters).''
MODEL = ''MODEL''
```

Listing 11.1 One Shot example

In this scenario if one domain is chosen it can be clearly stated in the prompt as it has shown to slightly increase accuracy [25]. So if Dermatology is chosen then the prompt would be:

```
rule_of_conduct = ''On the Following Dermatology
    Multiple choice Quiz choose correct response or
    responses. This is an example question: Which
    condition may include generalised pruritus as a
    symptom? A. Hodgkins disease B. Pityriasis rosea C.
     Diabetes mellitus D. Haemolytic jaundice E.
    Polycythaemia rubra vera and the Correct responses
    on that example are A-B-E.''
MODEL = ''MODEL''
```

Listing 11.2 One Shot example

Table 11.1 Benchmark card

MODEL	Description	Domain	Difficulty	COGNET-MD version
GPT-4	Add here related info	Full dataset	Production	1.0
Mistral	Add here related info	Dermatology	Specialty	1.0

Appendix

See Table 11.2.

Data Availability

Data and python code can be found at:
https://huggingface.co/datasets/DimitriosPanagoulias/COGNET-MD/,
https://huggingface.co/datasets/DimitriosPanagoulias/COGNET-MD/blob/main/COGNET-MD1.0.ipynb.

Table 11.2 COGNET-MD evaluation dataset

Questions	Correct-choices	Specialty
In which neurological condition is Gowers' sign or maneuver typically observed? A. Multiple sclerosis B. Myotonic dystrophy C. Huntington's disease D. Duchenne's muscular dystrophy E. Myasthenia gravis	D	Psychiatry
A husband and wife present at an emergency room. The wife informs that her husband has developed an unwavering belief of infidelity on her part, which she denies. His behaviors include tailing her, sniffing her clothes, scrutinizing her purse, and consistently accusing her. He does not fit the mood disorder criteria and denies having other psychotic symptoms. His medical history displays no substance abuse. What could be his diagnosis? A. Schizophrenia B. Major depressive disorder with psychotic features. C. Delirium D. Dementia E. Delusional disorder	E	Psychiatry
Which among the following options might prove beneficial for an old man suffering from a gravitational ulcer with adjacent eczema? A. The usage of a supportive elastic bandage B. Daily application of Betnovate-C ointment C. Povidine-iodine ointment on the ulcer D. Diuretic therapy for oedema reduction E. Lassar's paste application around the ulcer	A-C-E	Dermatology
Which condition may include generalised pruritus as a symptom? A. Hodgkin's disease B. Pityriasis rosea C. Diabetes mellitus D. Haemolytic jaundice E. Polycythaemia rubra vera	A-B-E	Dermatology

Abbreviations

The following abbreviations are used in this chapter:

LLM Large Language Model
AI Artificial Intelligence
MCQ Multiple-Choice Questionnaire

References

1. R. Agarwal, E. Karahanna, Time flies when you're having fun: Cognitive absorption and beliefs about information technology usage. MIS Q. 665–694 (2000). https://doi.org/10.2307/3250951
2. S.C. Bagheri, *Clinical Review of Oral and Maxillofacial Surgery-E-Book: Clinical Review of Oral and Maxillofacial Surgery-E-Book* (Elsevier Health Sciences, 2013)
3. A. Blandford, J. Wesson, R. Amalberti, R. AlHazme, R. Allwihan, Opportunities and challenges for telehealth within, and beyond, a pandemic. Lancet Glob. Health **8**(11), e1364–e1365 (2020)
4. J.L. Bolognia, J.V. Schaffer, K.O. Duncan, *Dermatology Essentials* (Elsevier, 2014)
5. J. Devlin, M.-W. Chang, K. Lee, K. Toutanova, Bert: pre-training of deep bidirectional transformers for language understanding (2018). arXiv preprint arXiv:1810.04805
6. Ditki Medical & Biological Sciences, Respiratory pathologies exam preparation (2024). https://ditki.com/course/usmle-comlex-step-2/respiratory-pathologies/exam
7. B. Fadem, *High Yield Endocrinology* (Lippincott Williams & Wilkins, 2010)
8. P. Fenichel, M.D. Harbison, Endocrinology in clinical practice. J. Clin. Endocrinol. **101**(2), 331–340 (2016)
9. M.B. First, *DSM-5 Handbook of Differential Diagnosis* (American Psychiatric Publishing, 2014)
10. G.O. Gabbard, *Gabbard's Treatments of Psychiatric Disorders*, 5th edn (American Psychiatric Publishing, 2014)
11. D.G. Gardner, D. Shoback, *Greenspan's Basic and Clinical Endocrinology*, 10th edn. (McGraw-Hill Education, 2015)
12. A. Gilson, C.W. Safranek, T. Huang, V. Socrates, L. Chi, R.A. Taylor, D. Chartash, et al., How does ChatGPT perform on the United States medical licensing examination? The implications of large language models for medical education and knowledge assessment. JMIR Med. Educ. **9**(1), e45312 (2023)
13. L.A. Goldsmith, S.I. Katz, B.A. Gilchrest, A.S. Paller, D.J. Leffell, K. Wolff, *Fitzpatrick's Dermatology in General Medicine*, 8th edn. (McGraw-Hill, 2012)
14. A. Grossman, *Clinical Endocrinology* (Blackwell Publishing, 2010)
15. N. Hawley, D. Johnson, K. Packer, M. Ditmyer, K. Kingsley, Dental students' preparation and study habits for the national board dental examination part I. J. Dental Educ. **73**(11), 1274–1278 (2009)
16. W.D. James, D. Elston, T. Berger, I. Neuhaus, *Andrews' Diseases of the Skin: Clinical Dermatology*, 12th edn. (Elsevier, 2019)
17. S. Khan, M. Naseer, M. Hayat, S.W. Zamir, F.S. Khan, M. Shah, Transformers in vision: a survey. ACM Comput. Surv. (CSUR) **54**(10s), 1–41 (2022)
18. A.L. McCann, E.D. Schneiderman, R.J. Hinton, E-teaching and learning preferences of dental and dental hygiene students. J. Dental Educ. **74**(1), 65–78 (2010)
19. S. Melmed, K.S. Polonsky, P. Reed Larsen, H. Kronenberg, *Williams Textbook of Endocrinology* 13th edn (Elsevier, 2016)

20. OpenAI, GPT-4 Technical Report (2023). ArXiv arXiv:2303.08774. https://api.semanticscholar.org/CorpusID:257532815
21. D.P. Panagoulias, M. Virvou, G.A. Tsihrintzis, Augmenting large language models with rules for enhanced domain-specific interactions: the case of medical diagnosis. Electronics **13**(2), 320 (2024)
22. D.P. Panagoulias, P. Papatheodosiou, A.P. Palamidas, M. Sanoudos, E. Tsoureli-Nikita, M. Virvou, G.A. Tsihrintzis, *COGNET-MD, an evaluation framework and dataset for large language model benchmarks in the medical domain* (2024). arXiv:2405.10893 [cs.CL]
23. D.P. Panagoulias, E. Tsoureli-Nikita, M. Virvou, G.A. Tsihrintzis, *Dermacen analytica: a novel methodology integrating multi-modal large language models with machine learning in tele-dermatology* (2024). arXiv:2403.14243 [cs.CL]
24. D.P. Panagoulias, M. Virvou, G.A. Tsihrintzis, Evaluating LLM—generated multimodal diagnosis from medical images and symptom analysis (2024). arXiv:2402.01730 [cs.CL]
25. D.P. Panagoulias, M. Virvou, G.A. Tsihrintzis, Knowledge space reduction via sequential language model integration, in *28th International Conference on Knowledge-Based and Intelligent Information & Engineering Systems (KES 2024)* (2025), pp. 1–
26. Pediatrics, The American Board of (2024). *Pediatric Pulmonology Certification*. https://www.abp.org
27. B.J. Sadock, V.A. Sadock, P. Ruiz, *Kaplan and Sadock's Comprehensive Textbook of Psychiatry*, 10th edn. (Wolters Kluwer, 2015)
28. C.L. Snoswell, G. Chelberg, K.R. De Guzman, H.H. Haydon, E.E. Thomas, L.J. Caffery, A.C. Smith, The clinical effectiveness of telehealth: a systematic review of meta-analyses from 2010 to 2019. J. Telemed. Telecare **29**(9), 669–684 (2023)
29. Society, European Respiratory, HERMES examinations in adult and paediatric respiratory medicine (2024). https://www.ersnet.org
30. S.M. Stahl, *Stahl's Essential Psychopharmacology: Neuroscientific Basis and Practical Applications*, 4th edn. (Cambridge University Press, 2013)
31. StatPearls, Pediatric pulmonology clerkship exam (2024). https://www.statpearls.com
32. A. Tasman, J. Kay, J.A. Lieberman, M.B. First, M. Maj, *Psychiatry*, 4th edn. (Wiley-Blackwell, 2015)

Chapter 12
Advances, Challenges and Contributions to Artificial Intelligence-Empowered Bio-Medical Applications

The research conducted in this book, set out to identify the shortcomings and opportunities of recent technologies in the medical domain, with the goal of proposing solutions and providing a basis for cost-effective and modern improvements that enhance the patient's journey and streamline the doctor's workflow. The evolving space of Artificial Intelligence and software development within the medical field often sees business scaling overshadowing important social parameters such as equity, safety, and validation.

12.1 Discussion of Advances in AI-Empowered Medical Applications

12.1.1 Discussion on AI-Empowered Software Engineering

The field of software engineering is undergoing a paradigm shift driven by advancements in Artificial Intelligence (AI). AI's integration into software engineering frameworks like the Rational Unified Process (RUP) has significantly transformed development methodologies, introducing higher levels of automation, intelligence, and adaptability.

RUP has traditionally provided a structured approach to software development through iterative and incremental processes [20]. While its strengths lie in its rigor and formalism, challenges such as adapting to dynamic requirements have been a persistent critique. However, the introduction of AI, particularly through clustering and recommender systems [17], has empowered RUP-based life cycles to dynamically adapt to user needs, enabling more tailored and responsive development cycles.

D. P. Panagoulias et al., *Artificial Intelligence-Empowered Bio-medical Applications*, Learning and Analytics in Intelligent Systems 49,
https://doi.org/10.1007/978-3-031-90174-4_12

Merging AI with RUP has not only streamlined requirements gathering and analysis but has also enhanced the traceability and optimization of workflows. The development of product configuration systems poses significant challenges, such as handling variability and ensuring user satisfaction. Shafiee et al. [19] compare RUP with Scrum, demonstrating how AI can mitigate challenges like requirement volatility and integration complexity. AI's capability to process large data sets, predict trends, and automate testing makes it an indispensable asset in this context. For instance, RUP's structured phases, namely Inception, Elaboration, Construction, and Transition (Anwar 2014)-can benefit from AI-driven decision-making tools that analyze historical project data for predictive insights.

The agility of RUP has been a topic of debate, with efforts to infuse Agile principles into RUP to make it more adaptive and user-centric [6]. AI further accelerates this transformation by automating iterative feedback mechanisms and enabling real-time adaptation to project changes. The use of pre-trained AI models and natural language processing tools like ChatGPT [34] demonstrates how AI can facilitate collaboration and communication among stakeholders, bridging the gap between formal RUP structures and Agile's flexibility.

AI-empowered microservices and reference architectures are transforming software design and deployment. Schäffer et al. [18] propose an AutomationML-based architecture that leverages AI for engineering configurators. This approach aligns with RUP's principles by enabling modular and scalable development while integrating automation to manage complexity. AI-driven automation enhances system scalability, resilience, and maintainability, crucial aspects of RUP's focus on quality assurance during the construction and transition phases.

The integration of AI into RUP is particularly impactful in domain-specific applications. For example, Panagoulias et al. [8, 9] present a patient-centered nutritional software tool, highlighting AI's role in personalizing user experiences. Such systems align with the user-centered principles of RUP while utilizing AI to improve usability, efficiency, and adaptability. Moreover, George A Tsihrintzis, Virvou, Sakkopoulos et al. [28] explore machine learning paradigms that bolster intelligent systems, making them capable of learning and evolving with minimal human intervention. These advancements enable software systems to better align with evolving user needs and business goals.

Virvou and Tsihrintzis [34] emphasize the role of human-AI collaboration in software engineering. While AI excels at automating tasks and generating insights, human expertise is essential for ethical decision-making, strategic planning, and managing AI-driven workflows. This collaboration aligns with RUP's governance models, which emphasize stakeholder involvement throughout the software lifecycle. Governance in AI-empowered development requires alignment between business goals and IT strategies [37]. Explainability is another critical dimension, especially in decision support systems. Dimitrios P Panagoulias, Sarmas et al. [12] propose methodologies for tailoring AI explainability, ensuring transparency and trustworthiness, which resonate with RUP's emphasis on documentation and stakeholder accountability.

AI's integration into software engineering processes, particularly through frameworks like RUP, is advancing significantly the field. By automating mundane tasks, improving adaptability, and enhancing decision-making, AI empowers developers to focus on creative and strategic aspects of software engineering. However, as Virvou and George A Tsihrintzis [34] note, the role of human expertise remains critical in steering AI's capabilities towards ethical and goal-oriented outcomes. AI-enabled software engineering holds the promise of creating systems that are not only efficient and scalable but also intelligent and user-focused. The synergy of RUP's structured approach and AI's transformative potential is paving the way for the next generation of intelligent software systems.

12.1.2 Discussion on Advances in Support Vector Machines (SVM) and Their Application to the Medical Domain

Support Vector Machines (SVM) are among the most robust and widely-used machine learning algorithms, particularly for classification tasks. Their capability to handle high-dimensional data, nonlinear boundaries, and complex classification problems makes them highly effective in applications across various fields. In the medical domain, the advancements in SVM and their integration with other technologies have significantly improved diagnostic accuracy, personalized medicine, and healthcare decision-making. One of the notable applications of SVM in the medical field is in breast cancer diagnosis. Agrios et al. [1] proposed a computer-aided diagnostic system for mammogram classification based on SVM. This approach highlights the algorithm's strength in handling complex medical imaging data, distinguishing between malignant and benign cases with high precision. The SVM's ability to manage nonlinear separable data through kernel functions ensures that subtle patterns in mammograms are effectively captured. This development aligns with the ongoing push towards early detection of diseases, where timely and accurate diagnosis can significantly improve treatment outcomes and reduce mortality rates.

A common challenge in medical datasets is the extreme class imbalance, where the number of samples in one class (e.g., diseased cases) is far fewer than the other (e.g., healthy cases). Traditional machine learning models often fail to perform well under such conditions. Sotiropoulos and Tsihrintzis (2017) addressed this issue by employing artificial immune system-based classification in extremely imbalanced problems. Their methods align well with SVM's capacity to work effectively in small sample spaces by adjusting the decision boundaries to prioritize minority classes. Further, Sotiropoulos et al. [21] compared one-class classifiers in handling class imbalances, demonstrating how hybrid approaches with SVM can provide robust classification performance in critical medical tasks.

The integration of SVM with decision support systems (DSS) has further expanded its utility in medical applications. Tsihrintzis et al. [26] included advancements in AI-empowered DSS and their ability to provide actionable insights. SVM-based DSS are particularly valuable in areas like radiology, pathology, and genomics, where the complexity and volume of data necessitate intelligent systems for accurate interpretation. Hybrid models that combine SVM with other machine learning techniques have gained traction in medical applications. For instance, Lampropoulos et al. [7] introduced cascade hybrid recommendation systems that use SVM alongside collaborative filtering techniques. While their study focused on e-commerce, the principles of hybrid modeling can be directly applied to medical recommendation systems, such as drug prescriptions or treatment plans.

The interpretability of machine learning models, including SVM, is critical in the medical field, where decisions often have life-altering consequences. Alogogianni et al. [2] emphasized the importance of interpretable machine learning techniques. Enhancements in SVM-based models to improve transparency and explainability make them more suitable for medical applications, fostering trust among practitioners and patients. Tsihrintzis et al. [24] explored broader advances in machine learning paradigms, including SVM, highlighting their role in data analytics and intelligent systems. In medical applications, these advancements have enabled the development of systems capable of analyzing large and heterogeneous datasets, such as genomic sequences and electronic health records.

Moreover, Tsihrintzis [27] emphasized the importance of surveys and reviews in tracking the evolution of AI technologies, including SVM, in intelligent decision-making systems. Such studies provide valuable insights into the algorithm's ongoing refinements and applications in the medical field. The integration of SVM with emerging technologies such as deep learning, artificial immune systems, and hybrid modeling frameworks offers exciting possibilities for medical applications. For example, combining SVM with clustering techniques Savvopoulos et al. [17] can enhance patient stratification in personalized medicine.

However, challenges remain, including the computational cost of large-scale SVM implementations, the need for better scalability in handling big medical data, and ensuring ethical and unbiased outcomes in AI-driven medical systems. Advances in SVM have significantly contributed to the medical domain, particularly in diagnostics, decision support systems, and addressing class imbalances. By leveraging SVM's strengths and integrating it with other AI techniques, researchers have developed innovative solutions for complex medical problems. The future of SVM in medicine lies in its hybridization with other models, improvements in interpretability, and its application to emerging medical challenges such as personalized medicine and predictive analytics.

12.1.3 Discussion on Personalization Techniques and Their Application to Medical Domains

Artificial Intelligence (AI)-driven personalization technologies have emerged as transformative tools across domains, including healthcare. These systems leverage adaptive learning, emotion recognition, explainable AI, and assistive technologies to deliver tailored solutions for medical education, patient care, and diagnostic processes. The following discussion explores their potential applications in the medical field using insights from the provided references.

AI-driven personalized learning platforms have made advancements to education, offering adaptable and user-centric training. Chrysafiadi et al. [5] developed a fuzzy-based intelligent tutoring system that evaluates user experience and learning outcomes. These systems can be tailored to train medical professionals by dynamically adapting to their skill levels and learning preferences. Chrysafiadi and Virvou [4] further demonstrated the efficacy of dynamically personalized e-training systems in computer programming. Applying these principles in medical education could provide scenario-based training, simulating clinical challenges and enhancing decision-making skills. The foundational work by Virvou (1999) on automatic reasoning and help for user errors in operating systems underscores the importance of intuitive, personalized interfaces in technology. These principles are directly applicable to AI-driven medical training systems, where user support and adaptability are critical for effective learning.

Emotion recognition systems play an important role in creating empathetic and patient-centered care environments. Tsihrintzis et al. [36] combined audio-lingual and visual-facial modalities for emotion recognition. Such technologies can be used in mental health applications to monitor patient emotions and provide real-time feedback to therapists. Stathopoulou et al. [22] integrated keyboard-stroke patterns to augment visual-facial emotion recognition, improving the system's accuracy. Similarly, Tsihrintzis et al. [25] emphasized the importance of complementary data modalities in enhancing affective computing systems. These innovations are particularly relevant for telemedicine, where non-verbal cues are crucial for understanding patient needs.

AI-driven personalization in e-learning platforms, such as LYRICEL by Panagoulias et al. [13], uses GPT and machine learning for culturally relevant education. Similar methods could be adapted for medical training to provide culturally sensitive and personalized learning materials for practitioners in diverse healthcare settings. The broader advancements in machine learning paradigms, as outlined by Virvou et al. [30], underscore the potential of AI to deliver customized and adaptive solutions across domains, including healthcare.

Explainable AI (XAI) ensures transparency and trust in AI-driven systems, which is crucial in medical applications where decisions directly impact patient outcomes. Panagoulias et al. [12] proposed tailored explainability methodologies, enabling medical professionals to understand and validate AI-driven decisions. Virvou [29] discussed the interplay between AI and user experience, emphasizing the importance of intuitive and interpretable systems. This balance ensures that advanced medical technologies remain accessible and user-friendly while maintaining their effectiveness.

Assistive technologies designed for personalized care enhance accessibility and independence for patients with disabilities. Bourbakis et al. [3] highlighted advancements in these systems, showcasing their potential to adapt to individual needs and improve patient outcomes. These technologies play a crucial role in inclusive healthcare, ensuring equitable access for all patients.

Interactive AI systems enhance engagement between stakeholders in healthcare, including patients and providers. Tsihrintzis et al. [23] explored AI systems such as ChatGPT for facilitating user interaction. These systems can improve communication in medical contexts, ensuring clarity and fostering trust between patients and healthcare providers.

AI-driven personalization technologies are reshaping the medical domain by offering tailored solutions to meet the unique needs of patients and practitioners. Adaptive learning, emotion recognition, explainable AI, and assistive systems are paving the way for more efficient, empathetic, and inclusive healthcare services. The integration of these technologies into medical practices holds immense potential for transforming how healthcare is delivered and experienced

12.1.4 Discussion on Advances in Large Language Models and ChatGPT and Their Application to the Medical Domain

Large Language Models (LLMs) and ChatGPT have contributed important advances to numerous domains through their ability to process natural language, generate human-like responses, and integrate multimodal information. In the medical field, these technologies show immense potential for enhancing diagnostics, treatment planning, telemedicine, and medical education. This discussion examines their applications, benefits, and challenges using insights from recent research. LLMs like GPT-4 and greater have demonstrated significant potential in medical diagnostics and treatment planning. Panagoulias et al. [10] evaluated ChatGPT's role in diagnosing, staging, and planning treatments for lung cancer. The study highlighted the model's capacity to synthesize patient history, imaging data, and clinical guidelines into actionable recommendations. Similarly, Panagoulias et al. [11] investigated ChatGPT's broader diagnostic capabilities, emphasizing its utility in providing evidence-based suggestions, particularly for complex cases. However, they also noted the necessity of domain-specific customization and continuous validation to ensure reliability and safety.

To address limitations in general-purpose LLMs, researchers have explored augmenting these models with rules and domain-specific knowledge. Panagoulias et al. [16] proposed a framework for enhancing LLMs in medical diagnosis by integrating rule-based systems. This approach improves the models' interpretability and ensures alignment with clinical guidelines. In tele-dermatology, Panagoulias et al. [15] introduced Dermacen Analytica, a methodology combining multimodal LLMs with machine learning. The integration enables accurate analysis of medical images alongside symptom narratives, facilitating remote consultations and early disease detection. Evaluating LLMs in medical contexts requires rigorous frameworks and datasets. Panagoulias et al. [14] developed COGNET-MD, a benchmarking framework for assessing LLM performance in medical domains. By focusing on accuracy, interpretability, and ethical considerations, this framework ensures that LLMs meet the high standards required in healthcare applications.

LLMs have significant implications for medical education, offering personalized learning experiences. Virvou and Tsihrintzis [33] evaluated ChatGPT's role in education, finding that it supports adaptive tutoring and interactive learning, fostering deeper understanding among students. These findings align with their subsequent work Virvou and Tsihrintzis [35], where they introduced the FEPER framework, emphasizing efficiency, pedagogical quality, and ethical requirements for AI-powered educational tools. In cultural and educational e-learning contexts, Panagoulias et al. [13] demonstrated the utility of rule-augmented AI platforms like LYRICEL. These methods could inspire similar applications in medical education, creating context-aware, adaptive curricula tailored to individual learners.

Human trust in AI systems is critical, particularly in high-stakes environments like healthcare. Tsihrintzis et al. [31] introduced VIRTSI, a trust dynamics model that evaluates human-AI collaboration using ChatGPT. The study revealed how trust fluctuates based on AI response accuracy, transparency, and the perceived impact of decisions, offering valuable insights for designing trustworthy medical AI systems. Expanding on this, Tsihrintzis [32] explored the role of consequences in shaping trust dynamics, highlighting the importance of aligning AI systems with ethical standards and user expectations.

Advances in multimodal LLMs, which integrate text, images, and other data types, hold immense promise for the medical field. Panagoulias et al. [16] explored these capabilities in generating multimodal diagnoses by combining image analysis with patient symptom narratives. The integration of these modalities enables more comprehensive and accurate assessments, particularly for complex conditions. Tsihrintzis et al. [26] underscored the broader implications of AI-empowered decision support systems, noting that LLMs could serve as foundational tools in future healthcare innovations. The application of LLMs and ChatGPT in the medical domain has shown new possibilities for diagnostics, treatment planning, education, and patient engagement. These models, when augmented with domain-specific knowledge and tailored methodologies, offer transformative potential. However, challenges such as trust, ethical considerations, and the need for robust benchmarking must be addressed to fully realize their benefits in healthcare.

12.2 Contributions to AI-Empowered Medical Applications

- Personalization of Technology via novel theoretical approaches: By focusing on the perceived usability and usefulness of technology as understood by its users (TAM), this book emphasizes tailored explainability functionalities. This approach ensures that the technology meets the unique needs of each user, thereby enhancing its overall effectiveness and acceptance.

- Stakeholder Identification: Regulators and experts have been identified as the main stakeholders who ensure safety and equity. This work provides design paradigms as part of novel and modern software development for AI-empowered systems, ensuring that these systems are both safe, equitable and scalable.
- Evaluation of Medical Systems: Various proposed, implemented, and open-sourced pipelines were developed to facilitate the evaluation of medical systems. This ensures that these systems are rigorously tested and validated, providing reliable and effective solutions for medical professionals and patients alike.

Through the implementation of three distinct systems and subsystems, an ecosystem of associated services was constructed. More specifically, this book has made these significant contributions to the field:

1. Classification engines and ML services as a basis of explainable decision-making systems and knowledge reduction tools.
2. NuhealtSoft: An AI-empowered medical application developed using the Rational Unified Process (RUP) and microservices architecture, encompasying those tools. This application places the needs of patients and stakeholders (regulators and doctors) at the center of a continuous development process focused on improvements. The application also incorporates chatbots in the domains of pathology and dermatology, enhancing its utility and user engagement.
3. Sequential Language Model Integration for Knowledge Space Reduction
4. PINXEL Framework: This framework employs targeted machine learning methods to acquire a user model for personalized, multi-layered explainability. It integrates a rule-based system that guides the degree of trustworthiness based on user perception and AI literacy level, ensuring that the system is both transparent and reliable.
5. COGNET-MD Toolkit: A novel toolkit consisting of expertly curated datasets and an intuitive framework for scoring and assessing AI-empowered medical systems. This toolkit addresses the limitations of multi-modal large language models (LLMs) by providing evaluation pipelines for assessing their accuracy in symptom evaluation, either unilateral or accompanied by associated images.

In summary, this book has provided a comprehensive approach to addressing key challenges in the medical domain through the use of AI and modern software development techniques. By focusing on personalization, stakeholder engagement, and rigorous evaluation, this work lays a solid foundation for the development of effective, equitable, and safe medical technologies. These innovative approaches have the potential to provide transformative solutions in the healthcare landscape, ensuring better efficient workflows for doctors.

References

1. A. Agrios, D.N. Sotiropoulos, G.A. Tsihrintzis, Computer aided breast cancer diagnosis: an SVM-based mammogram classification approach. Int. J. Comput. Intell. Stud. **7.2**, 138–163 (2018)
2. E. Alogogianni, M. Virvou, Interpretable machine learning for undeclared work prediction, in *2023 14th International Conference on Information, Intelligence, Systems and Applications (IISA)* (IEEE, 2023), pp. 1–8
3. N.G. Bourbakis, A. Esposito, G.A. Tsihrintzis, M. Virvou, L.C. Jain, Introduction to advances in assistive technologies, in *Advances in Assistive Technologies: Selected Papers in Honour of Professor Nikolaos G. Bourbakis*, vol. 3, pp. 1–7 (2022)
4. K. Chrysafiadi, M. Virvou, Dynamically personalized e-training in computer programming and the language C, in *IEEE Transactions on Education*, vol. 56.4, pp. 385–392 (2013)
5. K. Chrysafiadi, M. Virvou, G.A. Tsihrintzis, I. Hatzilygeroudis, Evaluating the user's experience, adaptivity and learning outcomes of a fuzzy-based intelligent tutoring system for computer programming for academic students in Greece, in *Education and Information Technologies*, vol. 28.6, pp. 6453–6483 (2023)
6. M. Hirsch, Making RUP agile, in *OOPSLA 2002 Practitioners Reports*, 1–ff (2002)
7. A.S. Lampropoulos, D.N. Sotiropoulos, G.A. Tsihrintzis, Cascade hybrid recommendation as a combination of one-class classification and collaborative filtering. Int. J. Artif. Intell. Tools **23.04**, 1460009 (2014)
8. D.P. Panagoulias, D.N. Sotiropoulos, G.A. Tsihrintzis, Nutritional biomarkers and machine learning for personalized nutrition applications and health optimization, in *Proceedings of the Twelfth IEEE International Conference on Information, Intelligence, Systems and Applications*, Chania, Greece, pp. 731–733, July 2021a
9. D.P. Panagoulias, D.N. Sotiropoulos, G.A. Tsihrintzis, Nutritional biomarkers and machine learning for personalized nutrition applications and health optimization (extended journal version). Intell. Decis. Technol. **15**, 645–653 (2021)
10. D. Panagoulias, F. Palamidas, M. Virvou, G.A. Tsihrintzis, Evaluating the potential of LLMs and ChatGPT on medical diagnosis and treatment, in *14th IEEE International Conference on Information, Intelligence, Systems, and Applications (IISA2023), Volos, Greece* (2023)
11. D.P. Panagoulias, F.A. Palamidas, M. Virvou, G.A. Tsihrintzis, Evaluation of ChatGPT-supported diagnosis, staging and treatment planning for the case of lung cancer, in *2023 20th ACS/IEEE International Conference on Computer Systems and Applications (AICCSA)* (IEEE, 2023), pp. 1–8
12. D.P. Panagoulias, E. Sarmas, V. Marinakis, M. Virvou, G.A. Tsihrintzis, H. Doukas, Intelligent decision support for energy management: a methodology for tailored explainability of artificial intelligence analytics, in *Electronics*, vol. 12.21, p. 4430 (2023)
13. D.P. Panagoulias, D. Sotiropoulos, K. Chrysafiadi, E.-A. Tsichrintzi, E. Sakkopoulos, G.A. Tsihrintzis, M. Virvou, LYRICEL: a rule-augmented artificial intelligence-empowered cultural E-learning with GPT and machine learning for song lyrics, in *2024 15th International Conference on Information, Intelligence, Systems and Applications (IISA)* (IEEE, 2024), pp. 1–5
14. D.P. Panagoulias, P. Papatheodosiou, A.P. Palamidas, M. Sanoudos, E. Tsoureli-Nikita, M. Virvou, G.A. Tsihrintzis, in *COGNET-MD, an Evaluation Framework and Dataset for Large Language Model Benchmarks in the Medical Domain* (2024). arXiv: 2405.10893 [cs.CL]
15. D.P. Panagoulias, E. Tsoureli-Nikita, M. Virvou, G.A. Tsihrintzis, *Dermacen Analytica: A Novel Methodology Integrating Multi-modal Large Language Models with Machine Learning in Tele-dermatology* (2024). arXiv: 2403.14243 [cs.CL]
16. D.P. Panagoulias, M. Virvou, G.A. Tsihrintzis, *Evaluating LLM – Generated Multimodal Diagnosis from Medical Images and Symptom Analysis* (2024). arXiv: 2402.01730 [cs.CL]
17. A. Savvopoulos, Virvou, M., Sotiropoulos, D.N., Tsihrintzis, G.A., Clustering for user modeling in recommender e-commerce application: a RUP-based intelligent software life-cycle, in *Knowledge-Based Software Engineering* (IOS Press, 2008), pp. 295–304

18. E. Schäffer, L.N. Penczek, M. Bartelt, M. Brossog, B. Kuhlenkötter, J. Franke, A microservice-And AutomationML-based reference architecture for an engineering configurator web platform. Procedia CIRP **103**, 274–279 (2021)
19. S. Shafiee, Y. Wautelet, L. Hvam, E. Sandrin, C. Forza, Scrum versus Rational Unified Process in facing the main challenges of product configuration systems development. J. Syst. Softw. **170**, 110732 (2020)
20. A.K. Shuja, J. Krebs, *IBM Rational Unified Process Reference and Certification Guide: Solution Designer (RUP)*. Pearson Education (2007)
21. D. Sotiropoulos, C. Giannoulis, G.A. Tsihrintzis, A comparative study of one-class classifiers in machine learning problems with extreme class imbalance, in *IISA 2014, the 5th International Conference on Information*. Intelligence Systems and Applications. (IEEE, 2014), pp. 362–364
22. I.-O. Stathopoulou, E. Alepis, G.A. Tsihrintzis, M. Virvou, On assisting a visual-facial affect recognition system with keyboard-stroke pattern information. Knowl.-Based Syst. **23.4**, 350–356
23. G.A. Tsihrintzis, E. Sarmas, V. Marinakis, D. Panagoulias, E.-A. Tsihrintzi, M. Virvou, Evaluating non-expert stakeholder interaction with artificial intelligence on energy urban domain using VIRTSI: the case of ChatGPT, in *2024 15th International Conference on Information, Intelligence, Systems and Applications (IISA)* (IEEE, 2024), pp. 1–10
24. G.A. Tsihrintzis, D.N. Sotiropoulos, L.C. Jain, *Machine Learning Paradigms: Advances in Data Analytics* (Springer, 2019)
25. G.A. Tsihrintzis, M. Virvou, E. Alepis, I.-O. Stathopoulou, Towards improving visual-facial emotion recognition through use of complementary keyboard-stroke pattern information, in *Fifth International Conference on Information Technology: New Generations (ITNG 2008)* (IEEE, 2008), pp. 32–37
26. G.A. Tsihrintzis, M. Virvou, H. Doukas, L.C. Jain, Introduction to advances in artificial intelligence-empowered decision support systems, in *Advances in Artificial Intelligence-Empowered Decision Support Systems: Papers in Honour of Professor John Psarras* (Springer, 2024), pp. 1–11
27. G.A. Tsihrintzis, M. Virvou, G. Phillips-Wren, Surveys in artificial intelligence-based technologies, in *Intelligent Decision Technologies*, vol. 13.4, pp. 393–394 (2019)
28. G.A. Tsihrintzis, M. Virvou, E. Sakkopoulos, L.C. Jain, *Machine Learning Paradigms: Applications of Learning and Analytics in Intelligent Systems*, vol. 1. (Springer, 2019)
29. M. Virvou, The emerging Era of Human-AI interaction: Keynote address, in *2022 13th International Conference on Information, Intelligence, Systems I and Applications (IISA)*, pp. 1–10 (2022). https://doi.org/10.1109/IISA56318.2022.9904422
30. M. Virvou, E. Alepis, G.A. Tsihrintzis, L.C. Jain, *Machine Learning Paradigms: Advances in Learning Analytics* (Springer, 2020)
31. M. Virvou, G.A. Tsihrintzis, E.-A. Tsichrintzi, VIRTSI: a novel trust dynamics model enhancing Artificial Intelligence collaboration with human users–Insights from a ChatGPT evaluation study, in *Information Sciences*, p. 120759 (2024)
32. M. Virvou, G.A. Tsihrintzis, A novel trust state-chart model for requirements engineering of trustful AI-Empowered software, in *2023 14th International Conference on Information, Intelligence, Systems and Applications (IISA)* (IEEE, 2023a), pp. 1–6
33. M. Virvou, G.A. Tsihrintzis, Is ChatGPT beneficial to education? A holistic evaluation framework based on intelligent tutoring systems, in *2023 14th International Conference on Information, Intelligence, Systems and Applications (IISA)* (IEEE, 2023b), pp. 1–8
34. M. Virvou, G.A. Tsihrintzis, Pre-made empowering artificial intelligence and ChatGPT: the growing importance of human AI-Experts, in *2023 14th International Conference on Information, Intelligence, Systems and Applications (IISA)* (IEEE, 2023c), pp. 1–8
35. M. Virvou, G.A. Tsihrintzis, Artificial intelligence chatbots and social robots in education: FEPER framework for efficiency, pedagogical and ethical requirements, in *2024 15th International Conference on Information, Intelligence, Systems and Applications (IISA)* (IEEE, 2024), pp. 1–8

36. M. Virvou, G.A. Tsihrintzis, E. Alepis, I.-O. Stathopoulou, K. Kabassi, Emotion recognition: empirical studies towards the combination of audio-lingual and visual-facial modalities through multi-attribute decision making. Int. J. Artif. Intell. Tools **21.02**, 1240001 (2012)
37. Y. Wautelet, Using the RUP/UML business use case model for service development governance: a business and IT alignment based approach, in *2020 IEEE 22nd Conference on Business Informatics (CBI)*, vol. 2 (IEEE, 2020), pp. 121–130

The manufacturer's authorised representative in the EU is Springer Nature Customer Service Centre GmbH, Europaplatz 3, 69115 Heidelberg, Germany. If you have any concerns regarding our products, please contact ProductSafety@springernature.com

Printed and bound by CPI Group (UK) Ltd, Croydon, CR0 4YY

30/09/2026

02225233-0005